Analytical Aspects in Environmental Pollution Monitoring

Analytical Aspects in Environmental Pollution Monitoring

Editors

Victoria Samanidou
Eleni Deliyanni
Dimitra Voutsa

MDPI

Basel • Beijing • Wuhan • Barcelona • Belgrade • Novi Sad • Cluj • Manchester

Editors

Victoria Samanidou
Chemistry- Laboratory of
Analytical Chemistry
University of Thessaloniki
Thessaloniki
Greece

Eleni Deliyanni
Chemistry
University of Thessaloniki
Thessaloniki
Greece

Dimitra Voutsa
Chremistry
Aristotle University of
Thessaloniki
Thessaloniki
Greece

Editorial Office
MDPI
St. Alban-Anlage 66
4052 Basel, Switzerland

This is a reprint of articles from the Special Issue published online in the open access journal *Molecules* (ISSN 1420-3049) (available at: www.mdpi.com/journal/molecules/special_issues/analy_environ_monitor).

For citation purposes, cite each article independently as indicated on the article page online and as indicated below:

Lastname, A.A.; Lastname, B.B. Article Title. *Journal Name* **Year**, *Volume Number*, Page Range.

ISBN 978-3-0365-8805-6 (Hbk)
ISBN 978-3-0365-8802-5 (PDF)
doi.org/10.3390/books978-3-0365-8802-5

Contents

About the Editors

Victoria Samanidou

Dr. Victoria Samanidou is Full Professor and Director of the Laboratory of Analytical Chemistry in the Department of Chemistry of Aristotle University of Thessaloniki, Greece. Her research interests focus on the development of sample preparation methods using sorptive extraction prior to chromatographic analysis. She has co-authored 218 original research articles in peer-reviewed journals and 67 reviews, 85 editorials/in view and 57 chapters in scientific books (H-index 44, 6748 citations)). She is an Editorial Board Member of more than 31 scientific journals and has been a Guest Editor for more than 32 Special Issues. She has peer-reviewed more than 760 manuscripts for 161 scientific journals.

In 2016, she was included in top 50 power list of women in Analytical Science, as proposed by Texere Publishers, while in 2021, she was included in the "The Analytical Scientist" 2021 Power List of Top 100 Influential People in Analytical Science. In 2023, she was included in the Power List 2023 in the category: Mentors and Educators. Since 2020, she is included in the 2% top world scientists in the field of Analytical Chemistry (career and single year) published in PLOS Biology based on citations from SCOPUS. She is also enlisted as one of the 50 scientists from Aristotle University of Thessaloniki Scientist in AD University Rankings 2023.

Eleni Deliyanni

Dr. Eleni A. Deliyanni was a Full Professor in the Laboratory of Chemical and Environmental Technology in the School of Chemistry of the Aristotle University of Thessaloniki (AUTh), Greece (retired in 2022).

From 20/12/85 to 30/3/86, she was part of the Technical University of München, Germany, for additional work on her PhD Thesis, as well as research work on Scanning Electron Microscopy. From 1/12/07 to 30/4/08, she worked at the Institute of Analytical Chemistry at "The City College of New York" of CUNY, Department of Chemistry and Biochemistry in New York, while from 1/1/10 to 30/5/10, she returned as a visiting professor at the same lab in New York. From 1/9/14 to 31/10/14, she visited the Environcentrum Company in Kosice, Slovakia within the frame of the collaborative program, "Water and soil clean-up from mixed contaminants".

Dr. Deliyanni has co-authored more than 100 original research and review articles in peer-reviewed journals with an H-index of 42, and her work has been cited more than 5200 times. She is a member of editorial boards of scientific journals like *Journal of Colloid and Interface Science*, Elsevier, while she has reviewed more than 550 manuscripts for more than 100 scientific journals. She has also been a Guest Editor for several Special Issues in scientific journals.

Her research interests include (nano)materials chemistry, materials characterization, environmental chemistry, nanotechnology, water and waste-water treatment, the desulfurization of gaseous streams or (bio)fuels, adsorption, catalysis, advanced oxidation processes (AOPs), activated (nano)porous carbons and graphene derivatives.

Dimitra Voutsa

Dr. Dimitra Voutsa is a full professor of Environmental Chemistry and the Direction of the Environmental Pollution Control Laboratory in the School of Chemistry, Aristotle University of Thessaloniki, Greece. Dr. D. Voutsa has authored and co-authored more than 100 research articles and reviews in international peer-reviewed scientific journals and chapters in books (H-index 43, >6900 citations). She has supervised many PhD, MSc and undergraduate theses. She has participated as a coordinator and a member of research teams in more 40 projects. She is a member of the advisory board in the scientific journal *Environmental Science Europe*. She has also participated in scientific committees aiming to achieve the critical evaluation of various environmental issues. She is the deputy top manager and quality manager of the "Environmental Pollution Control, Atmospheric Particles Measurements Team" accredited according to ISO/IEC 17025:2017 for the sampling and gravimetric determination of PM10 and PM2.5.

Preface

Environmental analysis is one of the most relevant scientific fields in chemistry that contributes to the monitoring of environmental pollution.

Green methodologies should be developed and applied in order to investigate the impact of pollution on the environment.

The scope of this Special Issue was to describe the state of the art in both analysis and sample pretreatment with regard to the demands of green perspectives for the examination of the current status of environmental pollution.

Analytical scientists were invited to present new strategies to fulfill all the requirements for reductions in waste, energy consumption, the use of solvents, etc.

The Special Issue includes fourteen papers, namely, 11 research articles, 2 review articles and 1 perspective.

The Guest Editors would like to thank all authors for their excellent contributions.

Victoria Samanidou, Eleni Deliyanni, and Dimitra Voutsa

Editors

molecules

MDPI

Article

A Turbidity-Compensation Method for Nitrate Measurement Based on Ultraviolet Difference Spectroscopy

Jing Dong [1,2], Junwu Tang [1,3,4], Guojun Wu [1,3,*] and Ruizhuo Li [1,2]

1 Xi'an Institute of Optics and Precision Mechanics, Chinese Academy of Sciences, Xi'an 710119, China
2 University of Chinese Academy of Sciences, Beijing 100049, China
3 Pilot National Laboratory for Marine Science and Technology (Qingdao), Qingdao 266237, China
4 Faculty of Information Science and Engineering, Ocean University of China, Qingdao 266100, China
* Correspondence: wuguojun@opt.ac.cn

Abstract: To solve the problem that turbidity in water has a significant effect on the spectra of nitrate and reduces the accuracy of nitrate detection, a turbidity-compensation method for nitrate measurement based on ultraviolet difference spectra is proposed. The effect of turbidity on the absorption spectra of nitrate was studied by using the difference spectra of the mixed solution and a nitrate solution. The results showed that the same turbidity had different effects on the absorbance of different concentrations of nitrate. The change in absorbance due to turbidity decreased with an increase in the nitrate concentration at wavelengths from 200 nm to 230 nm, although this change was constant when the wavelength was greater than 230 nm. On the basis of this characteristic, we combined the residual sum of squares (RSS) and interval partial least squares (iPLS) to select wavelengths of 230–240 nm as the optimal modeling interval. Furthermore, the turbidity-compensation model was established by the linear fitting of the difference spectra of various levels of turbidity. The absorption spectra of the nitrate were extracted by subtracting the turbidity-compensation curve from the original spectra of the water samples, and the nitrate concentration was calculated by using a partial least squares (PLS)-based nitrate-prediction model. The experimental results showed that the average relative error of the nitrate predictions was reduced by 50.33% to 1.33% by the proposed turbidity-compensation method. This indicated that this method can better correct the deviation in nitrate's absorbance caused by turbidity and improve the accuracy of nitrate predictions.

Keywords: nitrate; ultraviolet spectroscopy; turbidity compensation; difference spectrum; partial least squares

Citation: Dong, J.; Tang, J.; Wu, G.; Li, R. A Turbidity-Compensation Method for Nitrate Measurement Based on Ultraviolet Difference Spectroscopy. *Molecules* **2023**, *28*, 250. https://doi.org/10.3390/molecules28010250

Academic Editors: Dimitra Voutsa, Victoria Samanidou and Eleni Deliyanni

Received: 28 November 2022
Revised: 20 December 2022
Accepted: 22 December 2022
Published: 28 December 2022

1. Introduction

Nitrate is a key phytoplankton nutrient. The concentrations of nitrate in seawater can affect marine primary productivity and regulate the structure of the phytoplankton community [1]. The accurate monitoring and analysis of nitrate concentrations are essential for understanding biogeochemical cycles and preventing ecological imbalances caused by high nitrate concentrations [2].

Traditionally, the laboratory chemical method has been used to determine nitrate levels [3]. Water samples are collected by a survey ship and transported to a laboratory for analysis. However, this method cannot capture the continuous temporal and spatial variation in nitrate. Through a combination of flow-injection analysis (FIA) [4] and/or microfluidic technology [5], nitrate sensors based on wet-chemical methods can be used for in situ nitrate detection. However, the problems related to preserving the chemical reagent and waste discharge limit the wide application of these sensors [6].

Ultraviolet spectroscopy is a popular method used for determining concentrations of nitrate with the significant advantages of simplicity, rapidity, high accuracy, and lack of reagent. In addition, ultraviolet spectroscopy has also been increasingly used to measure other water-quality parameters [7–9]. Johnson and Coletti [10] developed an in situ ultraviolet spectrophotometer (ISUS) for the direct high-resolution measurement of nitrate, bromide, and bisulfide. Langergraber et al. [11] presented a submersible UV-vis spectrometer to monitor nitrate, chemical oxygen demand (COD), and total suspended solids (TSS). However, matrix effects can seriously interfere with the detection of nitrates in natural water, and one of the most critical interfering factors is turbidity caused by suspended particulate matter [12]. The scattering of light by suspended particles reduces the amount of light reaching the detector, which changes the magnitude and peak position of the nitrate's absorption spectrum, thereby affecting the calculation of the nitrate concentration [13]. Therefore, eliminating the interference of turbidity is necessary.

Various turbidity-compensation methods have been proposed to reduce or eliminate the interference of turbidity. There are two main strategies. One is to establish the relationship between before and after turbidity correction by the mapping function of the model to recover the pure absorbance spectra of the water-quality parameter for analysis. Examples include the multiplicative scatter correction (MSC) algorithm [9,11], the orthogonal signal correction (OSC) algorithm [14], and deep-learning methods [15]. These methods can suppress the interference of turbidity to a certain degree but cannot fundamentally eliminate its effect. The accuracy of the model depends on the size of the training sample, and it is often necessary to retrain the model in the face of new water bodies. Another turbidity-compensation method is to subtract the absorbance caused by turbidity from the original spectrum based on a single wavelength [16,17] or multiple wavelengths [18–21]. Because the multi-wavelength method provides more information than single-wavelength methods, it is more extensively used. On the basis of Mie's scattering theory, Chen et al. effectively predicted the extinction spectrum induced by particulate matter and managed to compensate for turbidity [18]. However, it is difficult to accurately estimate the negative refractive index of suspended particulate matter in natural water. Chen et al. proposed a compensation-curve method to correct the absorption spectrum of nitrate. The compensation-curve model was established by a mixed solution of 2 mg/L nitrate and 5–50 NTU turbidity [19]. Li et al. reconstructed an influence matrix of the turbidity-absorption spectra based on a compressed sensing algorithm and used it to eliminate the interference of turbidity in COD detection [20]. Cai et al. calculated a corrected-absorbance spectral curve by introducing a coefficient K_N, which describes the degree to which the absorption spectrum of a turbid mixed COD solution deviates from the superposition law [21]. These kinds of turbidity-interference studies have been conducted with constant nitrate concentrations, varying the turbidity to examine its impact on the nitrate's absorption spectrum. However, this ignores a problem: whether the same level of turbidity has the same effect on varying nitrate concentrations. If the turbidity-compensation model is established in this way, it inevitably affects the accuracy with which other nitrate concentrations not included in the modeled sample set are detected.

In this study, the effect of turbidity on the absorption spectra of nitrate at different wavelengths was studied. Moreover, a turbidity-compensation method based on difference spectra combined with linear fitting is proposed. The turbidity was determined by the spectral area, and the concentration of nitrate was finally determined using a multivariate-calibration diffmodel. The results demonstrated that the proposed method can significantly enhance the accuracy of nitrate predictions.

2. Results and Discussion

2.1. Analysis of the Interference of Turbidity

The absorption spectra of the nitrate and turbidity solutions were measured, and the curves are shown in Figure 1a,b, respectively. The main spectral-absorption range of the nitrate was 200–250 nm, and the absorbance of nitrate was almost zero after 250 nm. The absorption spectra of the turbidity solutions covered the entire ultraviolet band, which could cause spectral cross-sensitivity and interfere with the absorption spectra of the nitrate solutions.

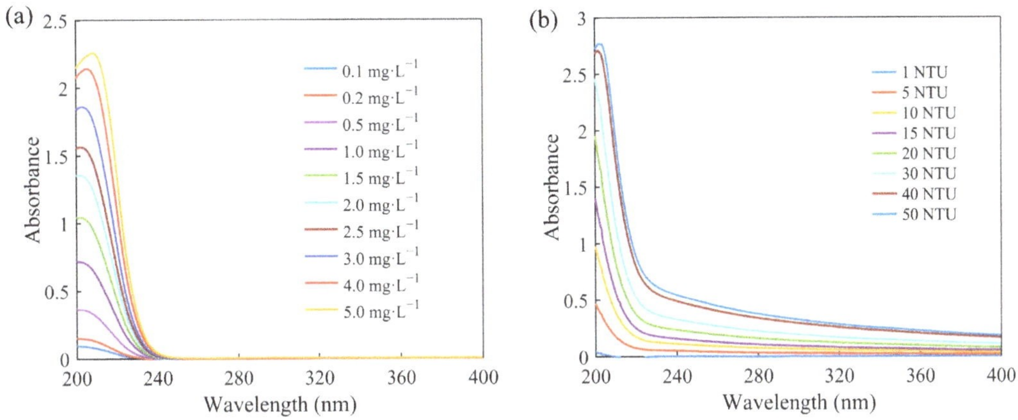

Figure 1. (**a**) Absorption spectra of solutions with different nitrate concentrations. (**b**) Absorption spectra of standard solutions with different levels of turbidity.

We prepared 15 groups of mixed solutions of nitrate and turbidity according to the sample concentrations and collected their absorption spectra. The spectra of the nitrate solutions were subtracted from the spectra of the mixed solutions to obtain their difference spectra.

The spectra of the mixed solutions and the difference spectra are shown in Figure 2. The spectra of the mixed solutions in Figure 2a were subjected to strong interference by the turbidity. As can be seen in Figure 2b,c, the amount of change in the absorbance caused by turbidity at different wavelengths varied. When the concentration of the nitrate was constant, the absorbance of the difference spectra increased with the increase in turbidity. However, when the turbidity was constant, the effect of the turbidity at different wavelengths changed as the nitrate concentration changed. At wavelengths of less than about 230 nm, the absorbance of the solutions with the same level of turbidity and different nitrate concentrations decreased with the increase in the nitrate concentration. At wavelengths greater than about 230 nm, the difference spectra almost overlapped. This means that the same turbidity had the same effect on the absorbance of the nitrate in this wavelength range.

Changes in absorbance are related to the effect of turbid particles in the solution on the nitrate molecules. The difference spectra reflect this effect, which includes two parts: one is the increase in absorbance caused by the absorption and scattering of suspended particles in turbid solutions, and the other is that suspended particles break the coplanar nature of nitrate molecules, causing steric hindrance and destroying the conjugate system, leading to a decrease in the absorbance of nitrate [19]. The superposition of these two parts is the change in the absorbance of nitrate caused by the inference of turbidity. Therefore, the superposition of the absorbance of turbidity solutions and nitrate solutions is usually greater than the absorbance of the mixed solution. Similar results were obtained in the research by Chen et al. [19].

Furthermore, the effect of turbidity was greatest at the wavelength of the absorption peak and decreased with the decrease in the absorbance. This issue was discussed by Hu et al., who concluded that the effect of turbidity was greatest at the central wavelength of the energy leap and became smaller with the decrease in the probability of energy leaps [13]. Therefore, the selection of the correct wavelength is necessary to account for the different effects of turbidity at different wavelengths.

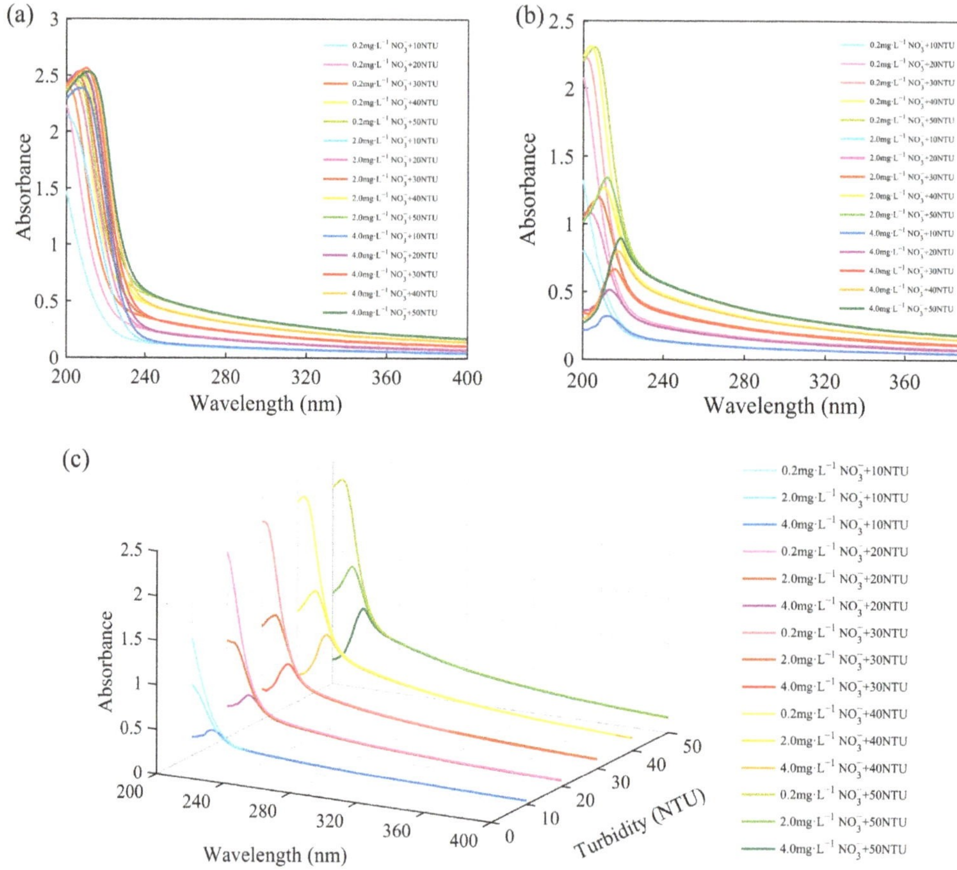

Figure 2. (**a**) Spectra of mixed solutions. (**b**) Difference spectra between the mixed solutions and standard nitrate solutions. (**c**) Difference spectra in three-dimensional coordinates.

We calculated the correlation coefficient between the difference spectra and the turbidity. In the band in which the difference spectra overlapped, the correlation coefficient was above 0.99. This demonstrates that the effect of turbidity on the absorbance of nitrate in this spectral band is proportional to the turbidity.

Based on this characteristic, a turbidity-compensation model was established by linear regression in a spectral interval. The effect of the turbidity on the nitrate was constant in this interval. Therefore, the mixed solutions with random levels of turbidity and random concentrations of nitrate were corrected accurately.

2.2. Interval Selection

The optimal interval is both the modeling interval of the turbidity-compensation model and the nitrate prediction model. Therefore, a suitable spectral interval for the turbidity-compensation model was chosen according to the characteristic effects of turbidity on the absorption spectra of nitrate. The interval was further optimized, and the interval with the highest accuracy in the nitrate-prediction model was taken as the final optimal interval.

It was shown in the previous section that the interval that is most suitable for establishing the turbidity-compensation model is the interval in which the difference spectra overlap. To extract the overlapping bands in the difference spectra, the difference spectra for a nitrate content of 0.2 mg/L were utilized as the reference spectra, and the evaluation method was the residual sum of squares (RSS) of the other difference spectra and the reference spectra. When the RSS was less than a specific threshold, this indicated that the difference in the spectra in this range was quite small, and the spectral curves essentially overlapped. The method of calculation is as follows

$$\mathrm{RSS} = \sum_{i=1}^{m}\sum_{j=1}^{n}\left(y_{ij} - y_{i0}\right)^2 \quad (1)$$

where y_{ij} is the jth difference spectrum at the ith level of turbidity, y_{i0} is the reference spectrum at the ith level of turbidity, n is the number of difference spectra apart from the reference spectrum, and m is the number of different turbidity levels.

Figure 3 shows the result of the RSS at different wavelengths. The RSS was less than 0.003 at wavelengths greater than 230 nm. Therefore, when this band was selected for turbidity-compensation modeling, the change in the absorbance of the mixed solution was only related to the turbidity and not to the nitrate concentration.

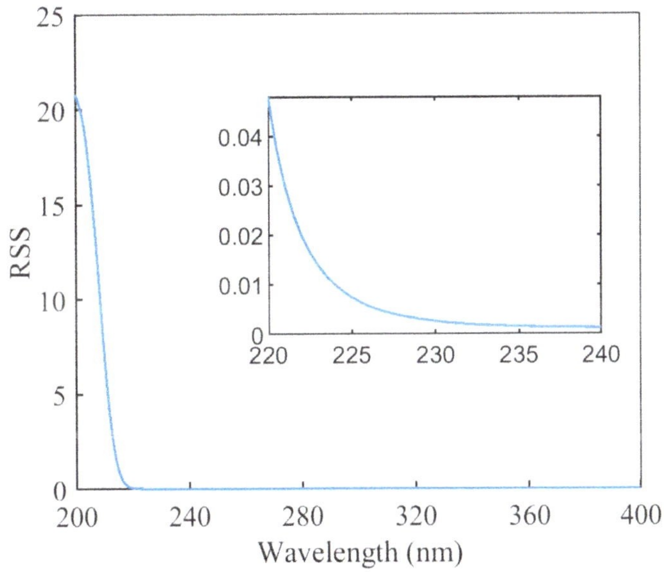

Figure 3. RSS at different wavelengths.

The wavelength range of 230–400 nm is still quite wide for establishing a nitrate-prediction model. This may include uninformative wavelengths, which reduce the accuracy of the prediction and increase the data-processing time.

The interval partial least squares (iPLS) method proposed by Norgaard et al. was used to optimize the wavelength range. The spectrum was divided into different intervals. The RMSECV for each interval was calculated, and the interval with the lowest RMSECV was chosen as the optimal wavelength interval. We divided the 230–400 nm wavelength band in the spectra data of the standard nitrate solutions into 17 intervals: 230–240, 240–250, . . . , 390–400 nm. The wavelength range with the lowest RMSECV was chosen by iPLS. For the range of 230–240 nm, the RMSECV value was the lowest, which was also lower than that of the model with the full spectral range. The result is shown in Figure 4 (which plots the first five intervals). Therefore, the modeling interval was finally determined to be 230–240 nm.

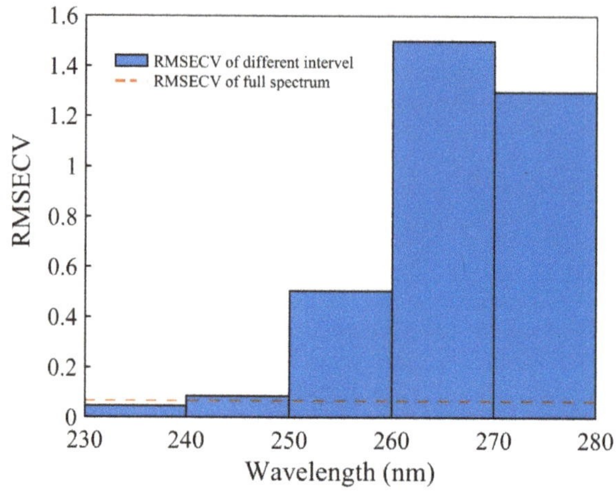

Figure 4. Interval optimization by interval-partial-least-squares method.

2.3. *Establishment of the Turbidity-Compensation Model*

The turbidity-prediction model was established by linear regression combined with the spectral area. According to the spectral characteristics of the nitrate and the turbidity, in the wavelength range of 250–400 nm, the absorbance of the nitrate solutions was almost zero and the absorption spectra of the mixed solutions were almost entirely attributable to the turbidity. In addition, the correlation coefficients of the absorbance and the turbidity at each wavelength in this band were greater than 0.9. The selection of this band can also eliminate the problem of the susceptibility of single-wavelength regression modeling to interference. The 15 spectral curves of the mixed solutions were divided into five groups, and those with the same level of turbidity were grouped together. Five spectral curves were obtained by averaging each group. We then integrated the five spectral curves between 250 and 400 nm and obtained their spectral areas. The integral values and turbidity values were used to build a turbidity-prediction model. The results of the modeling are shown in Figure 5. The linear-regression equation was $y = 60.51x - 2.708$, and the R^2 was 0.9989.

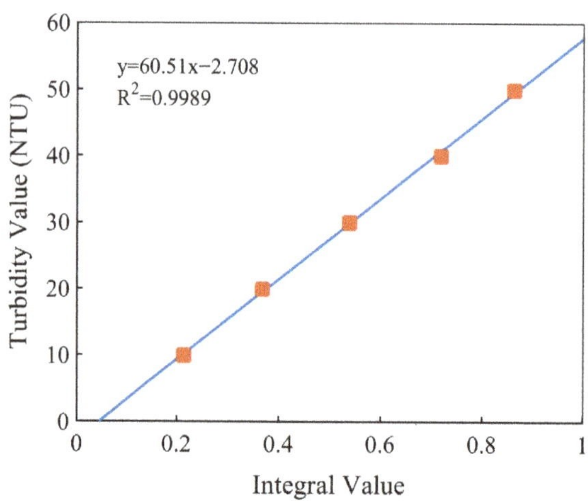

Figure 5. Fitting relationship between the integral values and the turbidity values.

The turbidity-compensation model was established by using the difference spectra at 230–240 nm. The mean values of the difference spectra with the same level of turbidity were calculated to obtain the reference compensation curve. The compensation curve of any turbidity value can be calculated by linear fitting, as described in Section 3.4. The regression parameters a_i (i = 1, 2, ... , λ) and b_i (i = 1, 2, ... , λ) for each wavelength were calculated as shown in Table 1. After calculating the turbidity of the unknown mixed solution using this turbidity-prediction model, we substituted the turbidity value into Formula (5) in Section 3.4 to obtain the turbidity-compensation curve. The spectra of nitrate can be extracted by subtracting the compensation curve from the mixed solution's spectra.

Table 1. Regression parameters in the wavelength range of 230–240 nm.

Parameter	The Values in the Wavelength Range of 230–240 nm (the Spectral Interval is 0.5 nm)
a_λ	(0.0127; 0.0125; 0.0124; 0.0123; 0.0122; 0.0121; 0.0120; 0.0119; 0.0118; 0.0117; 0.0116; 0.0115; 0.0114; 0.0114; 0.0113; 0.0112; 0.0111; 0.0111; 0.0110; 0.0109; 0.0109)
b_λ	(0.0333; 0.0327; 0.0322; 0.0317; 0.0311; 0.0305; 0.0303; 0.0298; 0.0293; 0.0289; 0.0285; 0.0280; 0.0277; 0.0275; 0.0276; 0.0275; 0.0272; 0.0268; 0.0267; 0.0265; 0.0263)

2.4. Establishment of the Nitrate-Prediction Model

The nitrate-prediction model was established by the PLS method. The nitrate samples were used as a calibration set. We used leave-one-out cross-validation to verify the model and RMSECV to evaluate the model's accuracy. The model showed good accuracy. The R^2 of the predicted values and the true values was 0.9996, and the RMSECV was 0.0462 mg/L. The plot of the predicted concentrations versus the actual concentrations is shown in Figure 6.

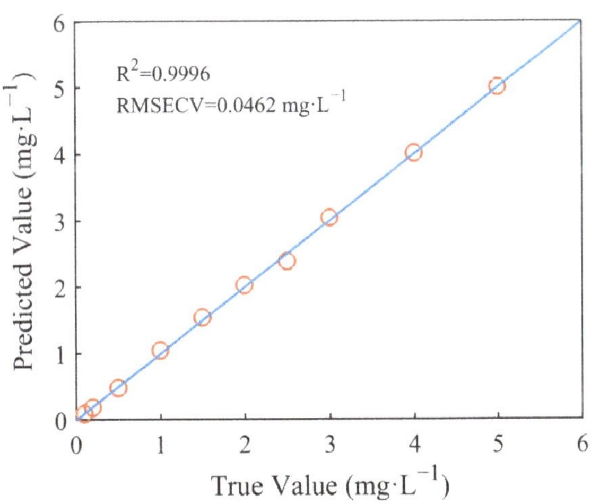

Figure 6. Relationship between the true values and the values of nitrate predicted by PLS.

2.5. Experiment with the Random Mixed-Solution Samples

To verify the turbidity-compensation method proposed, we prepared five mixed solutions with random concentrations of nitrate and levels of turbidity.

The absorption spectra of the five groups of mixed solutions were collected. The spectra of the mixed solutions were corrected by the turbidity-compensation model, after which the nitrate concentrations were calculated by substitution in the nitrate-prediction model. The results of the predictions before and after compensation are plotted in Figure 7, and the relative errors, RMSEP, and R^2 values before and after compensation are shown in Table 2.

Table 2. Prediction results of mixed samples before and after turbidity compensation.

Endpoint	Relative Error (100%)	RMSEP (mg L^{-1})	R^2
Uncompensated	50.33%	0.7845	0.7596
Compensated	1.33%	0.0359	0.9995

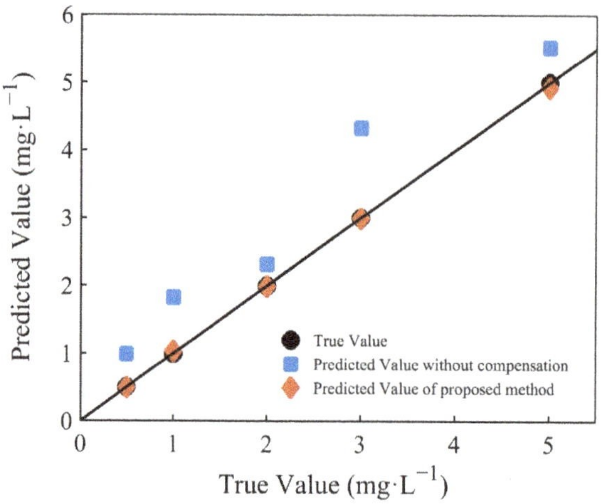

Figure 7. Comparison of the results of uncompensated and compensated predictions.

It can be seen that the results of the uncompensated prediction were quite different from the true values. As the turbidity of the mixed solution increased, the error in the prediction of the nitrate concentration became increasingly significant. After compensation, the predicted values were generally consistent with the true values. The average relative error decreased from 50.33% to 1.33%, and the RMSEP value was very small. The accuracy with which the nitrate was predicted was significantly improved after compensation for turbidity.

The use of an in situ nitrate sensor based on UV spectroscopy can help us to monitor environmental changes on a finer spatial and temporal scale, and it has wide prospects for application in freshwater systems and in the monitoring of sewage. The turbidity-compensation method proposed in this study can be used to improve the calibration and data-processing procedures of an in situ nitrate sensor to detect nitrate highly accurately in turbid water. Significantly, in this study, the water samples were prepared by using formazine turbidity particles, which generally conform to a normal distribution with a mean volume diameter of 2.5 μm and may not completely characterize the turbidity of some water bodies [22]. For example, in near-shore estuarine-water environments, in which larger particles are present and uniformity is poorer than in formazine turbidity solutions, the effect of compensation using this method may be affected. In addition, the presence of CDOM in water also affects the detection of nitrate due to its strong absorption in ultraviolet regions. Therefore, additional consideration should be given to eliminating the interference of CDOM to further improve the accuracy with which the concentration of nitrate in water is predicted.

3. Materials and Methods

3.1. Samples

The experiment was divided into two parts: research and verification of the method. The characteristics of the effect of turbidity on the absorption spectra of nitrate were investigated by single-solution samples of nitrate and turbidity and mixed-solution samples, and a turbidity-compensation model was established on the basis of the difference spectra of the nitrate-solution samples and the mixed-solution samples. The model was further validated with five sets of turbid mixed-solution samples of nitrate at arbitrary concentrations.

A number of samples with different nitrate solutions and levels of turbidity were created to investigate their UV-absorption characteristics at different wavelengths by measuring their absorption spectra, as follows. The data on the nitrate's absorption spectra were used as a modeling set for the model predicting the nitrate concentration. According to the standards for surface-water environmental quality in China (GB3838-2002), the nitrate-concentration range in the solution samples was set to 0.1–5 mg/L to meet the general requirements of detecting nitrate in surface water. The standard formazine solution is widely used worldwide for the determination of turbidity in water because of its good optical stability. Deionized water, a standard nitrate solution (analytical grade), and the standard formazine solution used for measuring turbidity (analytical grade) were used to prepare the water samples. The deionized water was supplied by a Milli-Q water-purification system (Millipore, Billerica, MA, USA). Nitrate solutions of 0.1, 0.2, 0.5, 1.0, 1.5, 2.0, 2.5, 3.0, 4.0, 5.0 mg/L were obtained by diluting standard nitrate solutions, and turbidity solutions of 1, 5, 10, 15, 20, 30, 40, 50 NTU were obtained by diluting standard solutions of formazine. The nitrate concentration was calculated as the concentration of nitrogen in the solution.

To investigate the effect of turbidity on nitrate's absorbance spectra and to develop a compensation model, 15 mixed-solution samples with different levels of nitrate and turbidity were prepared. In addition, to validate the compensation method, five different mixtures were produced with random concentrations of nitrate and levels of turbidity. The concentrations used in the samples are shown in Table 3.

Table 3. Composition of the mixed-solution samples with varying levels of nitrate and turbidity.

NO.	Nitrate ($mg \cdot L^{-1}$)	Turbidity (NTU)	NO.	Nitrate ($mg \cdot L^{-1}$)	Turbidity (NTU)
1	0.2	10	11	4	10
2	0.2	20	12	4	20
3	0.2	30	13	4	30
4	0.2	40	14	4	40
5	0.2	50	15	4	50
	/			Test Samples	
6	2	10	1	0.5	13
7	2	20	2	1	22
8	2	30	3	2	8
9	2	40	4	3	41
10	2	50	5	5	16

3.2. Measurements

A UV-vis spectrophotometer (UV-8000s, LASPEC, Shanghai, China) was used to measure the absorption spectra of different solutions from 200 to 400 nm. All the samples were measured in a 1-cm quartz cuvette. Deionized water was used as the reference. The scanning speed was set to medium and the spectral resolution was set to 0.5 nm.

3.3. Spectral-Subtraction Method

The effect of turbidity on the nitrate's absorption spectra was investigated using a difference spectra obtained by the spectral-subtraction method. An example of a difference spectrum is shown in Figure 8. We acquired the spectrum of mixed nitrate and turbidity solutions, as well as their individual spectra. By subtracting the nitrate spectrum from the spectrum of the mixed solution, the difference spectrum was obtained. The difference spectrum reflects the effect of turbidity on the absorption of nitrate. As can be seen in Figure 1, the appearance of turbidity caused an increase in the total absorbance, but this increase was not equal to the absorbance contributed by the turbidity. This phenomenon is related to molecular interactions in the solution. These are explained in more detail in the Section 2.

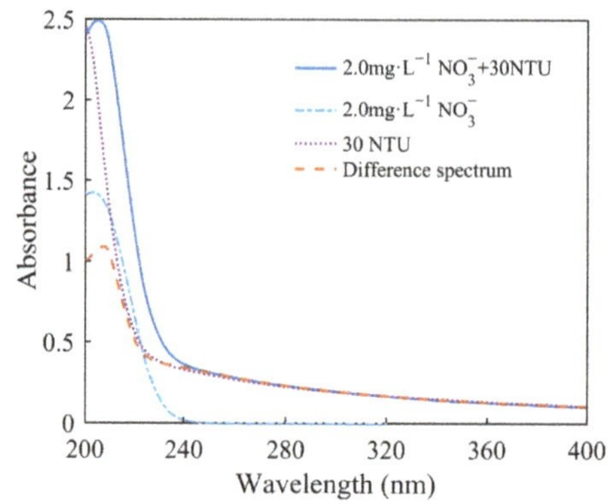

Figure 8. Spectra of the mixed solution, the nitrate solution, the turbidity solution, and their difference spectrum.

3.4. Turbidity-Compensation Method

According to the correlation coefficient between the difference spectra and the turbidity, it can be seen that an increase in the absorbance in a specific wavelength range is linearly correlated with the increase in turbidity. The correlation coefficient is calculated in Section 2.1. The change in the absorbance of nitrate as a result of a change in turbidity can be calculated by linear fitting of difference spectra of various levels of turbidity. The spectra of nitrate can then be extracted by subtracting the effect of turbidity from the mixed spectra. The specific steps of the turbidity-compensation method are described below.

Step 1: The absorption spectra of mixed solutions ($A_{mixture}$) and nitrate solutions ($A_{nitrate}$) are measured. The turbidity-prediction model is built through linear regression of the spectral integral at a selected wavelength range and the turbidity value of mixed solutions.

Step 2: The difference spectra of mixed solutions and nitrate solutions are obtained by spectral-subtraction method, which are known as the reference compensation curves (A_{ref}). The turbidity values of the difference spectra are known as the reference turbidity values (Tur_{ref}). These are calculated as shown in Equations (2) and (3), where n is the number of difference spectra and λ is the wavelength.

$$A_{ref} = A_{mixture} - A_{nitrate} = \begin{bmatrix} A_1^1 & \cdots & A_\lambda^1 \\ \vdots & \ddots & \vdots \\ A_1^n & \cdots & A_\lambda^n \end{bmatrix} \quad (2)$$

$$Tur_{ref} = \begin{bmatrix} T^1 \\ \vdots \\ T^n \end{bmatrix} \quad (3)$$

Step 3: The regression parameters a_i (i = 1, 2, ..., λ) and b_i (i = 1, 2, ..., λ) are determined by linear regression for each wavelength using A_{ref} and Tur_{ref}, as shown in Equation (4).

$$A_{ref}' = \begin{bmatrix} a_1 \\ \vdots \\ a_\lambda \end{bmatrix} \cdot Tur_{ref}' + \begin{bmatrix} b_1 \\ \vdots \\ b_\lambda \end{bmatrix} \quad (4)$$

Step 4: The absorbance effect of any level of turbidity (A_{tur}) can be calculated as shown in Equation (5), where Tur_{any} is the turbidity value of an unknown mixed solution. The turbidity value is calculated by the turbidity-prediction model.

$$A_{tur} = \begin{bmatrix} a_1 \\ \vdots \\ a_\lambda \end{bmatrix} \cdot Tur_{any} + \begin{bmatrix} b_1 \\ \vdots \\ b_\lambda \end{bmatrix} \tag{5}$$

Step 5: The spectrum of any mixed nitrate and turbidity solution after compensation ($A_{residual}$) is achieved by subtracting A_{tur} from the raw spectrum A_{any}, as shown in Equation (6).

$$A_{residual} = A_{any} - A_{tur} \tag{6}$$

3.5. Nitrate-Prediction Model

The nitrate-prediction model was constructed by using partial least squares (PLS), a multivariate calibration method. The principle of PLS is to extract the principal components of independent variables and then establish the regression of the dependent variables on these until accurate predictions can be achieved [23]. To improve the model's accuracy, the modeling-wavelength range was optimized through interval partial least squares (iPLS), which can help us focus on important spectral regions and eliminate interference from other regions [24]. In this method, the measured spectra were evenly divided into different subintervals. All samples were used for PLS modeling in each subinterval, after which the interval with minimum root mean square error of cross-validation (RMSECV) was selected as the optimal wavelength interval.

3.6. Process of Calculating the Nitrate Concentration

The process of calculating the nitrate concentration is shown in Figure 9. The turbidity-prediction model, turbidity-compensation model, and nitrate-prediction model were built in advance using the abovementioned methods. Firstly, the absorption spectrum of a mixed solution with an unknown concentration was measured. Secondly, the turbidity of the mixed solution was calculated by using the turbidity-prediction model. Thirdly, the turbidity-compensation model was applied to calculate the turbidity-compensation curve for any value of turbidity. The spectrum after turbidity compensation was obtained by subtracting the turbidity-compensation curve from the raw spectrum. Finally, the absorption spectrum after turbidity compensation was inputted into the nitrate prediction model, and the nitrate concentration was obtained.

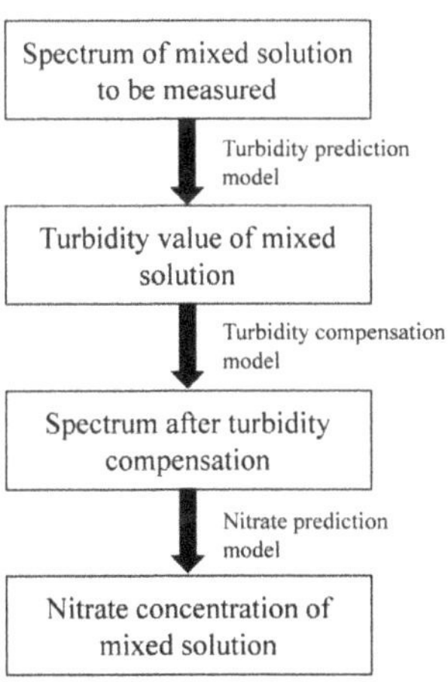

Figure 9. The process for calculating the nitrate concentration.

3.7. Model Validation

The whole model was evaluated according to the root mean square error of prediction (RMSEP), the correlation coefficient (R^2), and the relative error (RE). These are defined as follows (Equations (7)–(9))

$$\mathrm{RMSEP} = \sqrt{\frac{\sum_{i=1}^{m}(\hat{c}_i - c_i)^2}{m}} \tag{7}$$

$$R^2 = 1 - \frac{\sum_{i=1}^{m}(\hat{c}_i - c_i)^2}{\sum_{i=1}^{m}(c_i - \bar{c})^2} \tag{8}$$

$$\mathrm{RE} = \frac{\hat{c}_i - c_i}{c_i} \times 100\% \tag{9}$$

where $\hat{c}_i$ is the predicted value in the ith sample, c_i is the true value in the ith sample, $\bar{c}$ is the mean of the predicted value of all samples, and m is the number of prediction samples.

Leave-one-out cross-validation was used to validate the performance of the PLS models in predicting new data [25]. A total of N-1 elements out of n elements in the whole set were used as the training sets to build models for this analysis, and the remaining element acted as a test set to validate the model. After several rounds of cross-validation, the average validation results were obtained and used to estimate the overall prediction performance of the model. The root mean square error of cross-validation (RMSECV) was the mean value of the RMSEP obtained from the results of several rounds of cross-validation.

4. Conclusions

A method of turbidity compensation based on difference spectra was proposed for measuring nitrate. In this study, the two main aims were to (i) study the effect of turbidity on the absorption spectra of nitrate and (ii) to establish a process of compensating for turbidity, including the establishment of a turbidity-prediction model, a turbidity-compensation model, and a nitrate-prediction model. The effectiveness of the compensation method was further verified by experiments. The results showed that the turbidity-compensation method performed well, making the relative error of the predicted nitrate concentration decrease from 50.33% to 1.33%. The following conclusions can be drawn.

(i) The influence of turbidity on nitrate varies with wavelength. The change in absorbance caused by turbidity decreased with the increase in the nitrate concentration at wavelengths of 200–230 nm, and the same level of turbidity had the same effect on the absorbance of nitrate after 230 nm.

(ii) The nitrate-prediction model established in the optimal modeling interval selected by the data analysis of the difference spectra combined with the iPLS algorithm had good predictive accuracy. Meanwhile, arbitrary concentrations of nitrate and levels of turbidity in the water were suitable for the model because of the chosen wavelength.

(iii) The turbidity-compensation method based on the ultraviolet difference spectra can eliminate the interference of turbidity and improve the accuracy with which nitrate is detected in turbid water.

(iv) Although the proposed turbidity-compensation method was effective, there were still limitations in this study, such as the fact that CDOM was not considered when the method was applied to natural water. Further investigations will focus on fully understanding the effects of CDOM and studying the related compensation algorithms, which may improve the accuracy with which nitrate is predicted. Finally, experimental validation will be performed on natural-water samples.

Author Contributions: Conceptualization, J.D. and R.L.; methodology, J.D.; software, R.L.; validation, J.T.; writing—original draft preparation, J.D.; writing—review and editing, J.T. and G.W.; visualization, J.D.; supervision, G.W.; funding acquisition, G.W. All authors have read and agreed to the published version of the manuscript.

Funding: This research was funded by the National Key Research and Development Program of China, grant number 2022YFC3103900.

Institutional Review Board Statement: Not applicable.

Informed Consent Statement: Not applicable.

Data Availability Statement: The data presented in this study are available on request from the corresponding author.

Conflicts of Interest: The authors declare no conflict of interest.

References

1. Pajares, S.; Ramos, R. Processes and microorganisms involved in the marine nitrogen cycle: Knowledge and gaps. *Front. Mar. Sci.* **2019**, *6*, 739. [CrossRef]
2. Tyrrell, T. The relative influences of nitrogen and phosphorus on oceanic primary production. *Nature* **1999**, *400*, 525–531. [CrossRef]
3. Becker, S.; Aoyama, M.; Woodward, E.M.S.; Bakker, K.; Coverly, S.; Mahaffey, C.; Tanhua, T. GO-SHIP repeat hydrography nutrient manual: The precise and accurate determination of dissolved inorganic nutrients in seawater, using continuous flow analysis methods. *Front. Mar. Sci.* **2020**, *7*, 908. [CrossRef]
4. Wang, S.; Lin, K.N.; Chen, N.W.; Yuan, D.X.; Ma, J. Automated determination of nitrate plus nitrite in aqueous samples with flow injection analysis using vanadium (III) chloride as reductant. *Talanta* **2016**, *146*, 744–748. [CrossRef]
5. Nightingale, A.M.; Hassan, S.U.; Warren, B.M.; Makris, K.; Evans, G.W.; Papadopoulou, E.; Coleman, S.; Niu, X.Z. A droplet microfluidic-based sensor for simultaneous in situ monitoring of nitrate and nitrite in natural waters. *Environ. Sci. Technol.* **2019**, *53*, 9677–9685. [CrossRef]
6. Daniel, A.; Laës-Huon, A.; Barus, C.; Beaton, A.D.; Blandfort, D.; Guigues, N.; Knockaert, M.; Munaron, D.; Salter, I.; Woodward, E.M.S.; et al. Toward a harmonization for using in situ nutrient sensors in the marine environment. *Front. Mar. Sci.* **2020**, *6*, 773. [CrossRef]
7. Singh, P.; Singh, M.K.; Beg, Y.R.; Nishad, G.R. A review on spectroscopic methods for determination of nitrite and nitrate in environmental samples. *Talanta* **2019**, *191*, 364–381. [CrossRef]
8. Wang, H.; Ju, A.B.; Wang, L.Q. Ultraviolet spectroscopic detection of nitrate and nitrite in seawater simultaneously based on partial least squares. *Molecules* **2021**, *26*, 3685. [CrossRef]
9. Chen, X.W.; Yin, G.F.; Zhao, N.J.; Gan, T.T.; Yang, R.F.; Xia, M.; Feng, C.; Chen, Y.N.; Huang, Y. Simultaneous determination of nitrate, chemical oxygen demand and turbidity in water based on UV–Vis absorption spectrometry combined with interval analysis. *Spectrochim. Acta Part A* **2021**, *244*, 118827. [CrossRef]
10. Johnson, K.S.; Coletti, L.J. In situ ultraviolet spectrophotometry for high resolution and long-term monitoring of nitrate, bromide and bisulfide in the ocean. *Deep Sea Res. Part I* **2002**, *49*, 1291–1305. [CrossRef]
11. Langergraber, G.; Fleischmann, N.; Hofstaedter, F. A multivariate calibration procedure for UV/VIS spectrometric quantification of organic matter and nitrate in wastewater. *Water Sci. Technol.* **2003**, *47*, 63–71. [CrossRef]
12. Zielinski, O.; Voß, D.; Saworski, B.; Fiedler, B.; Körtzinger, A. Computation of nitrate concentrations in turbid coastal waters using an in situ ultraviolet spectrophotometer. *J. Sea Res.* **2011**, *65*, 456–460. [CrossRef]
13. Hu, Y.T.; Zhao, D.D.; Qin, Y.L.; Wang, X.P. An order determination method in direct derivative absorption spectroscopy for correction of turbidity effects on COD measurements without baseline required. *Spectrochim. Acta Part A* **2020**, *226*, 117646. [CrossRef]
14. Wang, C.X.; Li, W.X.; Huang, M.Z. High precision wide range online chemical oxygen demand measurement method based on ultraviolet absorption spectroscopy and full-spectrum data analysis. *Sens. Actuators B* **2019**, *300*, 126943. [CrossRef]
15. Zhang, H.M.; Zhou, X.; Tao, Z.; Lv, T.T.; Wang, J. Deep learning–based turbidity compensation for ultraviolet-visible spectrum correction in monitoring water parameters. *Front. Environ. Sci.* **2022**, *10*, 986913. [CrossRef]
16. Snazelle, T.T. *The Effect of Suspended Sediment and Color on Ultraviolet Spectrophotometric Nitrate Sensors*; US Geological Survey: Reston, VA, USA, 2016.
17. Bleyen, N.; Albrecht, A.; De Cannière, P.; Wittebroodt, C.; Valcke, E. Non-destructive on-line and long-term monitoring of in situ nitrate and nitrite reactivity in a clay environment at increasing turbidity. *Appl. Geochem.* **2019**, *100*, 131–142. [CrossRef]
18. Chen, X.W.; Yin, G.F.; Zhao, N.J.; Yang, R.F.; Xia, M.; Feng, C.; Chen, Y.N.; Dong, M.; Zhu, W. Turbidity compensation method based on mie scattering theory for water chemical oxygen demand determination by uv-vis spectrometry. *Anal. Bioanal. Chem.* **2021**, *413*, 877–883. [CrossRef]

19. Chen, Y.; He, L.; Cui, X.N.; Xiao, C.Y.; Zhang, J.; Zhang, C.; Yang, H.; Zhou, X.D.; Li, S.H. Study on Turbidity Compensation of Nitrate Nitrogen in Water Based on Ultraviolet Spectrum. *Spectrosc. Spectr. Anal.* **2020**, *40*, 472–477.
20. Li, F.X.; Tang, B.; Zhao, M.F.; Hu, X.Y.; Shi, S.G.; Zhou, M. Research on Correction Method of Water Quality Ultraviolet-Visible Spectrum Data Based on Compressed Sensing. *J. Spectrosc.* **2021**, *2021*, 6650630. [CrossRef]
21. Cai, J.C.; Deng, X.; Kan, C.X.; Zhao, S. Design of Water Quality COD Online Detection System Based on Ultraviolet-Visible Spectroscopy. *Laser Optoelectron.* **2022**, *59*, 347–354.
22. Ebie, K.; Yamaguchi, D.; Hoshikawa, H.; Shirozu, T. New measurement principle and basic performance of high-sensitivity turbidimeter with two optical systems in series. *Water Res.* **2006**, *40*, 683–691. [CrossRef] [PubMed]
23. Khatri, P.; Gupta, K.K.; Gupta, R.K. A review of partial least squares modeling (PLSM) for water quality analysis. *Model. Earth Syst. Environ.* **2021**, *7*, 703–714. [CrossRef]
24. Nørgaard, L.; Saudland, A.; Wagner, J.; Nielsen, J.P.; Munck, L.; Engelsen, S.B. Interval partial least-squares regression (iPLS): A comparative chemometric study with an example from near-infrared spectroscopy. *Appl. Spectrosc.* **2000**, *54*, 413–419. [CrossRef]
25. Mevik, B.H.; Cederkvist, H.R. Mean squared error of prediction (MSEP) estimates for principal component regression (PCR) and partial least squares regression (PLSR). *J. Chemom.* **2004**, *18*, 422–429. [CrossRef]

molecules

MDPI

Article

Simultaneous Dispersive Liquid–Liquid Microextraction and Determination of Different Polycyclic Aromatic Hydrocarbons in Surface Water

Zaual Temerdashev [1,*], Surendra Prasad [2], Tatiana Musorina [1], Tatiana Chervonnaya [1] and Zhanna Arutyunyan [1]

1 Department of Analytical Chemistry, Faculty of Chemistry and High Technologies, Kuban State University, Stavropolskaya St. 149, 350040 Krasnodar, Russia
2 School of Agriculture, Geography, Environment, Ocean and Natural Sciences (SAGEONS), Discipline of Biological and Chemical Sciences, The University of the South Pacific, Suva, Fiji
* Correspondence: temza@kubsu.ru

Abstract: Polycyclic aromatic hydrocarbons (PAHs) are a class of persistent organic pollutants of water, and their determination at trace levels in the aquatic ecosystems is essential. In this work, an ultrasound-assisted dispersive liquid–liquid microextraction (DLLME) procedure was suggested utilizing a binary dispersive agent for recovery of different molecular weight polycyclic aromatic hydrocarbons (PAHs) from waters. The detection was carried out by gas chromatography–mass spectrometry (GC-MS) as well as high-performance liquid chromatography with fluorescence and diode-array detection (HPLC-FD/PDA). The method was optimized for the extraction of analytes with respect to the mixture composition, ratios of components, ultrasonication time and centrifugation parameters. The analytical schemes for PAHs extraction from water samples using different ratios of extraction and dispersive solvents are reported. The mixture consisting of chloroform and methanol was applied for the extraction of PAHs containing two or three fused aromatic rings; the mixture of chloroform and acetonitrile is suitable for PAHs containing more than four aromatic rings. The mixture of chloroform:acetone + acetonitrile was applied in the universal scheme and allowed for the simultaneous extraction of 20 PAHs with different structures. The developed sample preparation schemes were combined with GC-MS and HPLC-FD/PDA, which allowed us to determine the analytes at low concentrations (from 0.0002 μg/L) with the recoveries exceeding 80% and relative standard deviations of about 8%. The developed methods for the determination of 20 PAHs were applied to the analysis of water samples from the Karasun Lake (Krasnodar), Azov Sea (Temryuk) and Black Sea (Sochi).

Keywords: dispersive liquid–liquid microextraction; polycyclic aromatic hydrocarbons; priority pollutants; water samples

Citation: Temerdashev, Z.; Prasad, S.; Musorina, T.; Chervonnaya, T.; Arutyunyan, Z. Simultaneous Dispersive Liquid–Liquid Microextraction and Determination of Different Polycyclic Aromatic Hydrocarbons in Surface Water. *Molecules* 2022, 27, 8586. https://doi.org/10.3390/molecules27238586

Academic Editors: Victoria Samanidou, Eleni Deliyanni and Dimitra Voutsa

Received: 1 November 2022
Accepted: 2 December 2022
Published: 5 December 2022

1. Introduction

Polycyclic aromatic hydrocarbons (PAHs) are related to eco-toxicants due to their mutagenic and carcinogenic nature [1]. Therefore, the United States Environmental Protection Agency (US-EPA) has recommended the routine monitoring of 16 PAHs in environmental samples. Dibenz[a,h]anthracene, benz[e]pyrene, benz[a]anthracene, benz[b]fluoranthene, benz[k]fluoranthene and inden[1,2,3-c,d]pyrene are the most toxic PAHs. However, less dangerous PAHs can have a synergetic effect when present in a complex mixture. Toxic equivalency factors (TEF) in combination with the maximum permissible concentration (MPC) of individual PAHs in water express the toxicity of complex mixtures in the USA and European Union (EU). TEF for different priority PAHs is expressed as benz[a]pyrene equivalents, and the most toxic substances have the TEF value of 1 [2].

Besides PAHs' differentiation based on toxicity, it is essential to consider their distribution in the water bodies [3,4]. PAHs can be conditionally categorized into three groups based on their water solubility (S) and vapor pressure (P) [5]. The first group corresponds to PAHs containing two to three fused aromatic rings with P of 10^{-2}–10^{-3} Pa and water solubility from 2 to 35 mg/L. The second group includes PAHs containing three or four aromatic rings with P of 10^{-2}–10^{-3} Pa and solubility between 0.1 and 2 mg/L. The third group consists of PAHs with more than four aromatic rings, vapor pressure less than 10^{-5} Pa and solubility below 0.01 mg/L (Table 1). Thus, it can be assumed that lighter PAHs, such as naphthalene, acenaphthene, fluorene, etc., will be present in the water bodies in dissolved form [6]. On the other hand, the high molecular weight toxic PAHs containing more than four aromatic rings can aggregate on the surface of different particles and precipitate into the sediments [7]. Consequently, it is necessary to consider the transformation and distribution of PAHs in waters [8,9]. With the increase in the molecular weight, the octanol/water partition coefficient (Kow) changes from 3.2 (light PAHs) to 7.1 (heavy PAHs), which shows the tendency of PAHs to accumulate in the lipophilic phase. Naphthalene usually evaporates from the surface of water bodies during 50–200 h depending on natural factors. The evaporation of biphenyl from the surface can take from 7.5 days for surface water to 14 days for groundwater, while the evaporation of acenaphthylene varies with seasonal and climatic factors and can range from 42 to 120 days. In addition, since the degradation of PAHs in water depends on several factors, such as light intensity, time, climatic conditions and microorganisms, the correlation between degradation speed and the number of aromatic rings of PAHs has not been established. It should be considered that this parameter can vary for almost all the components in a wide range from several hours to months and in some cases can even reach years. Thus, fast and precise methods are required for the determination and ecological monitoring of different molecular weight PAHs in water samples within a wide linear range including trace levels and allowing one to account for the degradation of these substances.

Table 1. Some physicochemical properties of polycyclic aromatic hydrocarbons.

		PAHs	Mr	S, mg/L	P, Pa	K_{ow}	Half-Life in the Environment, h
Light	2 to 3 rings	Naphthalene	128	31–34	37–42	3.2–3.8	16–6193
		2-Methylnaphthalene	142	20.0–27.3	6.3–10.7	3.9–4.1	54–9840
		Biphenyl	154	7.0–7.8	1.3–6.9	3.2–4.3	36–336
		Acenaphthylene	152	3.4–16.1	0.89–1.1	3.7–4.1	1020–1440
		Acenaphthene	154	3.9–3.8	0.21–3.1	3.9–4.5	3–4896
	3 to 4 rings	Fluorene	166	1.6–1.8	0.08–0.79	3.7–4.3	768–2880
		Phenanthrene	178	1.0–1.2	0.02–0.11	4.5–4.6	3–9600
		Anthracene	178	0.04–0.08	5.7×10^{-4}–0.1	4.2–5.3	1–22,080
		Pyrene	202	0.1–0.2	1.7×10^{-4}–0.2	4.8–5.5	1–91,200
		Fluoranthene	202	0.21	1.3×10^{-4}–0.1	4.5–5.2	21–21,120
Heavy	more than 4 rings	Benz[a]anthracene	228	0.0090–0.0094	3.9×10^{-7}–2.5×10^{-4}	5.0–5.9	0.5–32,640
		Chrysene	228	0.0015–0.0170	8.4×10^{-7}–2.3×10^{-4}	5.5–5.9	0.5–48,000
		Triphenylene	228	0.040	3.9×10^{-7}–1.2×10^{-2}	4.8–6.3	–
		Benz[b]fluoranthene	252	0.0011–0.0015	5.0×10^{-8}–6.7×10^{-5}	5.8	9–29,280
		Benz[k]fluoranthene	252	0.0008–0.0011	1.3×10^{-8}–6.7×10^{-5}	5.9–7.2	4–102,720
		Benz[a]pyrene	252	0.0016	7.5×10^{-7}–1.1×10^{-4}	6.0–8.0	0.4–25,440
		Benz[e]pyrene	252	0.0001–0.0073	7.4×10^{-7}–1.8×10^{-5}	5.7–7.4	–
		Inden[1,2,3,-c,d]pyrene	278	0.0002–0.0004	1.3×10^{-8}–1.3×10^{-7}	6.7–8.2	–
		Dibenz[a,h]anthracene	276	0.0006–0.0025	3.7×10^{-10}–2.5×10^{-7}	5.8–7.1	6–45,120
		Benz[g,h,i]perylene	276	0.0001–0.0008	1.3×10^{-8}–6.7×10^{-7}	6.2–7.1	14,160–31,200

Mr—Molecular weight. S—Water solubility at 25 °C. P—Vapor pressure. K_{ow}—Octanol/water partition coefficient.

As a rule, chromatographic methods, such as gas chromatography–mass spectrometry (GC-MS) and high-performance liquid chromatography (HPLC) with a fluorescence detector (FD), or their combination in complicated cases, are used to quantify PAHs in water [10–12]. However, the sensitivity of GC-MS and HPLC-FD can be insufficient for the determination of trace PAH levels in complex matrices without proper sample pretreatment. To solve this problem, it is necessary to apply an effective sample preparation technique providing high recoveries and concentration factors of the analytes [11,13,14].

There exist numerous solid-phase extraction methods for the recovery and concentration of organic pollutants from environmental samples [15,16]. However, liquid–liquid extraction (LLE) methods are more widely used for these purposes due to their availability and simplicity. Combining these methods with ultrasound treatment accelerates mass transfer and consequently increases effectiveness of the processes [17]. A procedure consisting of ultrasound-assisted LLE with *n*-hexane and GC-MS detection has been suggested by us for the extraction and quantification of PAHs from different types of water samples [18]. The procedure provided high recoveries (90%) of PAHs; however, this method is time- and labor-consuming and has high organic solvent consumption.

Dispersive liquid–liquid microextraction (DLLME) seems to be a perspective technique [19–21] free from the above-mentioned disadvantages. In this technique, the chlorinated solvents are used as extractants, while acetone, acetonitrile, methanol, etc., are used as dispersive agents. Cloudy solution formation provides a high surface area between the extraction mixture and sample, which results in quick and effective extraction of PAHs from water samples [22–24]. The separation of extractant is usually performed by centrifugation [22]. In addition, shaking and ultrasound-assisted extraction techniques were suggested to improve the extraction efficiency [24,25].

In recent years, environmentally friendly solvents have also become attractive for the development of "green" chemistry. In this way, surfactants, low-toxic brominated and other derivates of hydrocarbons [26], deep eutectic solvents (DES) [27,28], ionic liquids (ILs) [29], etc., have been utilized in the developed methods to decrease toxicity of systems. On the other hand, the large-scale use of ionic liquids results in significant material costs and difficulties in understanding their mechanism of action due to the lack of studies. High viscosity of ILs can lead to a decrease in the mass transfer efficiency [30], while the hygroscopicity and absorption of moisture from atmospheric air limit the application of ILs and DES for the extraction of analytes from water samples and make these solvents poorly compatible with chromatographic systems [31]. It should also be noticed that ionic liquids are non-biodegradable [32], which complicates their application for "green" chemistry purposes. The use of deep eutectic solvents in DLLME is complicated by their hydrophilicity. In this case, sample pretreatment is more cumbersome, since it is necessary to apply a multi-step extraction procedure of analytes or properly choose hydrophobic components [28,33]. It can be assumed that a wide application of ionic liquids and deep eutectic solvents in DLLME with chromatographic detection is quite problematic due to insufficient study of their properties. To avoid these shortcomings, it is necessary to conduct a supplementary investigation and develop a conventional DLLME procedure for efficient extraction of PAHs from water samples. The combination of chlorinated solvents and conventional dispersive agents, such as acetone and acetonitrile, provides high extraction efficiency of individual PAHs, e.g., 96% for naphthalene group PAHs, 98% for PAHs with three and four aromatic rings and 95% for PAHs containing more than four aromatic rings [34–38]. However, the behavior of co-existing PAHs with diverse molecular weights in the extraction procedures with conventional solvents was not discussed in these works. The difference in volatility and solubility of particular PAH groups can influence their behavior in the extraction systems. It is essential to consider the problem of the simultaneous microextraction of individual PAHs with significantly different solubilities and vapor pressures from waters. Moreover, no data are available on the investigation of PAHs behavior in the DLLME extraction mixtures and applying a binary solvent mixture as a dispersive agent for these purposes.

In this work, the peculiarities of the dispersive liquid–liquid microextraction and concentration of different molecular weight PAHs with individual conventional solvents (acetone, methanol, acetonitrile) and their binary mixtures as dispersive agents followed by GC-MS and HPLC-FD/PDA detection were investigated.

2. Experiment

2.1. Reagent and Standards

In total, 17 polycyclic aromatic hydrocarbons (PAHs)—naphthalene (Naph), 2-methylnaphthalene (2-MN), biphenyl (Biph), acenaphthylene (Acy), acenaphthene (Ace), fluorene (Flu), phenanthrene (Phe), anthracene (Anth), fluoranthene (Fluor), pyrene (Pyr), benz[b]fluoranthene (B[b]F), benz[k]fluoranthene (B[k]F), benz[a]anthracene (B[a]A), chrysene (Chry), benz[a]pyrene (B[a]P), dibenz[a,h]anthracene (D[a,h]A) and benz[g,h,i]perylene (B[g,h,i]P)—in acetonitrile with a concentration of 100 or 200 μg/mL of each PAH were purchased from Ecros (St. Petersburg, Russia). Triphenylene (Triph) (analytical standard, 98.8% purity), benz[e]pyrene (B[e]P) and inden[1,2,3,-c,d]pyrene (I [1,2,3,-c,d]P) in cyclohexane were obtained from Sigma-Aldrich, USA (100 μg/mL). Methanol (gradient grade, ≥99.8%) was purchased from Avantor Performance Materials Poland S.A. (Poland). Acetonitrile (gradient grade, ≥99.9%), dichloromethane (ACS reagent, ≥99.5%), tetrachloroethane (ACS reagent, ≥99.0%), carbon tetrachloride (≥99.5%) and acetone (suitable for HPLC, ≥99.8%) were purchased from Sigma-Aldrich, USA, and Merck, Germany. Chloroform (ACS reagent, ≥99.8%) was obtained from Acros Organic, Belgium. Ultrapure water was obtained from a Milli-Q system (Millipore, Bedford, MA, USA).

2.2. Samples

Real surface water samples with different salinity and matrices were used to assess the applicability of the developed method for the analysis of real samples. Lake water was collected from the local Lake Karasun (Krasnodar, Russia). Sea water samples were obtained from the Azov Sea (Temruk, Russia) and the Black Sea (Sochi, Russia). The blank water was used for the model samples. All water samples were stored in brown glass bottles and kept at 4 °C prior to analysis.

2.3. Instruments

In this work, two chromatographic systems were used for the determination of PAHs: a gas chromatograph with a split/splitless injector coupled to a quadrupole mass spectrometer (Shimadzu GCMS-QP2020, Shimadzu, Japan) and a Shimadzu LC-30 Nexera high-performance liquid chromatograph (Shimadzu, Japan) with an SPD-M30A photodiode array detector (PDA) and an RF-20A/20Axs fluorescence detector (FD) equipped with an autosampler. A Liston C 2201 centrifuge (Russia) was used for phase separation. The compounds were identified using Wiley8 and NIST-17.1 mass spectral libraries as well as the retention times of the individual PAHs standards.

2.4. GC-MS Analysis

The injection port temperature was 250 °C, and a split ratio of 1:10 was applied. The oven temperature program was as follows: the initial temperature was 60 °C, then it increased from 60 °C (held for 1 min) to 170 °C at the rate of 15 °C/min, from 170 °C (held for 3 min) to 280 °C at the rate of 10 °C/min, from 280 °C(held for 8 min) to 290 °C at the rate of 10 °C/min and held constant at 290 °C for 25 min. The ion source temperature was maintained at 250 °C. Ultra-pure helium (99.995%) was used as a carrier gas at a constant linear velocity of 30 cm/s. Selected ion monitoring (SIM) mode was used to achieve higher sensitivity and minimize the influence of matrix. Separation of 20 PAHs including difficult to separate isomers, such as phenanthrene/anthracene, chrysene/triphenylene/benz[a]anthracene, benz[b]fluoranthene/benz[k]fluoranthene and benz[a]pyrene/benz[e]pyrene, was carried out on a 5 ms Zebron capillary column (60 m × 0.25 mm, 0.25 μm). The integration of target analyte peaks on the chromotogram was performed by the GCMSsolution software Version 4.45.

2.5. HPLC-FD/PDA Analysis

To separate the PAHs, a Kinetex 3.5 μm PAH column (150 × 4.5 mm) was used. Deionized water and acetonitrile were used as eluents in the gradient elution mode as follows: the initial eluent composition was 50% acetonitrile for 3 min; 3–10 min: a linear ramp to 100% acetonitrile with a plateau at 100% acetonitrile for 8 min; 18–18.5 min: a decrease to 50% acetonitrile; 18.5–20 min: a plateau at 50% acetonitrile for 1.5 min; the flow rate was 1 mL/min. Since acenaphthylene does not fluoresce, PDA was used for its detection at the wavelength of 254 nm. Detection was performed by programming the excitation and emission wavelengths to obtain better sensitivity and minimize interferences. The excitation/emission wavelength pairs (nm) were as follows: 0.01–7.50 min: 280/325 nm; 7.50–8.90 min: 265/380 nm; 8.90–9.70 min: 290/420 nm; 9.70–20 min: 300/500 nm. The column was thermostated at 35 °C. The total HPLC run time was 20 min. The software LabSolutions Version 5.73 was used for the integration of the target analyte peaks on the chromatogram.

2.6. Extraction Procedure

Based on the literature data [36,37] and results obtained in our laboratory, the extractant for DLLME was selected. Chloroform, dichloromethane, dichloroethane and carbon tetrachloride were tested as extraction solvents. Acetonitrile was used as a dispersive solvent in the initial experiments, because it provided stable and reproducible extraction systems. To evaluate the effect of extraction solvent type, a series of sample solutions were tested using 1 mL of acetonitrile and 50, 100, 150, 200 and 500 μL of chlorinated solvents for water sample volumes of 10 mL. Experiments were carried out by spiking model water samples with PAH standard solutions at different concentration levels, i.e., low (0.0002 μg/L), medium (0.1 μg/L) and high (1 μg/L). The experiments were carried out in three replicates. Based on the obtained results, three procedures for the extraction of various PAHs from water have been developed.

Extraction of light PAHs. In a glass centrifuge tube, 10 mL of water was added, and the extraction mixture consisting of 150 μL of chloroform and 400 μL of methanol was quickly injected by using a 2 mL syringe. Then, the tube was shaken for 1 min and ultrasonicated for 2 min at 35 kHz. To separate the sample and extractant (chloroform) phases, the tube was centrifuged for 2 min at 2600 rpm.

Extraction of heavy PAHs. In a glass centrifuge tube, 10 mL of water was added, and the extraction mixture (150 μL of chloroform and 1.5 mL of acetonitrile) was quicky introduced into the sample by using a 2 mL syringe. Then, the tube was shaken for 1 min and centrifuged for 5 min at 3600 rpm for phase separation.

Extraction of 20 PAHs. In a glass centrifuge tube, 10 mL of water was quickly spiked with the extraction mixture, i.e., 150 μL of chloroform and 1.0 mL of acetonitrile + acetone (1:1) binary dispersive agent, by using a 2 mL syringe. Then, the extraction system was shaken for 1 min and ultrasonicated for 6 min at 35 kHz. For the phase separation, the tube was centrifuged for 2 min at 2600 rpm.

3. Results and Discussion

3.1. Estimation of Effectiveness of DLLME Extraction Solvents for the Recovery of PAHs from Water

As shown in Figure 1, the systems containing dichloroethane or carbon tetrachloride as extractants provided the lowest recoveries. Moreover, the extractant phase was not separated in the case of dichloromethane. Satisfactory analyte recoveries were achieved by using chloroform, which agreed with the results of our previous study, i.e., the application of DLLME for the extraction of PAHs from soils and bottom sediments [38].

At this stage, the use of chloroform volumes less than 150 μL was established to result in the insufficient extractant phase separation (Figure 2a). With further increase in chloroform volume, a decrease in the sensitivity was observed. A stable extraction system and sufficient recoveries were obtained by using 150 μL of chloroform (Figure 2b) and, consequently, this extractant volume was chosen as optimum.

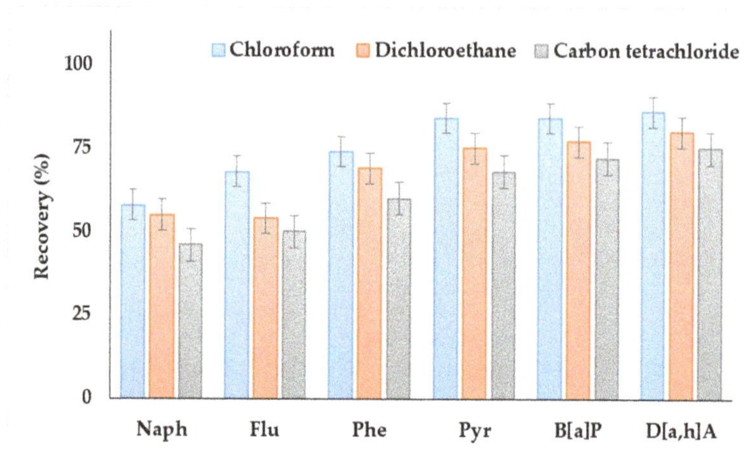

Figure 1. Comparison of polycyclic aromatic hydrocarbon (PAH) recoveries from water samples by using various extraction solvents.

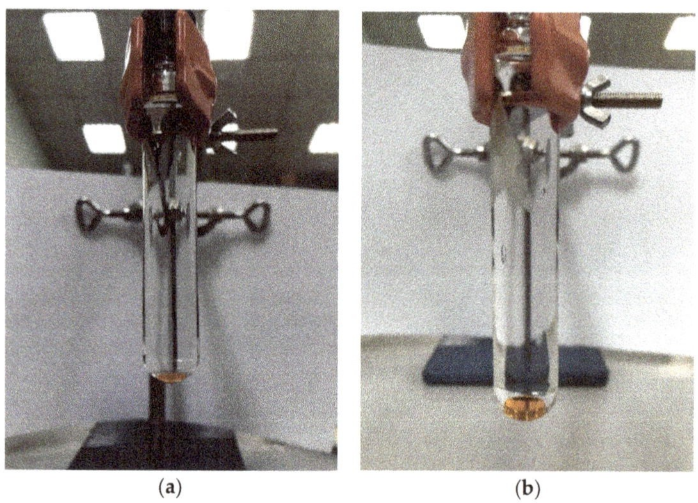

Figure 2. Stability of the extraction system for the recovery of PAHs using chloroform in volume: (**a**) 100 µL; (**b**) 150 µL.

When the ratio of a tested sample to solvent volume is selected, several requirements should be considered: the volume of water has to be sufficient for the effective concentration of analytes and the quantitative separation of chloroform drop from the extraction system. The last condition limits the maximum sample volume. According to these limitations, the sample volume of 10 mL was chosen in the following extraction steps.

Based on the obtained results, chloroform volume of 150 µL and 10 mL of water sample were used in further experiments.

3.2. Selection of Dispersive Solvent and It Is Optimal Volume

The selection of appropriate combinations and ratios of dispersive and extraction solvents is essential because it affects the formation of a stable and effective extraction system. Acetonitrile, acetone, methanol and their binary mixtures like acetone + acetonitrile, methanol + acetone and methanol + acetonitrile were chosen as dispersive agents. The extraction mixtures with a constant volume of the extractant (150 μL of chloroform) were used in the following investigation, and volumes of dispersive solvents were between 0.2 and 1.7 mL. The criteria for selecting the volume of the dispersive solvent were the formation of a cloudy solution and minimum dissolution of the extractant in the dispersive solvents in water phase (the necessary and sufficient conditions). The extraction systems that were satisfactory reproducible in terms of droplet volume in replicate experiments were chosen for further tests (Table S1). Based on experimental results, the optimal volumes of individual dispersive solvents were selected, and for acetonitrile it was 1.5 mL, for methanol it was 0.4 mL and for acetone it was 1.0 mL. Among the binary solvent mixtures, stable systems were obtained only when acetonitrile volume was equal or exceeded the other solvent volume (methanol or acetone). In the case of acetone and methanol mixture, utilization of excessive acetone volumes allowed us to obtain stable extraction systems. Thus, it was 0.5 + 0.5 mL for mixtures containing acetone and acetonitrile or acetone and methanol, and 0.75 + 0.75 mL of the mixture with methanol and acetonitrile. These dispersive solvent volumes were used in the following stage. The extraction capability of the developed solvent mixtures by PAHs was assessed by analysis of water samples spiked with PAHs at two concentration levels (5 and 50 ng/mL).

The obtained results demonstrated several peculiarities of PAHs extraction when individual solvents were used as dispersive solvents (Figure 3). Acetonitrile was preferable for the extraction of heavy and several light PAHs (fluorene, phenanthrene, anthracene, fluoranthene, pyrene); for these compounds, the recoveries were up to 89%. However, the recoveries of naphthalene group PAHs (acenaphthene, acenaphthylene, biphenyl and 2-methylnaphthalene) were less than 60%. Application of methanol allowed us to increase the recoveries of naphthalene group PAHs to 78–85% and provided high extraction efficiency for fluorene, phenanthrene, anthracene, fluoranthene and pyrene (80–89%). Acetone showed the worst results as a dispersive solvent among the solvents. In this case, the maximum recoveries were 75–76% for benz[b]fluoranthene, benz[k]fluoranthene and inden[1,2,3–c,d]pyrene, while the recoveries were less than 65% for other compounds; therefore, acetone was not used in the further studies.

Thus, the results demonstrated the possibility of highly efficient extraction of particular groups of PAHs by individual dispersive solvents, i.e., acetonitrile for the extraction of PAHs containing three or more aromatic rings and methanol for light PAHs. However, the individual dispersive agents did not provide simultaneous efficient extraction of a large list of PAHs. The application of binary mixtures as the dispersive agents was studied to solve this problem. The binary mixtures of acetonitrile with methanol or acetone provided high extraction recovery of PAHs containing more than three aromatic rings. The recoveries were comparable in both cases and ranged between 82 and 89%. However, the mixtures were ineffective for the extraction of naphthalene group PAHs; for example, the recoveries of naphthalene by methanol + acetonitrile and acetone + acetonitrile were 43 and 64%, respectively. The application acetone + methanol mixture resulted in obtaining the lowest extraction recoveries of all compounds.

It can be stated that the individual dispersive solvents provide potentially high extraction recoveries for light PAHs or the analytes containing three or more aromatic rings, and it is possible to develop a universal effective scheme for the simultaneous recovery of different molecular weight PAHs by applying the binary mixture consisting of acetone and acetonitrile.

To increase PAHs recoveries, the influence of ultrasound treatment and centrifugation parameters on the developed extraction mixtures was investigated.

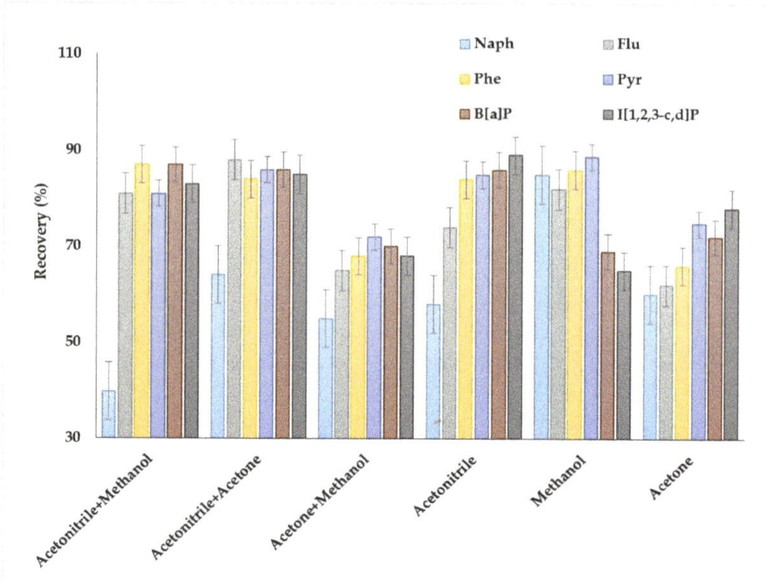

Figure 3. Effect of the dispersive agent compositions on the extraction recovery of PAHs.

3.3. Ultrasonication Effect on Extraction of PAHs from Water

The ultrasound treatment in DLLME provides decreased size of solvent drops and, consequently, increases the extraction efficiency [39]. To study this influence, water samples spiked with PAHs at 50 ng/L (concentration of analytes in the extract was 2.5 ng/mL) were ultrasonicated from 2 to 8 min. Ultrasonication of the extraction mixture containing acetonitrile as a dispersive solvent (mixture I) resulted in decreased recoveries of compounds (Figure 4a). The observed effect could be related to the reverse mass transfer of the analytes from chloroform to acetonitrile-water phase. Ultrasonication showed a positive effect on the extraction of analytes, when mixtures with methanol (mixture II) and the binary dispersive agent (mixture III) were used, and the optimal treatment times were 2 min (Figure 4b) and 6 min (Figure 4c), respectively. The effect of ultrasonication on the extraction of representative PAHs by the mixture containing binary dispersive agent is shown in Figure 4c; similar results were obtained for other investigated compounds.

3.4. Effect of Centrifugation on Extraction of PAHs from Water

The selection of centrifugation parameters is also important for the formation of the extractant drop. To study the effects of centrifugation speed and time on the recoveries of PAHs, the following conditions were tested: 2600 rpm (2 min), 3000 rpm (2 min), 3000 rpm (5 min), 3000 rpm (10 min) and 3600 rpm (5 min). Water samples spiked with PAHs at 50 ng/L (concentration of analytes in the extract was 2.5 ng/mL) were used in the experiments (3 replicates).

The results of experiments showed that best conditions for mixture I were 3000 and 3600 rpm for 5–10 min. In this case, the recoveries exceeded 99% for each PAH. However, the mixtures II and III had different behavior at long centrifugation times and high speed, and the extraction recoveries of PAHs were decreased. The optimized centrifugation conditions for mixtures II and III were 2600 rpm and 2 min. The optimization of ultrasonication and centrifugation conditions allowed us to increase the extraction efficiency of analytes by using the three suggested mixtures.

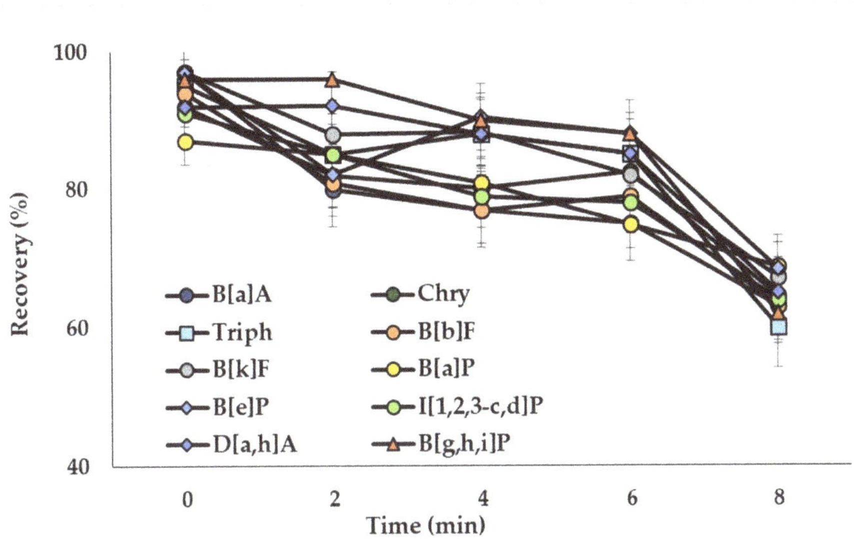

(a)

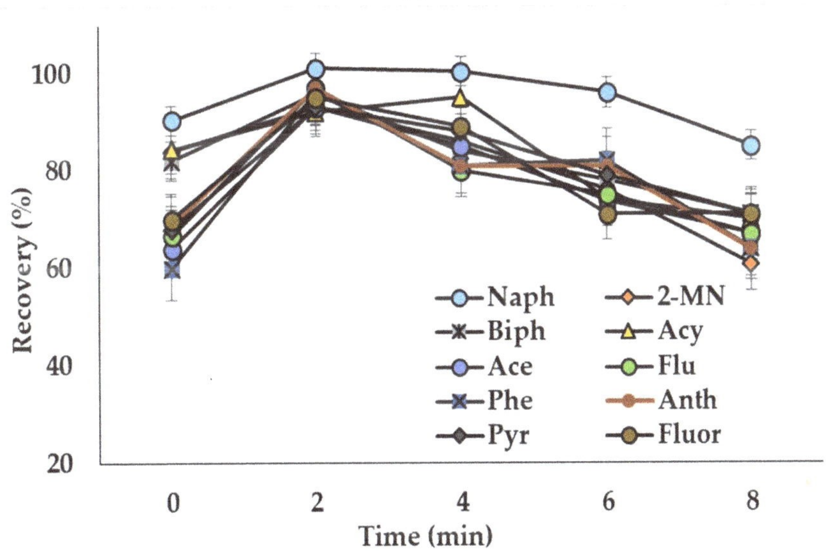

(b)

Figure 4. *Cont.*

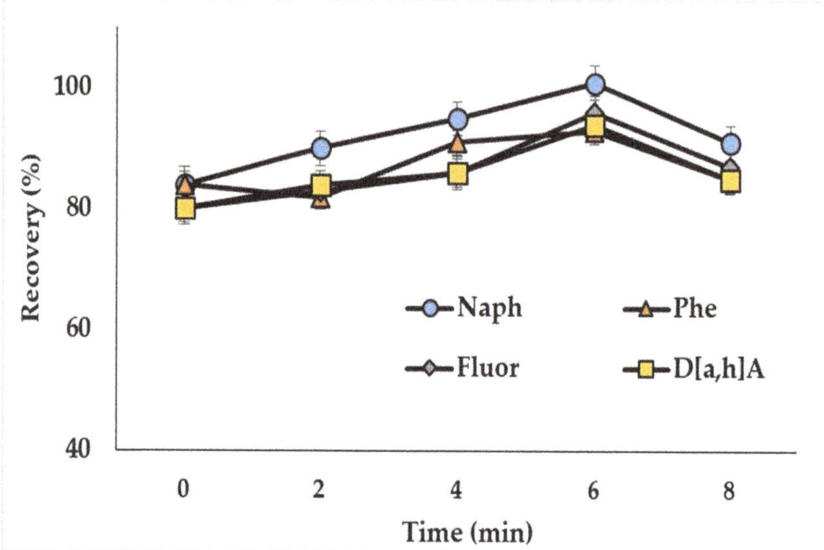

(c)

Figure 4. Effect of ultrasonication time on the recoveries of PAHs with extraction mixture: (**a**) chloroform and acetonitrile (mixture I); (**b**) chloroform and methanol (mixture II); (**c**) chloroform and acetone + acetonitrile (mixture III). Concentration of the analytes: 2.5 ng/mL.

Based on obtained results, three extraction schemes were suggested for the sample preparation of water by using DLLME with physical effects: procedure A for the extraction of PAHs containing more than three fused aromatic rings; procedure B for the extraction of naphthalene group PAHs and PAHs containing no more than four aromatic rings; a universal procedure C for the extraction different PAHs. For procedure A, in the first step, mixture I was added to the water sample. Next, the samples were shaken and centrifuged for 5 min at 3600 rpm to separate the extractant. For procedures B and C, after the addition of mixture II or III to the water sample, shaking and ultrasonication of the systems were conducted for 2 or 6 min, respectively. Centrifugation was carried out for 2 min at 2600 rpm to separate the extractant. Under the optimized conditions, procedure A provided the recoveries of heavy PAHs of 95–102% depending on the analyte (from fluorene to benz[g,h,i]perylene); the recoveries of light PAHs (from naphthalene to pyrene) were 95–101% by using procedure B; the recoveries between 91 and 99% for PAHs were obtained by procedure C (Table 2).

The universal DLLME procedure with the binary mixture (acetone + acetonitrile) as the dispersive agent was validated, because it provided high recoveries in the simultaneous extraction of a large list of PAHs.

Table 2. The recoveries of PAHs by using different sample preparation procedures.

PAHs	Recovery (%)		
	Procedure A	Procedure B	Procedure C
Naph	74 ± 5	101 ± 4	91 ± 5
2-MN	75 ± 4	99 ± 4	92 ± 3
Biph	75 ± 5	99 ± 4	92 ± 4
Acy	74 ± 5	98 ± 3	92 ± 5
Ace	76 ± 4	100 ± 3	93 ± 4
Flu	95 ± 3	100 ± 8	92 ± 5
Phe	97 ± 3	99 ± 4	93 ± 3
Anth	97 ± 4	95 ± 4	93 ± 4
Pyr	98 ± 4	101 ± 7	95 ± 3
Fluor	99 ± 4	98 ± 4	92 ± 4
B[a]A	100 ± 6	82 ± 6	98 ± 4
Chry	98 ± 3	82 ± 6	98 ± 3
Triph	98 ± 3	80 ± 5	95 ± 4
B[b]F	100 ± 4	79 ± 4	96 ± 4
B[k]F	98 ± 3	78 ± 4	97 ± 4
B[a]P	98 ± 4	76 ± 3	97 ± 4
B[e]P	102 ± 5	73 ± 3	98 ± 4
I [1,2,3-c,d]P	101 ± 4	71 ± 3	98 ± 3
D[a,h]A	98 ± 3	72 ± 3	98 ± 3
B[g,h,i]P	99 ± 3	72 ± 3	99 ± 3

3.5. Determination of Different PAHs by Dispersive Liquid–Liquid Microextraction and GC-MS and HPLC-FD/PDA Detection

The combination of DLLME with GC-MS is the simplest option, because chloroform can be directly injected into GC-MS after centrifugation. Previously optimized conditions of the GC-MS detection of PAHs in soils [38] were used in this work. The separation of isomeric pairs, such as phenanthrene/anthracene, chrysene/triphenylene/benz[a]anthracene, benz[b]fluoranthene/benz[k]fluoranthene and benz[a]pyrene/benz[e]pyrene, was achieved by using the temperature program: from 60 °C to 290 °C at the three stages on a specialized capillary column (Figure 5a). Application of the selected ion monitoring (SIM) mode allowed us to increase sensitivity and reliability of PAHs determination in complex matrices of natural waters.

However, to determine lower concentrations of PAHs in waters, the application of HPLC-FD/PDA is advisable. The separation PAHs was achieved by using the gradient elution and time program of FD (Figure 5b).

The chromatographic conditions described in EPA 8310 method and work [40] were used as initial to optimize HPLC-FD determination of PAHs. However, when these conditions were applied for the separation of 20 PAHs at the flow rate of 1.5 mL, peaks overlapped; at the flow rate of 0.5 mL, the HPLC performance was insufficient. The flow rate of 1 mL/min under gradient elution conditions and timing program of the fluorescence detector allowed us to solve the problem. Since acenaphthylene does not fluoresce, the PDA detector was used at the wavelength of 254 nm for the determination of this analyte.

The combination of DLLME with HPLC-FD/PDA requires evaporation of chloroform and re-extraction of the residue in a solvent compatible both with the chromatographic system and PAHs; acetonitrile was selected for this purpose. It is well known that naphthalene group PAHs with high vapor pressure can evaporate during this process [5]. In our work, naphthalene, 2-methylnaphthalene, biphenyl, acenaphthene, acenaphthylene and fluorene losses did not exceed 8–11%, while for other PAHs, the losses were lower than 4%, when chloroform was evaporated under the stream of nitrogen (water samples were spiked with PAHs at low, medium and high concentration levels, 3 replicates).

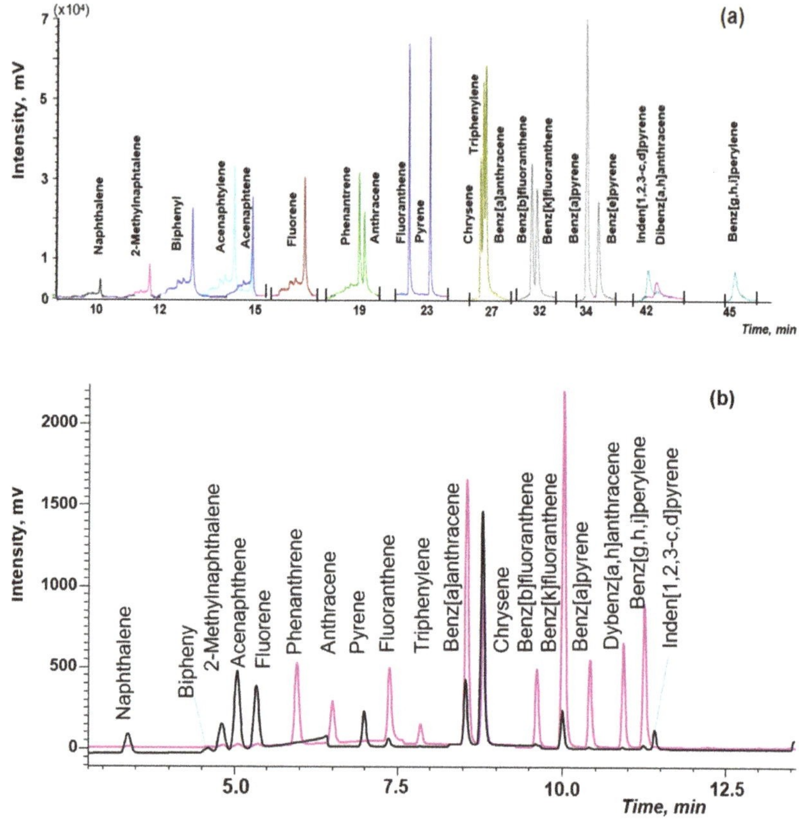

Figure 5. The chromatogram of the standard PAHs solution (50 ng/mL) by: (**a**) gas chromatography and mass spectrometry (GC-MS); (**b**) high-performance liquid chromatography and fluorescence detector (HPLC-FD).

3.6. Analytical Performance and Real Water Analysis

The developed method was validated in terms of linearity, limit of detection (LOD), limit of quantitation (LOQ) and precision. As can be seen in Table S2, good linearity of calibration curves for GC-MS and HPLC-FD/PDA detection was established, and the determination coefficients (R^2) were >0.99. The LODs for all PAHs were calculated as the signal/noise (S/N) ratios of 3, and they were in the range of 0.10–0.20 ng/L for HPLC-FD/PDA and between 10 and 20 ng/L for GC-MS. To check the intra-day assay precision of the developed method, 16 replicate samples with concentrations of individual PAHs of 30 ng/L (HPLC-FD/PDA) and 100 ng/L (GC-MS) were extracted with the DLLME method with binary solvents (acetone + acetonitrile) as dispersive agents and analyzed within one day. The inter-day assay precision was assessed at 30 ng/L (HPLC-FD/PDA) and 100 ng/L (GC-MS) spiking concentration during 10 consecutive days (Table S2). The relative standard deviations (RSDs) for the target PAHs were in the ranges of 3.1–6.5% and 3.7–7.8% (intra-day, HPLC-FD/PDA and GC-MS, respectively) and 4.3–7.0% and 5.4–8.2% (inter-day, HPLC-FD/PDA and GC-MS, respectively). It may be concluded that the combination of the universal sample preparation scheme with the chromatographic detection method can be selected depending on the required sensitivity of PAHs determination. The combination of DLLME with GC-MS is advisable for the determination of PAHs at concentrations above 10 ng/L in waters with complex matrices due to the increased reliability of PAHs identification. For trace concentrations of PAHs (0.10–0.20 ng/L), DLLME should be combined with HPLC-FD/PDA.

The validated method for the determination of PAHs was applied to different types of water samples. Table 3 shows the results of real sample analyses. To evaluate the matrix effect of the validated method, the spike-recovery test was performed by spiking 20 PAHs at three concentration levels (0.2, 10 and 750 ng/L) into the tap water sample and at two concentration levels of 10 and 750 ng/L into lake (salinity of 0.5 ‰) and seawater samples (the Azov Sea salinity of 6–18 ‰ and the Black Sea salinity of 22 ‰). As can be seen in Table 3, the recoveries for the PAHs in real water samples were in the range from 88 to 103%, and the RSDs were in the range of 3.1–7.8%, demonstrating good precision. Satisfactory repeatability of measurements proved that the sample matrices and salinity of water negligibly affected the determination of PAHs; consequently, the developed method can be applied to the analysis of different types of waters. Furthermore, the developed method allows for determining naphthalene, 2-methylnaphthalene, biphenyl and acenaphthene in waters at trace levels.

Table 3. Results of HPLC-FD/PDA determination of PAHs in real water samples (n = 2, p = 0.95).

Analyte	Sample												
	Tap Water				Lake Water			Sea Water [1]			Sea Water [2]		
	Mean (ng/L)	ER + RSDs (%) Spicked Amount (ng/L):			Mean (ng/L)	ER + RSDs (%) Spicked Amount (ng/L):		Mean (ng/L)	ER + RSDs (%) Spicked Amount (ng/L):		Mean (ng/L)	ER + RSDs (%) Spicked Amount (ng/L):	
		0.2	10	750		10	750		10	750		10	750
Naph	<0.20	91 ± 7.7	92 ± 7.5	95 ± 7.5	<0.20	88 ± 7.5	91 ± 7.1	<0.20	91 ± 7.4	95 ± 7.7	<0.20	93 ± 7.1	95 ± 7.2
2-MN	<0.15	92 ± 6.2	93 ± 6.3	93 ± 6.0	<0.15	90 ± 7.2	92 ± 7.5	<0.15	92 ± 7.7	94 ± 7.8	<0.15	92 ± 7.5	97 ± 7.7
Biph	<0.15	93 ± 6.2	92 ± 6.9	94 ± 6.8	<0.15	85 ± 6.8	86 ± 6.1	<0.15	95 ± 6.5	96 ± 6.8	<0.15	92 ± 7.6	95 ± 7.8
Acy	<0.15	94 ± 6.5	95 ± 6.4	95 ± 6.3	<0.15	87 ± 5.8	89 ± 6.5	<0.15	96 ± 6.5	99 ± 6.6	12	93 ± 6.1	96 ± 6.3
Ace	<0.15	93 ± 4.1	94 ± 3.8	95 ± 3.6	<0.15	87 ± 3.7	91 ± 3.5	<0.15	89 ± 4.0	92 ± 4.2	<0.15	94 ± 4.3	97 ± 4.5
Flu	<0.15	93 ± 5.0	95 ± 4.7	98 ± 4.5	<0.15	92 ± 4.7	91 ± 4.8	1.6	99 ± 4.5	95 ± 4.3	<0.15	92 ± 4.1	95 ± 4.3
Phe	<0.15	92 ± 3.9	99 ± 3.7	98 ± 3.5	3.1	87 ± 3.3	90 ± 2.9	1.4	96 ± 3.5	94 ± 3.7	1.3	99 ± 3.5	98 ± 3.6
Anth	<0.15	93 ± 5.1	95 ± 4.9	93 ± 4.7	0.46	98 ± 3.8	95 ± 3.7	<0.15	92 ± 3.6	95 ± 3.8	<0.15	95 ± 3.8	97 ± 4.1
Pyr	<0.15	95 ± 3.8	98 ± 3.5	95 ± 3.4	2.2	101 ± 3.1	98 ± 3.2	<0.15	93 ± 3.5	96 ± 3.4	14	94 ± 3.3	98 ± 3.2
Fluor	<0.15	96 ± 4.1	98 ± 4.0	96 ± 4.0	5.9	93 ± 3.8	95 ± 3.6	0.72	96 ± 3.7	98 ± 3.5	0.19	97 ± 3.7	99 ± 3.5
B[a]A	<0.10	95 ± 3.8	97 ± 3.7	95 ± 3.5	<0.10	98 ± 3.5	96 ± 3.6	<0.10	94 ± 3.2	95 ± 3.4	<0.10	97 ± 3.6	101 ± 3.8
Chry	<0.10	96 ± 4.5	98 ± 4.0	96 ± 4.1	<0.10	99 ± 4.1	97 ± 3.9	0.11	91 ± 4.0	96 ± 4.2	0.27	102 ± 4.0	98 ± 3.9
Triph	<0.10	95 ± 4.6	95 ± 4.4	95 ± 4.1	<0.10	97 ± 4.0	95 ± 4.1	0.84	89 ± 4.5	95 ± 4.3	5.7	96 ± 4.4	98 ± 4.6
B[b]F	<0.10	94 ± 4.7	95 ± 4.4	94 ± 4.2	0.78	92 ± 4.1	95 ± 4.2	<0.10	91 ± 4.3	94 ± 4.5	0.19	94 ± 4.1	99 ± 4.4
B[k]F	<0.10	95 ± 4.1	96 ± 4.0	95 ± 4.0	0.35	103 ± 3.9	99 ± 3.8	<0.10	103 ± 3.5	98 ± 3.7	0.1	98 ± 3.6	101 ± 3.8
B[a]P	<0.10	95 ± 4.0	96 ± 3.8	95 ± 3.7	<0.10	96 ± 3.8	97 ± 4.0	<0.10	93 ± 3.5	95 ± 3.7	<0.10	95 ± 3.3	99 ± 3.5
B[e]P	<0.10	95 ± 4.5	97 ± 4.4	95 ± 4.2	1.8	92 ± 4.1	95 ± 4.3	<0.10	92 ± 4.5	96 ± 4.0	0.18	95 ± 4.4	98 ± 4.6
I [1,2,3-c,d]P	<0.10	96 ± 3.5	98 ± 3.3	96 ± 3.2	<0.10	98 ± 3.1	101 ± 3.0	<0.10	96 ± 3.5	99 ± 3.2	<0.10	97 ± 3.7	99 ± 3.9
D[a,h]A	<0.10	97 ± 3.7	95 ± 3.5	97 ± 3.2	3	103 ± 3.2	99 ± 3.3	<0.10	96 ± 3.6	97 ± 3.7	0.27	96 ± 3.5	99 ± 3.7
B[g,h,i]P	<0.10	96 ± 3.4	95 ± 3.5	96 ± 3.3	0.21	105 ± 3.2	101 ± 3.1	<0.10	96 ± 3.5	98 ± 3.7	0.12	92 ± 3.3	97 ± 3.5

[1]—The Azov Sea; [2]—The Black Sea; ER—extraction recovery; RSDs—relative standard deviations; mean—the average resulted of analytes concentration in the real water samples.

4. Estimation of the Extraction Effectiveness of PAHs from Waters with Different DLLME Types

The proposed method for the extraction of 20 PAHs and DLLME concentration with the binary dispersive agent and HPLC-FD/PDA detection was compared with the previously reported methods (Table 4). For comparison, the following procedures were selected: DLLME with tetrachloroethylene as an extractant and acetone as a dispersive agent [19], low-density solvent-based DLLME (LDS-DLLME) with acetonitrile as a dispersive agent and hexane as an extractant [41], vortex-assisted DLLME (VSA-DLLME) technique with dichloromethane as an extractant and vortex mixing for the dispersion [24] and solidification of deep eutectic solvent DLLME (DLLME–SFDES) using deep eutectic solvent as an extractant and ultrasonication for the dispersion [42]. It should be noted that all the procedures have comparable recoveries; however, the proposed procedure is more universal regarding different PAHs. Meanwhile, the developed method has lower LODs for most analytes in comparison with other methods [19,24,41,42]. The suggested approach for the optimization of analyte extraction recoveries covers the differences between physicochemical properties of particular PAH groups more comprehensively. It allows us to improve the analysis scheme and choose the optimal parameters for the simultaneous recovery and concentration of components followed by their trace quantification.

Table 4. The LODs (ng/L) and recoveries (%) of PAHs using various DLLME techniques.

PAHs	n-DLLME [a] GC-FID * [19]		VSA-DLLME [b] GC-MS [24]		LDS-DLLME [c] GC-MS [41]		SFDES–DLLME [d] HPLC-FD [42]		UA-DLLME [f] HPLC-FD/PDA **	
	LOD	ER	LOD	ER	LOD	ER	LOD	ER	LOD	ER
Naph	10	99	2	82	58	85	6.6	103	0.07	91
2-MN	n/d	n/d	n/d	n/d	n/d	n/d	n/d	n/d	0.05	92
Biph	n/d	n/d	n/d	n/d	n/d	n/d	n/d	n/d	0.05	92
Acy	10	97	2	82	43	93	n/d	n/d	0.05	92
Ace	7	82	2	83	34	92	n/d	n/d	0.05	93
Flu	8	92	2	85	23	98	1.2	97	0.05	92
Phe	9	99	2	82	28	93	3.4	83	0.05	93
Anth	9	95	2	83	26	99	0.7	103	0.05	93
Pyr	10	91	3	84	37	96	0.9	107	0.05	95
Fluor	10	111	2	83	38	102	4.3	93	0.05	92
B[a]A	10	103	3	76	46	83	n/d	n/d	0.03	98
Chry	10	94	3	81	49	83	n/d	n/d	0.03	98
Triph	n/d	n/d	n/d	n/d	n/d	n/d	n/d	n/d	0.03	95
B[b]F	n/d	n/d	2	77	n/d	n/d	n/d	n/d	0.03	96
B[k]F	n/d	n/d	3	74	n/d	n/d	n/d	n/d	0.03	97
B[a]P	15	102	3	84	n/d	n/d	n/d	n/d	0.03	97
B[e]P	20	102	n/d	n/d	n/d	n/d	n/d	n/d	0.03	98
I [1,2,3-c,d]P	n/d	n/d	5	78	n/d	n/d	n/d	n/d	0.03	98
D[a,h]A	n/d	n/d	5	78	n/d	n/d	n/d	n/d	0.03	98
B[g,h,i]P	30	101	5	74	n/d	n/d	n/d	n/d	0.03	99

LOD—limit of detection; ER—extraction recovery of PAHs at concentration of each analyte: [a] 5 µg/L; [b] 0.2 µg/L; [c] 0.5 µg/L; [d] 0.3 µg/L; [f] 0.2 µg/L. n/d—no data; *FID—flame ionization detector; ** developed method HPLC-FD/PDA determination of PAHs.

The developed method is fast and precise, which is proved by accuracy values lower than 8% that are reproducible with the relative standard deviation values of 4.3–7.0%, thus allowing researchers to use it in analytical laboratories for environmental monitoring.

5. Conclusions and Future Perspectives

In the present paper, different water sample preparation schemes were developed for the determination of PAHs based on DLLME concentration and HPLC-FD/PDA or GC-MS detection. Extraction solvent mixtures containing individual dispersive solvents (methanol or acetonitrile) provided high recoveries (>95%) for light PAHs or the analytes containing three or more aromatic rings. However, this sample preparation scheme is not universal when PAHs with different molecular weights are simultaneously determined. DLLME with acetone and acetonitrile as a binary dispersing solvent in combination with ultrasonication allowed us to achieve higher recoveries of a large list of PAHs from water samples (>99%), as they are compatible with chromatographic detection methods. With the relative standard deviation of 3.1–7.8%, GC-MS provided detection and quantification limits of 3.0–6.0 and 10–20 ng/L, respectively. In the case of HPLC-FD/DAD determination of PAHs, the detection and quantification limits were 0.03–0.07 and 0.10–0.20 ng/L, respectively. Analysis of water samples with different mineralization has shown that salinity has negligible effect on PAH recoveries.

The developed analysis schemes allowed us to determine PAHs with different structures in water samples within a wide concentration range. The analysis schemes are suitable for routine environmental monitoring, do not require highly qualified personnel and minimize the volumes of used chloroform, which agrees with "green" analytical chemistry.On the other hand, despite the mentioned advantages of the DLLME procedure for the concentration of the analytes, its shortcoming is the use of a toxic chlorinated solvent, even in small volumes. The aim of subsequent studies is to find biodegradable solvents with comparable analyte concentration factors.

Supplementary Materials: The following supporting information can be downloaded https://www.mdpi.com/article/10.3390/molecules27238586/s1, Table S1: The results obtained for different extraction systems containing chloroform for DLLME; Table S2: Analytical performance of the DLLME with binary solvents as dispersive agents for the determination of PAHs in waters.

Author Contributions: Conceptualization and supervision and data curation and funding acquisition, Z.T.; validation and formal analysis and writing—original draft preparation, T.M. and T.C.; investigation and visualization, Z.A.; writing—review and editing, S.P. All authors have read and agreed to the published version of the manuscript.

Funding: This research was funded by the Russian Foundation for Basic Research, grant no. 19-43-230003 p_a, and was carried out with the use of the scientific equipment of the Center for Collective Use "Ecological and Analytical Center" at the Kuban State University.

Institutional Review Board Statement: Not applicable.

Informed Consent Statement: Not applicable.

Data Availability Statement: The data presented in this study are available on request from the corresponding author.

Conflicts of Interest: The authors declare no conflict of interest.

References

1. Achten, C.; Andersson, J.T. Overview of Polycyclic Aromatic Compounds (PAC). *Polycycl. Aromat. Compd.* **2015**, *35*, 177–186. [CrossRef] [PubMed]
2. Environmental Protection Agency. *Health Effects Assessment for Polycyclic Aromatic Hydrocarbons (PAH)*; EPA 540/1-86-013; Environmental Criteria and Assessment Office: Cincinnati, OH, USA, 1984; pp. 1–61.
3. Jesus, F.; Pereira, L.J.; Campos, I.; Santos, M.; Ré, A.; Keizer, J.; Nogueira, A.; Gonçalves, J.M.F.; Abrantes, N.; Serpa, D. A review on polycyclic aromatic hydrocarbons distribution in freshwater ecosystems and their toxicity to benthic fauna. *Sci. Total Environ.* **2022**, *820*, 153282. [CrossRef] [PubMed]
4. Zheng, B.; Wang, L.; Lei, K.; Nan, B. Distribution and ecological risk assessment of polycyclic aromatic hydrocarbons in water, suspended particulate matter and sediment from Daliao River estuary and the adjacent area, China. *Chemosphere* **2016**, *149*, 91–100. [CrossRef] [PubMed]
5. Mackay, D.; Shiu, W.Y.; Lee, S.C. *Handbook of Physical-Chemical Properties and Environmental Fate for Organic Chemicals*; Taylor & Francis Group: Boca Raton, FL, USA, 2006; pp. 1–925.
6. Mezzanotte, V.; Anzano, M.; Collina, E.; Marazzi, F.A.; Lasagni, M. Distribution and Removal of Polycyclic Aromatic Hydrocarbons in Two Italian Municipal Wastewater Treatment Plants in 2011–2013. *Polycycl. Aromat. Compd.* **2015**, *36*, 213–228. [CrossRef]
7. Semenov, M.Y.; Marinaite, I.I.; Golobokov, L.P.; Khuriganova, O.I.; Khodzher, T.V.; Semenov, Y.M. Source apportionment of polycyclic aromatic hydrocarbons in Lake Baikal water and adjacent air layer. *Chem. Ecol.* **2017**, *33*, 977–990. [CrossRef]
8. Huang, Y.; Sui, Q.; Lyu, S.; Wang, J.; Huanga, S.; Zhaoc, W.; Wang, B.; Xu, D.; Kong, M.; Zhang, Y.; et al. Tracking emission sources of PAHs in a region with pollution-intensive industries, Taihu Basin: From potential pollution sources to surface water. *Environ. Pollut.* **2020**, *264*, 114674. [CrossRef]
9. Igwe, J.C.; Ukaogo, P.O. Environmental effects of polycyclic aromatic hydrocarbons. *J. Nat. Sci. Res.* **2015**, *5*, 2224–3186.
10. Lawal, A.T. Polycyclic Aromatic Hydrocarbons. A Review. *Cogent Environ. Sci.* **2017**, *3*, 1339841. [CrossRef]
11. Ahad, M.E.J.; Macdonald, R.; Parrot, J.; Yang, Z.; Zhang, Y.; Siddique, T.; Kuznetsova, A.; Rauert, C.; Galarneau, E.; Studabaker, W.; et al. Polycyclic aromatic compounds (PACs) in the Canadian environment: A review of sampling techniques, strategies and instrumentation. *Environ. Pollut.* **2020**, *266*, 114988. [CrossRef] [PubMed]
12. Mojiri, A.; Zhou, J.L.; Ohashi, A.; Ozakia, N.; Kindaichi, T. Comprehensive review of polycyclic aromatic hydrocarbons in water sources, their effects and treatments. *Sci. Total Environ.* **2019**, *696*, 133971. [CrossRef]
13. Zhang, Q.; Liu, P.; Li, S.; Zhang, X.; Chen, M. Progress in the analytical research methods of polycyclic aromatic hydrocarbons (PAHs). *J. Liq. Chromatogr. Relat. Technol.* **2020**, *43*, 425–444. [CrossRef]
14. Temerdashev, Z.; Musorina, T.; Chervonnaya, T.; Arutyunyan, Z.V. Possibilities and Limitations of Solid-Phase and Liquid Extraction for the Determination of Polycyclic Aromatic Hydrocarbons in Environmental Samples. *J. Anal. Chem.* **2021**, *76*, 1357–1370. [CrossRef]

15. Umar, K.; Yaqoob, A.A.; Ibrahim, M.N.M.; Parveen, T.; Safian, M.T. Environmental applications of smart polymer composites. Smart Polym. Nanocompos. *Biomed. Environ. Appl.* **2020**, *15*, 295–320.
16. Safian, M.T.; Umar, K.; Parveen, T.; Yaqoob, A.A.; Ibrahim, M.N.M. Chapter Eight-Biomedical applications of smart polymer composites. In *Smart Polymer Nanocomposites: Biomedical and Environmental Applications*; Woodhead Publishing Series in Composites Science and Engineering; Elsevier Inc.: Cambridge, MA, USA, 2021; pp. 183–204.
17. Hassan Mohd, F.W.; Raoov, M.; Kamaruzaman, S.; Sanagi Mohd, M.; Yoshida, N.; Hirota, Y.; Nishiyama, N.; Yahaya, N. Dispersive liquid–liquid microextraction combined with dispersive solid-phase extraction for gas chromatography with mass spectrometry determination of polycyclic aromatic hydrocarbons in aqueous matrices. *J. Sep. Sci.* **2018**, *41*, 3751–3763. [CrossRef] [PubMed]
18. Temerdashev, Z.A.; Musorina, T.N.; Kiseleva, N.V.; Eletskii, B.D.; Chervonnaya, T.A. Gas chromatography–mass spectrometry determination of polycyclic aromatic hydrocarbons in surface water. *J. Anal. Chem.* **2018**, *73*, 1154–1161. [CrossRef]
19. Rezaee, M.; Assadi, Y.; Hosseini, M.-R.M.; Aghaee, E.; Ahmadi, F.; Berijani, S. Determination of organic compounds in water using dispersive liquid–liquid microextraction. *J. Chromatogr. A* **2006**, *1116*, 1–9. [CrossRef]
20. Berijani, S.; Assadi, Y.; Anbia, M.; Milani-Hosseini, M.R.; Aghaee, E. Dispersive liquid–liquid microextraction combined with gas chromatography-flame photometric detection: Very simple, rapid and sensitive method for the determination of organophosphorus pesticides in water. *J. Chromatogr. A* **2006**, *1123*, 1–9. [CrossRef]
21. Rezaeia, F.; Bidari, A.; Birjandi, A.P.; Hosseini, M.R.M.; Assadi, Y. Development of a dispersive liquid–liquid microextraction method for the determination of polychlorinated biphenyls in water. *J. Hazard. Mater.* **2008**, *158*, 621–627. [CrossRef] [PubMed]
22. Campillo, N.; Viñas, P.; Šandrejová, J.; Andruch, V. Ten years of dispersive liquid-liquid microextraction and derived techniques. *Appl. Spectrosc. Rev.* **2017**, *52*, 267–415. [CrossRef]
23. Wang, Q.; Chen, R.; Shatner, W.; Cao, Y.; Bai, Y. State-of-the-art on the technique of dispersive Liquid-liquid microextraction. *Ultrason. Sonochem.* **2019**, *51*, 369–377. [CrossRef]
24. Zheng, J.; Liu, B.; Ping, J.; Chen, B.; Wu, H.; Zhang, B. Vortex- and Shaker-Assisted Liquid–Liquid Microextraction (VSA-LLME) Coupled with Gas Chromatography and Mass Spectrometry (GC-MS) for Analysis of 16 Polycyclic Aromatic Hydrocarbons (PAHs) in Offshore Produced Water. *Water Air Soil Pollut.* **2015**, *226*, 318–331. [CrossRef]
25. Kokosa, J.M. Chapter 16-Dispersive Liquid-Liquid Microextraction. Liquid-Phase Extraction. In *Handbooks in Separation Science*; Elsevier: Amsterdam, The Netherlands, 2020; pp. 473–497. [CrossRef]
26. Tan, Y.H.; Chai, M.K.; Wong, L.S. A review on extraction solvents in the dispersive liquidliquid microextraction. *Malays. J. Anal. Sci.* **2018**, *22*, 166–174. [CrossRef]
27. Farajzadeh, M.A.; Afshar, M.M.R.; Aghanassab, M. Deep eutectic solvent-based dispersive liquid–liquid microextraction. *Anal. Methods.* **2016**, *8*, 2576–2583. [CrossRef]
28. Makos, P.; Przyjazny, A.; Boczkaj, G. Hydrophobic deep eutectic solvents as "green" extraction media for polycyclic aromatic hydrocarbons in aqueous samples. *J. Chromatogr. A* **2018**, *1570*, 28–37. [CrossRef]
29. Rykowska, I.; Ziemblińska, J.; Nowak, I. Modern approaches in dispersive liquid-liquid microextraction (DLLME) based on ionic liquids: A review. *J. Mol. Liq.* **2018**, *259*, 319–339. [CrossRef]
30. Jordana, A.; Gathergood, N. Biodegradation of ionic liquids—A critical review. *Chem. Soc. Rev.* **2015**, *44*, 8200–8237. [CrossRef] [PubMed]
31. Ma, C.; Laaksonen, A.; Liu, C.; Lu, X.; Ji, X. The peculiar effect of water on ionic liquids and deep eutectic solvents. *Chem. Soc. Rev.* **2018**, *47*, 8685–8720. [CrossRef]
32. Flieger, J.; Flieger, M. Ionic Liquids Toxicity—Benefits and Threats. *Int. J. Mol. Sci.* **2020**, *21*, 6267. [CrossRef] [PubMed]
33. Shishov, A.; Pochivalov, A.; Nugbienyo, L.; Andruch, V.; Bulatov, A. Deep eutectic solvents are not only effective extractants. *Trends Analyt. Chem.* **2020**, *129*, 115956. [CrossRef]
34. Mohammadi, A.; Jahani, S.M.-M.; Kamankesh, M.; Jazaeri, S.; Eivani, M.; Esmaeili, S.; Abdi, S. Determination of Polycyclic Aromatic Hydrocarbons in Edible Oil Using Fast and Sensitive Microwave-assisted Extraction and Dispersive Liquid–Liquid Microextraction Followed by Gas Chromatography-Mass Spectrometry. *Polycycl. Aromat. Compd.* **2020**, *40*, 705–713. [CrossRef]
35. Kamankesh, M.; Mohammadi, A.; Hosseini, H.; Tehrani, Z.M. Rapid determination of polycyclic aromatic hydrocarbons in grilled meat using microwave-assisted extraction and dispersive liquid–liquid microextraction coupled to gas chromatography–mass spectrometry. *Meat Sci.* **2015**, *103*, 61–67. [CrossRef] [PubMed]
36. Mahmoudpour, M.; Mohtadinia, J.; Ansarin, M.; Nemati, M. Dispersive Liquid–Liquid Microextraction for HPLC-UV Determination of PAHs in Milk. *J. AOAC Int.* **2016**, *99*, 527–533. [CrossRef] [PubMed]
37. Mahmoudpour, M.; Mohtadinia, J.; Mousavi, M.-M.; Ansarin, M.; Nemati, M. Application of the Microwave-Assisted Extraction and Dispersive Liquid–Liquid Microextraction for the Analysis of PAHs in Smoked Rice. *Food Anal. Methods.* **2017**, *10*, 277–286. [CrossRef]
38. Temerdashev, Z.A.; Musorina, T.N.; Chervonnaya, T.A. Determination of polycyclic aromatic hydrocarbons in soil and bottom sediments by gas chromatography–mass spectrometry using dispersive liquid–liquid microextraction. *J. Anal. Chem.* **2020**, *75*, 1000–1010. [CrossRef]
39. Avino, P.; Notardonato, I.; Perugini, L.; Russo, M.V. New protocol based on high-volume sampling followed by DLLME-GC-IT/MS for determining PAHs at ultra-trace levels in surface water samples. *Microchem. J.* **2017**, *133*, 251–257. [CrossRef]

40. Li, L.; Han, D.; Wang, M.; Han, Y.; Yan, H. Molybdenum disulfide–hypercrosslinked polymer composite as an adsorbent for determination of polycyclic aromatic hydrocarbons in environmental water coupled with HPLC–FLD. *Microchim. Acta* **2020**, *187*, 242. [CrossRef] [PubMed]
41. Guo, L.; Tan, S.; Li, X.; Lee, H.K. Fast Automated Dual-Syringe Based Dispersive Liquid-Liquid Microextraction Coupled with Gas Chromatography–Mass Spectrometry for the Determination of Polycyclic Aromatic Hydrocarbons in Environmental Water Samples. *J. Chromatogr. A* **2016**, *1438*, 1–9. [CrossRef] [PubMed]
42. Yousefi, S.M.; Shemirani, F.; Ghorbanian, S.A. Hydrophobic Deep Eutectic Solvents in Developing Microextraction Methods Based on Solidification of Floating Drop: Application to the Trace HPLC/FLD Determination of PAHs. *Chromatographia* **2018**, *81*, 1201–1211. [CrossRef]

molecules MDPI

Article

Parabens Permeation through Biological Membranes: A Comparative Study Using Franz Cell Diffusion System and Biomimetic Liquid Chromatography

Ilaria Neri [1], Sonia Laneri [1], Ritamaria Di Lorenzo [1], Irene Dini [1], Giacomo Russo [1,2,*] and Lucia Grumetto [1,*]

1 Department of Pharmacy, School of Medicine and Surgery, University of Naples Federico II, Via D. Montesano, 49, I-80131 Naples, Italy; ilaria.neri@unina.it (I.N.); slaneri@unina.it (S.L.); ritamaria.dilorenzo@unina.it (R.D.L.); irdini@unina.it (I.D.)
2 School of Applied Sciences, Sighthill Campus, Edinburgh Napier University, 9 Sighthill Ct, Edinburgh EH11 4BN, UK
* Correspondence: g.russo@napier.ac.uk (G.R.); grumetto@unina.it (L.G.)

Abstract: Parabens (PBs) are used as preservatives to extend the shelf life of various foodstuffs, and pharmaceutical and cosmetic preparations. In this work, the membrane barrier passage potential of a subset of seven parabens, i.e., methyl-, ethyl-, propyl- isopropyl, butyl, isobutyl, and benzyl paraben, along with their parent compound, p-hydroxy benzoic acid, were studied. Thus, the Franz cell diffusion (FDC) method, biomimetic liquid chromatography (BLC), and in silico prediction were performed to evaluate the soundness of both describing their permeation through the skin. While BLC allowed the achievement of a full scale of affinity for membrane phospholipids of the PBs under research, the permeation of parabens through Franz diffusion cells having a carbon chain > ethyl could not be measured in a fully aqueous medium, i.e., permeation enhancer-free conditions. Our results support that BLC and in silico prediction alone can occasionally be misleading in the permeability potential assessment of these preservatives, emphasizing the need for a multi-technique and integrated experimental approach.

Keywords: parabens; investigative toxicology; skin; Franz cell; lipophilicity; chromatography approach; immobilized artificial membrane

Citation: Neri, I.; Laneri, S.; Di Lorenzo, R.; Dini, I.; Russo, G.; Grumetto, L. Parabens Permeation through Biological Membranes: A Comparative Study Using Franz Cell Diffusion System and Biomimetic Liquid Chromatography. *Molecules* **2022**, *27*, 4263. https://doi.org/10.3390/molecules27134263

Academic Editors: Victoria Samanidou, Eleni Deliyanni and Dimitra Voutsa

Received: 7 June 2022
Accepted: 28 June 2022
Published: 1 July 2022

Publisher's Note: MDPI stays neutral with regard to jurisdictional claims in published maps and institutional affiliations.

1. Introduction

Parabens (PBs) are a group of C-4 esterified molecules of hydroxybenzoic acid (pHBA) with a broad antimicrobial and antifungal spectrum, commonly used since the 1920s [1] as preservatives in foodstuffs, pharmaceuticals, and personal care products (PCPs) to increase their shelf life [2,3]. PBs having a higher value of n-octanol–water partition coefficient (log P ranging from 1.96 to 3.57) show low water solubility, making the shorter-chained PBs better suited for their application. These compounds can leak into the environment because they are massively employed in the industry and released mainly through wastewater treatment discharges [2]. PBs can be easily absorbed by the human body, both orally and after dermal or respiratory exposure [4]. These are stable in the bloodstream for 24h and then are metabolized, mainly by carboxylesterases, releasing nonspecific pHBA and p-hydroxy hippuric acid and, less frequently, are excreted to various molecules as free, oxidized, or conjugated in the microsomes [5]. Some non-metabolized parabens bioaccumulate in the various compartments of the human body [6,7]. In addition, these toxicants can cross the placenta barrier, leading to potential fetal exposure [8].

PBs act by altering mitochondria and membrane transport, with their long-chain ester group showing higher antimicrobial properties. Moreover, the longer-chained PBs, such as Butyl 4-hydroxybenzoate (BuP) and Propyl 4-hydroxybenzoate (PrP), show higher endocrine disruptive properties than short-chain analogs, such as Methyl 4-hydroxybenzoate

(MP) [9,10]. In recent decades, numerous studies aimed at assessing the effects of PBs on the endocrine system, and currently, an ample body of scientific literature, indicate that these can exert estrogenic activity and can therefore be considered as endocrine disrupting compounds (EDCs) featuring a non-steroidal chemical structure [11–15]. Indeed, several studies confirm the hypothesis that PBs can act as EDCs that can modulate the functions of the endocrine system [16–18]. Some of them, MeP, EtP, PrP, and BuP, competitively bind the ERs [13]) and impact ER-dependent gene expression [19,20], interfering in the normal functioning of natural endocrine hormones.

Moreover, PBs can affect reproduction by altering fertility and reproductive functions in male rodents after repetitive oral exposure by causing epigenetic hypermethylation of sperm DNA, which may impact transcription regulation and be transmitted to the offspring [21]. Although their acute toxicity is low [22], a gap in knowledge subsists about their health effects after chronic exposure, which currently prevents an accurate risk assessment. This work aimed to explore the passage of seven parabens and their parent compound, pHBA, through the skin, the most prevalent route of human exposure [23], by measuring (a) their permeability using a traditional Franz diffusion cell system and (b) their affinity for membrane phospholipids by immobilized artificial membrane (IAM) liquid chromatography (LC).

Figure 1 shows the chemical structures of analytes we investigated: Methyl 4-hydroxybenzoate (MP), Ethyl 4-hydroxybenzoate (EP), Propyl 4-hydroxybenzoate (PrP), Butyl 4-hydroxybenzoate (BuP), Isopropyl 4-hydroxybenzoate (iPrP), Benzyl 4-hydroxybenzoate (BzP), Isobutyl 4-hydroxybenzoate (iBuP), and pHBA. Some PBs, i.e., MP, EP, BuP, PrP, and pHBA, are authorized in PCPs, up to 0.4% for a single ester and 0.8% in mixtures. Isopropyl-, isobutyl-, phenyl-, benzyl-, and pentylparabens have been banned from use in PCPs in Europe since 2014 [24].

Immobilized artificial membrane stationary phases consist of phosphatidylcholine (PC) analogs covalently bound to silica, aiming to mimic biological cell membranes closely. The degree of affinity between analytes and the IAM stationary phases is regarded as phospholipophilicity and indicated by the logarithm of coefficient of analytical retention achieved at or extrapolated to 100% aqueous conditions (log kw IAM). Indeed, previous studies by our research team demonstrated a good correlation between data obtained with traditional in vitro methods (intestinal tissues and passage through cell bilayers) and those achieved with this analytical technique [25,26], which was proven to be effective in studying the membrane permeability potential of target compounds both in vivo and in situ [27,28]. Analogously, we decided to explore the possible relationships between BLC and transdermic passage in order to verify if the lipophilicity can predict the PBs' permeation.

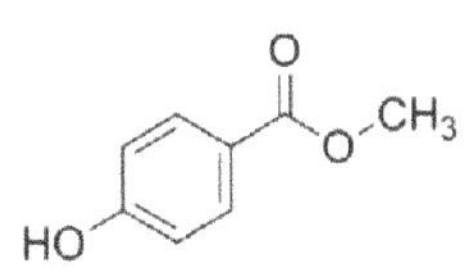

Methyl 4-hydroxybenzoate (MP)

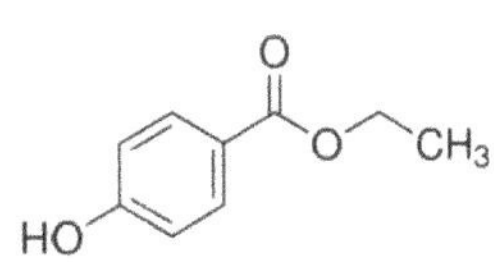

Ethyl 4-hydroxybenzoate (EP)

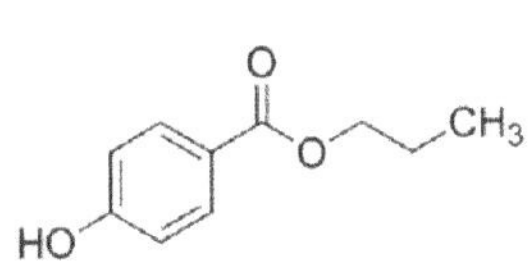

Propyl 4-hydroxybenzoate (PrP)

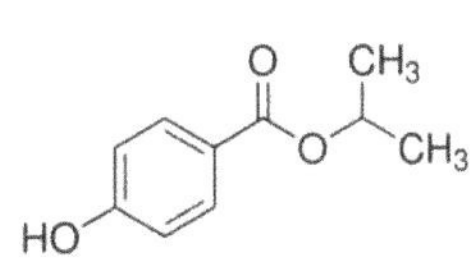

Isopropyl 4-hydroxybenzoate (iPrP)

Figure 1. *Cont.*

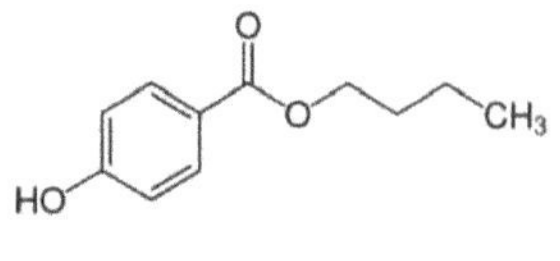

Butyl 4-hydroxybenzoate (BuP)

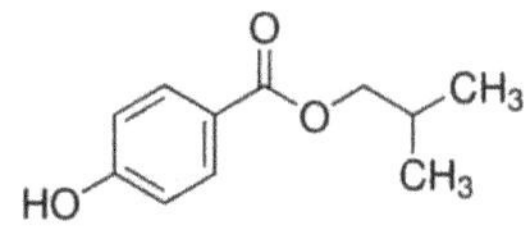

Isobutyl 4-hydroxybenzoate (iBuP)

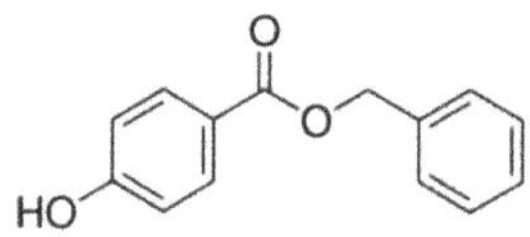

Benzyl 4-hydroxybenzoate (BzP)

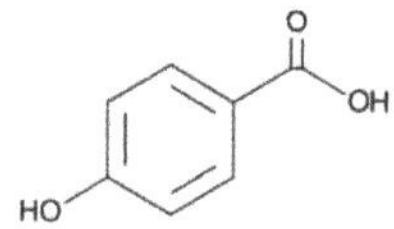

4- Hydroxybenzoic Acid (pHBA)

Figure 1. Chemical structure of parabens under investigation.

2. Results and Discussion

This work aimed at studying the membrane affinity and the dermal absorption of seven parabens—some of which, i.e., MP, EP, BuP, PrP, are allowed in PCPs—and their parent compound, pHBA. Table 1 summarizes chromatographic parameters applied to determine the analytes in the acceptor chambers of the FDC, the calibration curve parameters, and relevant physicochemical properties. Chromatographic conditions used in HPLC-UV methods for the quantification of all eight chemicals in receptor fluid solution were found sensitive (up to 2.5 μg mL^{-1}) to their detection, and a sound linear relationship ($r^2 > 0.9256$) was found between the instrument response and the applied concentration values spanning in the range 0.052–0.886 μg mL^{-1} (LOD values in Table S1).

Table 1. Chromatographic and physicochemical parameters of the eight chemicals.

Compound	S (mg L^{-1})	Molecular Weight (g/mol)	Log P	Range (μg mL^{-1})	Mobile Phase Composition ACN:H_2O	Rt (min)	Slope	Intercept	r^2
pHBA	5.00×10^3	138.12	1.58	20–40	40:60 *	6.1	43879	11,792	0.9992
MP	2.50×10^3	152.16	1.96	5–40	50:50	8.3	402.09	1120.1	0.9988
EP	8.85×10^2	166.18	2.47	5–40	50:50	10.9	3503.5	1344.3	0.9861
PrP	5.00×10^2	180.21	1.96	2.5–20	50:50	15.7	1059.8	1647.2	0.9973
iPrP	5.00×10^2	180.21	1.96	2.5–20	60:40	10.6	795.67	2048.8	0.9975
BuP	2.07×10^2	194.23	3.57	2.5–20	60:40	14.5	2275.3	2863.3	0.9256
iBuP	2.07×10^2	194.23	3.57	2.5–20	50:50	23.0	1271.4	1831.0	0.9984
BzP	0.92×10^2	228.25	3.56	2.5–20	60:40	14.5	723.07	2675.1	0.9976

S = Solubility in water; Log P were taken from PubCHEM; * the aqueous phase was phosphate buffer pH 3.0.

Permeation results, as assessed by Franz cell diffusion study, are shown in Table 2 along with the percentage permeation consistent with skin permeation values.

Table 2. Permeability coefficient and maximum flux ± standard deviation of the three compounds able to cross the skin. Median of % permeation, with 95% interval confidence given in parentheses.

Compound	Maximum Flux (μg/cm^2/h)	Kp (cm/h)	Permeation (%) (Median)
pHBA	12.68 ± 4.08	0.012	0.997 (0–1.22)
MP	76.23 ± 26.60	0.305	9.148 (4.94–19.70)
EP	2.35 ± 0.71	9.4×10^{-4}	0.110 (0–0.11)
PrP	Nd	nd	nd
iPrP	Nd	nd	nd
BuP	Nd	nd	nd
iBuP	Nd	nd	nd
BzP	Nd	nd	nd

MP and pHBA exhibited a maximum passage through the porcine skin within 30 min from the beginning of the experiment, while EP passage was realized within 120 min.

Furthermore, as shown in Table 2, the EP Kp value was lower than MP's. In fact, MP was found to have a maximum flux value of 76.23 ± 26.60 (μg/cm^2/h), which is significantly higher than that of EP (maximum flux of 2.35 ± 0.71 μg/cm^2/h). Thus, PBs with a shorter alkyl chain present higher permeation, MP > EP, and show that the skin permeation degree of PBs can be considered moderate, according to the definition of Marzulli et al. [29]. Conversely, starting from PrP, analogs with longer chains were not detected in receptor fluids at any time. The pHBA showed an intermediate passage profile between MP and EP, probably due to its free ionizable carboxyl acid group with a pKa of 4.01 (SwissADME), which is partially ionized at the experimental pH (5.0).

Over the years, in vitro Franz diffusion experiments have evolved into one of the most important methods for researching transdermal absorption [30]. However, these procedures have some shortcomings, including their being time-consuming and poorly inter-lab reproducible. For this reason, an increased demand for higher throughput, more inter-lab reproducible methods emerged. Chromatography performed on stationary phases emulating the asset of biological membranes cannot directly describe the permeation of the analytes but can accidentally assist in the pharmacokinetic (PK) profiling of an array of target compounds and potentially offer usefulness in surrogating permeation data [31].

Table 3 shows experimental log kw IAM performed on both columns, IAM.PC.MG and IAM.DD2.

Table 3. Experimental phospholipophilicity values measured on both the IAM.PC.MG and IAM.PC.DD2 stationary phases.

Compound	$\log k_w$ $^{IAM.MG}$	$\log k_w$ $^{IAM.DD2}$
pHB	−0.955	−1.054
MP	0.615	1.306
EP	0.956	1.751
PrP	1.339	2.267
iPrP	1.227	1.914
BuP	1.818	2.812
iBuP	1.708	2.448
BzP	2.292	2.735

A good relationship (r^2 = 0.898) was found between log kwIAM values determined on the IAM.PC.MG and IAM.PC.DD2 stationary phases, as can be seen in Figure 2, suggesting that the role of the end-capping in the analytical retention is only marginal.

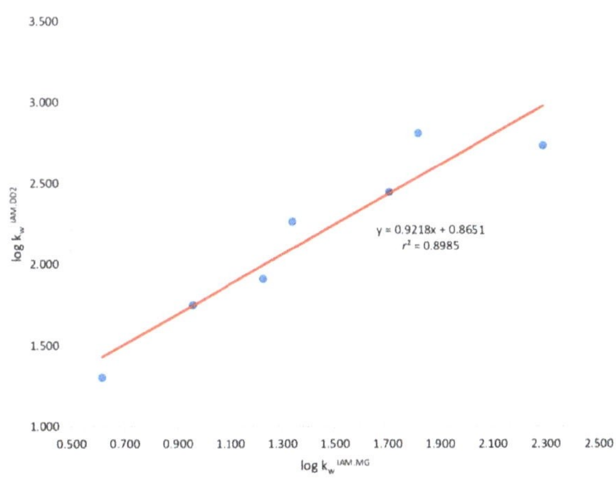

Figure 2. Relationship with phospholipophilicity data achieved on IAM.PC.MG and IAM.PC.DD2 stationary phases.

In the Supplementary Materials (Table S2 and Figure S1), we also show the in silico calculated clogkw IAM.MG vs. clogkw IAM.DD2 values having a similar trend but are slightly higher than the experimental values. These are achieved via a user-friend Web service (available at https://nova.disfarm.unimi.it/vegaol/logkwiam.htm (accessed on 27 June 2022) able to predict the phospholipophilicity of any molecule included in the PubChem collection [28]. Table 4 shows Kp values achieved by SwissADME® software (Swiss Institute of Bioinformatics© 2022, Lausanne, Switzerland) compared to those for the PBs able to cross the skin in our experimental conditions.

Table 4. Logarithm of experimental and calculated permeability coefficient (Kp). Log Kp * = Determined in the present study; Log Kp ** = SwissADME calculator.

Compound	Log Kp * cm/s	Log Kp ** cm/s
pHBA	−5.48	−6.02
MP	−4.07	−5.84
EP	−6.58	−5.56
PrP	Nd	−5.24
iPrP	Nd	−5.40
BuP	Nd	−4.95
iBuP	Nd	−5.05
BzP	Nd	−4.93

In Figure 3, we report the relationship between log kwIAM.MG (a) and log kwIAM.DD2 (b) values achieved for all chemicals under investigation but pHBA, already reported in Table 3, versus Log Kp values achieved from SwissADME® software. Indeed, pHBA is the only acidic analyte of the dataset. According to the pH piston hypothesis formulated by Avdeef [32], after some experimental observations about the partitioning of drugs into liposomes, negatively charged molecules interact with the positively charged ammonium groups of PC moieties located at the outer side of the membranes, engage more externally, and remain on the surface. In contrast, cations interact with the negatively charged phosphate moieties that are located inside. Consistently, pHBA, as a carboxylate at experimental pH 5.0, is less retained by both IAM phases than a neutral isolipophilic molecule and therefore exhibited a log kw IAM value < 1. When we compare IAM data with in silico permeation values achieved by the software SwissADME®, a good relationship is observed if pHBA is left out (Figure 3). Indeed, phospholipophilicity, which effectively describes the affinity of chemicals to the membrane [33], can be used in combination with other methods to shed light on the molecular features of these preservatives driving their toxic effects and specifically their endocrine-disrupting properties.

Similar to SwissADME® software, many in silico methods share the objective of predicting the penetration of molecules through the skin, and these normally exploit quantitative structure–property relationships (QSPRs), models based on diffusion mechanisms, or a combination of both [34]. Mathematical tools consist of equations to predict skin permeability based on physicochemical properties, such as molecular weight and the octanol-water partition coefficient (log Kow) [35–39]. The Kp data achieved by SwissADME® contradict the experimental permeation data. Indeed, according to the in silico data, the Kp values would be rather similar for the whole dataset, and there is only a slight difference in permeation rate starting from MP up to the higher chain analog. Instead, in our FDC experiment, only three compounds could cross the pig skin membrane.

We should highlight that membrane crossing is a complex phenomenon where multiple passage pathways that cannot be necessarily accounted for by either in silico or IAM methods can occur. The routes of permeation of the molecules across the skin includes the transcellular route where the molecule encounters the low lipid regions in the cytoplasm of the corneocytes; the intercellular route of dense lipids and fatty acids featuring both hydrophilic and lipophilic regions, making this route very resistant to permeation; and the shunt pathway or appendageal route with diffusion into the sweat gland, hair follicles, and sebaceous gland [40]. Indeed, the diffusion of the target molecule through the lipoidal layer is regarded as transcellular passage, and that is the mechanism that most compounds of pharmaceutical/environmental concern seem to exploit. However, smaller hydrophilic molecules can overcome the skin barrier by passing through the intercellular space between the cells, therefore using the intercellular route [41]. The relative contribution in the overall passage depends mainly (but not exclusively) upon the molecular volume. It is reasonable to assume that the contribution of the intercellular passage for MP is greater than that of EP for the relatively lower molecular weight and, therefore, molecular volume.

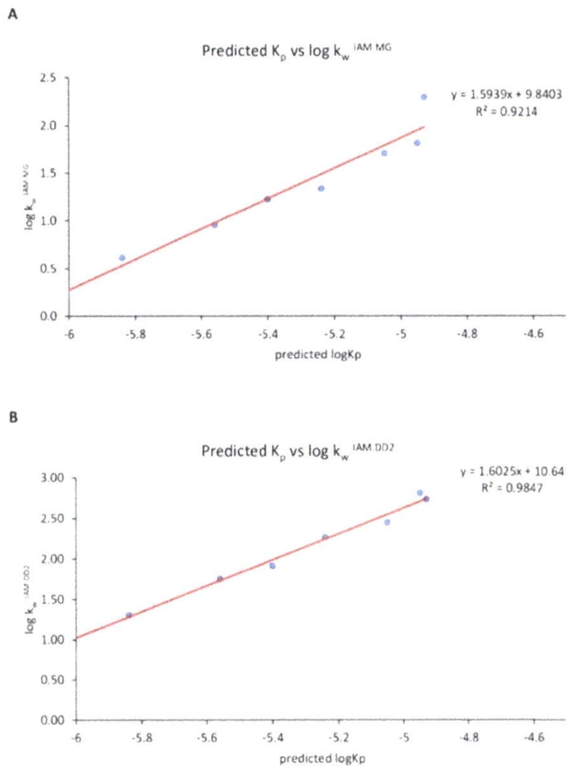

Figure 3. Relationship between Log Kp values achieved from SwissADME® software and log kw IAM.MG (**A**) and logkw IAM.DD2 (**B**) values achieved by IAM chromatography. A good relationship is observed, demonstrating that IAM experimental data of affinity for phospholipid agree with skin permeation data.

Our FDC data substantiate that the permeation potential of the target preservatives could also be affected by solubility, which decreases at increasing chain length, and so, the absolute amount exposed to the membrane for PBs with chain length > propyl is fairly inferior to that of other PBs [42]. The excipients contained in various formulations may potentially affect the skin crossing [43–45] by

- Decreasing the polarity of the medium and increasing the relative solubility of our target preservatives, and
- Disrupting the packing of lipids into biological membranes and increasing their leakiness.

For these reasons, most of these excipients can act as enhancers to increase skin barrier permeability. Indeed, in the Caon et al. study [42], PBs are dissolved in ethanol, a well-known promoter of skin passage [46], at various percentages (0.1% paraben dissolved in an ethanol/PBS mixture—20:80 and 50:50), while other studies introduced PBs in donor chambers as cream (O/W emulsion) formulations [47] or as commercial body lotions [48]. For instance, Hatami et al. evaluated the passage of MP and BuP from an emulsion using the FDC system, observing that MP has lower permeation and lower extent than BuP, adding PBs in the oily phase of the formulation [49]. Conversely, our experiments were run in a fully aqueous phase at pH 5.0, enhancer-free conditions, with saturated solutions of each PB, and thus, our results are partially consistent with those other previously mentioned studies, since only MP and EP and not PrP and BuP were found in receptor fluids. We underline that the design of the FDC experiments was underpinned to study the skin passage of the PBs per se, not as ingredients of a formulation because even if the employed formulations contain enhancer excipients, a 100% aqueous vehicle is the most appropriate standard state to assess the permeation [39].

Lipophilicity is a driving force of the transmigration of xenobiotics through the membrane, but the presence of other active/excipients can reduce surface tension, minimizing its impact on the overall permeation.

For this reason, the surrogate models such as in silico prediction and BLC should only be used as preliminary screening tools for chemicals to be followed by diffusion studies exploiting natural skin model, as these can fail in the estimation of molecule's permeation.

3. Materials and Methods

3.1. Chemicals

Methyl 4-hydroxybenzoate (MP), Ethyl 4-hydroxybenzoate (EP), Propyl 4-hydroxybenzoate (PrP), and Butyl 4-hydroxybenzoate (BuP) were purchased from Merck & Co. (Poole, UK). Isopropyl 4-hydroxybenzoate (iPrP) was purchased from Fluorochem (Hadfield, UK), Benzyl 4-hydroxybenzoate (BzP) was purchased from Sigma Aldrich (Milan, Italy), and Isobutyl 4-hydroxybenzoate (iBuP) and 4-Hydroxybenzoic acid (pHBA) were purchased from J&K Scientific (San Jose, CA, USA). The purity of all PBs was equal to or higher than 98%. Milli-Q water was produced in-house, and its conductivity was 0.055 $\mu S\ cm^{-1}$ at 25 °C (resistivity equals 18.2 MΩ·cm). Phosphate buffered saline (PBS) was freshly made in-house by adding disodium hydrogen phosphate and sodium chloride (J.T. Baker, Phillipsburg, NJ, USA), and potassium chloride (Sigma Aldrich), by diluting with Milli-Q water to 1 L. The pH (7.00) was determined by a pH meter. Acetonitrile (ACN) and methanol (MeOH) were HPLC grade and purchased from Sigma Aldrich (Milan, Italy).

3.2. Tissue Preparation

For the ex vivo skin penetration studies, the Organization for Economic Co-operation and Development (OECD) guideline recommends pig ear skin as a suitable skin surrogate to mimic human percutaneous penetration [24]. The skin samples were excised from a male pig ear, post-sacrifice, obtained from a local abattoir (Avellino, Italy) within three hours from animal death. The outer part of the ear was used.

The integrity of the skin area used for the permeation experiment was examined by measuring non-invasive trans-epidermal water loss (TEWL) using an In-vitro Tewameter® VT 310® (Courage + Khazaka electronic GmbH, Köln, Germany), and only the skin samples with TEWL values < 10 $g/m^2/h$ were used in this study. Subcutaneous fat was removed, and skin samples were kept at room temperature and hydrated in saline solution (0.9% NaCl) for 5 min. Then, skin specimens were put on a filter paper (Fisherbrand™ Grade 601, Fisher Scientific, Leicestershire, UK) and cut into pieces before assembly in the Franz Diffusion Cell (FDC).

3.3. Skin Permeation

Skin absorption tests were performed using Franz-type vertical static diffusion cells (FDC Ø 9 mm, 5 mL receptor compartment, SES GmbH-Analyse System, Bechenheim, Germany) with an effective diffusion area of 0.6 cm^2 and receptor volume of 5.0 mL [50]. The total thickness of the skin was washed three times with 1.0 mL of water and mounted between the two halves of the cell with the stratum corneum facing the donor compartment.

The receptor compartment was filled with 5.0 mL of PBS pH 7.4 and checked to ensure there were no air bubbles between the receptor fluid (PBS) and the skin. PBS is a buffer solution that is particularly valuable because it mimics the ion concentration, osmolarity, and pH of human body fluids [51]. The donor compartment was filled with 1 mL of a saturated solution (concentration values are listed in Table 1) of each analyte dissolved in PBS pH 5.0. Two cells were added as blanks filled with only PBS pH 5.0 as donor solutions. The experiments were repeated in triplicate for a total of 24 exposed cells and two blanks. The FDC was maintained at a constant temperature of 37 ± 0.5 °C through thermostatic bath circulation, and the receptor medium was constantly kept under magnetic stirring throughout the experiments using a Teflon cylindrical magnetic stirring bar (dimension 10 × 3 mm). At fixed time intervals ($\frac{1}{2}$, 1, 2, 3, 4, 5, 6, and 24 h), aliquots were collected from the receptor chambers and immediately replaced with the same volume of fresh PBS medium. The collected samples were analyzed by high-performance liquid chromatography (HPLC) in triplicate to quantify the amounts of each PB.

3.4. *Chromatographic Analysis*

PBs' quantification was obtained by the regression equation obtained from standard curves prepared on the same day and performed in concentration levels according to the solubility of each PB to achieve a concentration close to saturated solution. Analytical methods were set up ad hoc to analyze chemicals of interest in the receptor's fluids. Analyses were performed at room temperature (22 ± 2 °C). HPLC (LC-20 AD VP; Shimadzu Corp., Kyoto, Japan), equipped with an ultraviolet (UV)–visible detector (Shimadzu Model SPD10 AV) set at λ 254 nm, was used. The stainless-steel column was a reversed-phase Supelco Ascentis C18 (250 × 4.6 mm, 5.0 μm i.d.) with a Supelguard Ascentis C18 guard column (both from Supelco, Bellefonte, PA, USA). All mobile phases were vacuum filtered through 0.45 μm nylon membranes (Millipore, Burlington, MA, USA). Overall analyses were conducted at 0.5 mL min^{-1}. Data acquisition and integration were accomplished by Cromatoplus 2011 software (Messina, Italy). Methanol injections were made every five runs to assess that no carryover occurred.

3.5. *IAM Chromatography*

Phospholipophilicity was experimentally determined using two IAM analytical columns, i.e., IAM.PC.MG (150 × 4.6 mm, 10 μm p.s., pore size: 300 Å) and IAM.PC.DD2 (150 × 4.6 mm, 10 μm p.s., pore size: 300 Å), both from Regis Chemical Company (Morton Grove, IL, USA). The analyses on the IAM.PC.MG were conducted on a LC-20 AD VP (from Shimadzu Corp., Kyoto, Japan), equipped with an ultraviolet (UV)–visible detector (Shimadzu Model SPD10 AV) set at λ = 254 nm. For the determinations on the IAM.PC.DD2, an alliance HPLC system (Waters Corporation, Milford, MA USA) with 2690/2695 separation module, 2996 PDA detector, 717 plus autosampler, and 2487 dual λ absorbance detector, was used at 254 nm wavelength instead. The methods were set, and chromatograms were recorded by Waters Empower® 3 software (Milford, MA USA). On both columns, employed eluents were 0.1 M phosphate buffer at pH 5.0 and acetonitrile at various percentages with a flow rate of 1.0 mL min^{-1}. All samples were dissolved in acetonitrile (ca. 10^{-4} M), and chromatographic analyses were carried out at 22 ± 2 °C. The two different IAM stationary phases differ from each other in the end-capping of residual amino groups of the silica-propylamine core, which support C10 and C3 alkyl chains being end-capped by both decanoic and propionic anhydrides (IAM.PC.DD2), while IAM.PC.MG supports hydroxy groups being end-capped by methyl glycolate. IAM parameters performed on both stationary phases have already been found to be strongly interrelated, as assessed in our previous studies [25–28,31]. The affinity of the chemicals for both IAM.PC.MG and IAM.PC.DD2 was measured as a retention factor extrapolated at 100% of the aqueous phase (kw IAM) by performing a polycratic extrapolation method [52]. Since all the compounds under our investigation required at least the addition of acetonitrile to the mobile phase to elute within 20 min, at least three different mobile phases containing acetonitrile in percentages (φ), ranging from 10% to 30% (*v*/*v*), were employed. All kw IAM values average at least three individual measurements, and the final values are reported as decimal logarithms.

3.6. Experimental Permeability Calculation

The skin permeability was calculated as follows: PBs flux (J) was calculated from the quantity of each PB, which permeated through the membrane, divided by the insert membrane surface and the time duration (μg/cm^2*h). The permeability coefficient (Kp (cm/h)) was determined from J and the drug concentration in the donor phase (Cd (μg/cm^3)):

$$J = Qr/A \times h \tag{1}$$

$$Kp = J/Cd \tag{2}$$

The ratio of each PB total amount in the receptor fluid compared to the applied amount was calculated to determine the total absorption rate:

$$\text{Total absorption (\%)} = \text{Amount receptor fluid/Amount total} \times 100 \tag{3}$$

3.7. Skin Permeation Calculator

For in silico determination of log Kp, the SwissADME® engine (freely accessible at http://www.swissadme.ch (accessed on 27 June 2022) was used, based on a linear model adapted from Potts and Guy, who found that Kp linearly correlated with molecular size and lipophilicity [36].

3.8. Statistical Analysis

Calculations were made using Microsoft® Excel® of Office 365. All data sets are shown as the mean ± standard deviation (SD), with statistically significant differences determined by a *t*-test with probability (p) values < 0.05.

4. Conclusions

We studied the affinity of PBs for biomembranes, which plays a crucial role in their membrane barrier passage that strongly affects their toxicokinetic. The study of the permeation of PBs in a formulation vehicle was not in the scope of this work, as whichever experimental design would have never mirrored the plethora of cosmetic and pharmaceutical formulations in which PBs can be found. The actual issue of PBs' toxicity is their occurrence, even at low regulated doses, in many consumer goods of large consumption that can create continuous and multiple sources of exposure for humans and generate a chronic toxicity issue.

Our results demonstrate that caution is required when using in silico and chromatographic data, with the aim of surrogating in vitro permeation values. Indeed, IAM data reflect the affinity of compounds for phospholipids, but not necessarily their permeation through complex barrier. In silico data are based on the application of mathematical models that are often based in full or in part on molecular lipophilicity. However, their performance can unfortunately be substandard when passage is modulated by other features.

Supplementary Materials: The following supporting information can be downloaded at: https://www.mdpi.com/article/10.3390/molecules27134263/s1, Figure S1: Relationship between calculated phospholipophilicity data (clog kw IAM.MG and clog kw IAM.DD2) for both stationary phases achieved by Web service (available at https://nova.disfarm.unimi.it/vegaol/logkwiam.htm.); Table S1: Chromatographic Limit of Detection (LOD) of the seven analyzed parabens and their metabolite; Table S2: Phospholipophilicity calculated for both stationary phases (clog kw IAM.MG**, clog kw IAM.DD2**) in silico, via a user-friendly tool Web service (available at https://nova.disfarm.unimi.it/vegaol/logkwiam.htm.), able to predict phospholipophilicity of any molecule included in the PubChem collection.

Author Contributions: Conceptualization: G.R. and L.G.; Data curation: G.R.; Formal analysis: I.N. and R.D.L.; Methodology: G.R. Software: I.N. and R.D.L.; Supervision: L.G. and G.R.; Validation: I.N.; Visualization: I.D.; Writing—original draft preparation: L.G.; Writing—review and editing: S.L. All authors have read and agreed to the published version of the manuscript.

Funding: This research received no external funding.

Institutional Review Board Statement: Not applicable.

Informed Consent Statement: Not applicable.

Data Availability Statement: Not applicable.

Acknowledgments: Ashna Moidu and Eloise Debret are gratefully acknowledged for their participation in the IAM.LC experiments.

Conflicts of Interest: The authors declare no conflict of interest.

References

1. Oliveira, M.M.; Martins, F.; Silva, M.G.; Correia, E.; Videira, R.; Peixoto, F. Use of Parabens (Methyl and Butyl) during the Gestation Period: Mitochondrial Bioenergetics of the Testes and Antioxidant Capacity Alterations in Testes and Other Vital Organs of the F1 Generation. *Antioxidants* **2020**, *9*, 1302. [CrossRef] [PubMed]
2. Nowak, K.; Ratajczak-Wrona, W.; Górska, M.; Jabłońska, E. Parabens and their effects on the endocrine system. *Mol. Cell. Endocrinol.* **2018**, *474*, 238–251. [CrossRef] [PubMed]
3. Liao, C.; Liu, F.; Kannan, K. Occurrence of and dietary exposure to parabens in foodstuffs from the United States. *Environ. Sci. Technol.* **2013**, *47*, 3918–3925. [CrossRef] [PubMed]
4. Wei, F.; Mortimer, M.; Cheng, H.; Sang, N.; Guo, L.H. Parabens as chemicals of emerging concern in the environment and humans: A review. *Sci. Total Environ.* **2021**, *778*, 146150. [CrossRef] [PubMed]
5. Moos, R.K.; Angerer, J.; Dierkes, G.; Brüning, T.; Koch, H.M. Metabolism and elimination of methyl, iso- and n-butyl paraben in human urine after single oral dosage. *Arch. Toxicol.* **2016**, *90*, 2699–2709. [CrossRef] [PubMed]
6. Li, C.; Cui, X.; Chen, Y.; Liao, C. Paraben concentrations in human fingernail and its association with personal care product use. *Ecotoxicol. Environ. Saf.* **2020**, *202*, 110933. [CrossRef] [PubMed]
7. Greige-Gerges, H.; Kaissi, R.; Magdalou, J.; Jraij, A. Reviewing the binding of a series of parabens to human serum albumin. *Biopharm. Drug Dispos.* **2013**, *34*, 186–194. [CrossRef]
8. Andersen, M.H.G.; Zuri, G.; Knudsen, L.E.; Mathiesen, L. Placental transport of parabens studied using an ex-vivo human perfusion model. *Placenta* **2021**, *115*, 121–128. [CrossRef]
9. Oishi, S. Effects of butylparaben on the male reproductive system in rats. *Toxicol. Ind. Health* **2001**, *17*, 31–39. [CrossRef]
10. Crovetto, S.I.; Moreno, E.; Dib, A.L.; Espigares, M.; Espigares, E. Bacterial toxicity testing and antibacterial activity of parabens. *Toxicol. Environ. Chem.* **2017**, *99*, 858–868. [CrossRef]
11. Goodson, W.H., 3rd; Luciani, M.G.; Sayeed, S.A.; Jaffee, I.M.; Moore, D.H., 2nd; Dairkee, S.H. Activation of the mTOR pathway by low levels of xenoestrogens in breast epithelial cells from high-risk women. *Carcinogenesis* **2011**, *32*, 1724–1733. [CrossRef] [PubMed]
12. Harvey, P.W.; Everett, D.J. Parabens detection in different zones of the human breast: Consideration of source and implications of findings. *J. Appl. Toxicol.* **2012**, *32*, 305–309. [CrossRef] [PubMed]
13. Byford, J.R.; Shaw, L.E.; Drew, M.G.; Pope, G.S.; Sauer, M.J.; Darbre, P.D. Oestrogenic activity of parabens in MCF7 human breast cancer cells. *J. Steroid Biochem. Mol. Biol.* **2002**, *80*, 49–60. [CrossRef]
14. Prusakiewicz, J.J.; Harville, H.M.; Zhang, Y.; Ackermann, C.; Voorman, R.L. Parabens inhibit human skin estrogen sulfotransferase activity: Possible link to paraben estrogenic effects. *Toxicology* **2007**, *232*, 248–256. [CrossRef] [PubMed]
15. Hu, P.; Chen, X.; Whitener, R.J.; Boder, E.T.; Jones, J.O.; Porollo, A.; Chen, J.; Zhao, L. Effects of parabens on adipocyte differentiation. *Toxicol. Sci.* **2013**, *131*, 56–70. [CrossRef] [PubMed]
16. Nohynek, G.J.; Borgert, C.J.; Dietrich, D.; Rozman, K.K. Endocrine disruption: Fact or urban legend? *Toxicol. Lett.* **2013**, *223*, 295–305. [CrossRef]
17. Rutkowska, A.; Rachoń, D.; Milewicz, A.; Ruchała, M.; Bolanowski, M.; Jędrzejuk, D.; Bednarczuk, T.; Górska, M.; Hubalewska-Dydejczyk, A.; Kos- Kudła, B.; et al. Polish Society of Endocrinology Position statement on endocrine disrupting chemicals (EDCs). *Endokrynol. Pol.* **2015**, *66*, 276–285. [CrossRef]
18. Taxvig, C.; Vinggaard, A.M.; Hass, U.; Axelstad, M.; Boberg, J.; Hansen, P.R.; Frederiksen, H.; Nellemann, C. Do Parabens Have the Ability to Interfere with Steroidogenesis? *Toxicol. Sci.* **2008**, *106*, 206–213. [CrossRef]
19. Vo, T.T.B.; Jung, E.-M.; Choi, K.-C.; Yu, F.H.; Jeung, E.-B. Estrogen receptor α is involved in the induction of Calbindin-D9k and progesterone receptor by parabens in GH3 cells: A biomarker gene for screening xenoestrogens. *Steroids* **2011**, *76*, 675–681. [CrossRef]
20. Lillo, M.A.; Nichols, C.; Perry, C.; Runke, S.; Krutilina, R.; Seagroves, T.N.; Miranda-Carboni, G.A.; Krum, S.A. Methylparaben stimulates tumor initiating cells in ER+ breast cancer models. *J. Appl. Toxicol.* **2017**, *37*, 417–425. [CrossRef]

21. Park, C.J.; Nah, W.H.; Lee, J.E.; Oh, Y.S.; Gye, M.C. Butyl paraben-induced changes in DNA methylation in rat epididymal spermatozoa. *Andrologia* **2012**, *44* (Suppl. S1), 187–193. [CrossRef] [PubMed]
22. Cherian, P.; Zhu, J.; Bergfeld, W.F.; Belsito, D.V.; Hill, R.A.; Klaassen, C.D.; Liebler, D.C.; Marks, J.G., Jr.; Shank, R.C.; Slaga, T.J.; et al. Amended Safety Assessment of Parabens as Used in Cosmetics. *Int. J. Toxicol.* **2020**, *39* (Suppl. S1), 5S–97S. [CrossRef]
23. Guo, Y.; Kannan, K. A Survey of Phthalates and Parabens in Personal Care Products from the United States and Its Implications for Human Exposure. *Environ. Sci. Technol.* **2013**, *47*, 14442–14449. [CrossRef] [PubMed]
24. EU, Regulation (EC) No 1004/2014 of the European Parliament and of the Council on cosmetic products. *Off. J. Eur. Union* **2014**. No 1004/2014. Available online: http://data.europa.eu/eli/reg/2014/1004/oj (accessed on 27 June 2022).
25. Grumetto, L.; Russo, G.; Barbato, F. Immobilized Artificial Membrane HPLC Derived Parameters vs PAMPA-BBB Data in Estimating in Situ Measured Blood-Brain Barrier Permeation of Drugs. *Mol. Pharm.* **2016**, *13*, 2808–2816. [CrossRef] [PubMed]
26. Russo, G.; Capuozzo, A.; Barbato, F.; Irace, C.; Santamaria, R.; Grumetto, L. Cytotoxicity of seven bisphenol analogues compared to bisphenol A and relationships with membrane affinity data. *Chemosphere* **2018**, *201*, 432–440. [CrossRef]
27. Grumetto, L.; Russo, G.; Barbato, F. Relationships between human intestinal absorption and polar interactions drug/phospholipids estimated by IAM-HPLC. *Int. J. Pharm.* **2015**, *489*, 186–194. [CrossRef]
28. Russo, G.; Grumetto, L.; Barbato, F.; Vistoli, G.; Pedretti, A. Prediction and mechanism elucidation of analyte retention on phospholipid stationary phases (IAM-HPLC) by in silico calculated physico-chemical descriptors. *Eur. J. Pharm. Sci.* **2017**, *99*, 173–184. [CrossRef]
29. Marzulli, F.N.; Brown, D.W.C.; Maibach, H.I. Techniques for studying skin penetration. *Toxicol. Appl. Pharmacol.* **1969**, *14*, 76–83. [CrossRef]
30. Bartosova, L.; Bajgar, J. Transdermal drug delivery in vitro using diffusion cells. *Curr. Med. Chem.* **2012**, *19*, 4671–4677. [CrossRef]
31. Grumetto, L.; Russo, G. cΔlog k(w) (IAM): Can we afford estimation of small molecules' blood-brain barrier passage based upon in silico phospholipophilicity? *ADMET DMPK* **2021**, *9*, 267–281. [CrossRef]
32. Avdeef, A.; Box, K.J.; Comer, J.E.; Hibbert, C.; Tam, K.Y. pH-metric logP 10. Determination of liposomal membrane-water partition coefficients of ionizable drugs. *Pharm. Res.* **1998**, *15*, 209–215. [CrossRef] [PubMed]
33. Taillardat-Bertschinger, A.; Carrupt, P.A.; Barbato, F.; Testa, B. Immobilized artificial membrane HPLC in drug research. *J. Med. Chem.* **2003**, *46*, 655–665. [CrossRef] [PubMed]
34. Moss, G.P.; Dearden, J.C.; Patel, H.; Cronin, M.T.D. Quantitative structure–permeability relationships (QSPRs) for percutaneous absorption. *Toxicol. Vitr.* **2002**, *16*, 299–317. [CrossRef]
35. Barratt, M.D. Quantitative structure-activity relationships for skin permeability. *Toxicol. Vitr.* **1995**, *9*, 27–37. [CrossRef]
36. Potts, R.O.; Guy, R.H. Predicting Skin Permeability. *Pharm. Res.* **1992**, *9*, 663–669. [CrossRef] [PubMed]
37. Wilschut, A.; ten Berge, W.F.; Robinson, P.J.; McKone, T.E. Estimating skin permeation. The validation of five mathematical skin permeation models. *Chemosphere* **1995**, *30*, 1275–1296. [CrossRef]
38. Mitragotri, S. A theoretical analysis of permeation of small hydrophobic solutes across the stratum corneum based on Scaled Particle Theory. *J. Pharm. Sci.* **2002**, *91*, 744–752. [CrossRef]
39. Lee, P.H.; Conradi, R.; Shanmugasundaram, V. Development of an in silico model for human skin permeation based on a Franz cell skin permeability assay. *Bioorg. Med. Chem. Lett.* **2010**, *20*, 69–73. [CrossRef]
40. Fore, J. A review of skin and the effects of aging on skin structure and function. *Ostomy/Wound Manag.* **2006**, *52*, 24–35.
41. Yu, Y.-Q.; Yang, X.; Wu, X.-F.; Fan, Y.-B. Enhancing Permeation of Drug Molecules Across the Skin via Delivery in Nanocarriers: Novel Strategies for Effective Transdermal Applications. *Front. Bioeng. Biotechnol.* **2021**, *9*, 646554. [CrossRef]
42. Caon, T.; Costa, A.C.; de Oliveira, M.A.; Micke, G.A.; Simões, C.M. Evaluation of the transdermal permeation of different paraben combinations through a pig ear skin model. *Int. J. Pharm.* **2010**, *391*, 1–6. [CrossRef]
43. Björklund, S.; Pham, Q.D.; Jensen, L.B.; Knudsen, N.Ø.; Nielsen, L.D.; Ekelund, K.; Ruzgas, T.; Engblom, J.; Sparr, E. The effects of polar excipients transcutol and dexpanthenol on molecular mobility, permeability, and electrical impedance of the skin barrier. *J. Colloid Interface Sci.* **2016**, *479*, 207–220. [CrossRef] [PubMed]
44. Björklund, S.; Engblom, J.; Thuresson, K.; Sparr, E. Glycerol and urea can be used to increase skin permeability in reduced hydration conditions. *Eur. J. Pharm. Sci.* **2013**, *50*, 638–645. [CrossRef] [PubMed]
45. Osborne, D.W. Diethylene glycol monoethyl ether: An emerging solvent in topical dermatology products. *J. Cosmet. Dermatol.* **2011**, *10*, 324–329. [CrossRef] [PubMed]
46. Di Lorenzo, R.; Bernardi, A.; Grumetto, L.; Sacchi, A.; Avagliano, C.; Coppola, S.; de Giovanni di Santa Severina, A.F.; Bruno, C.; Paparo, L.; Laneri, S.; et al. Phenylalanine Butyramide Is a New Cosmetic Ingredient with Soothing and Anti-Reddening Potential. *Molecules* **2021**, *26*, 6611. [CrossRef]
47. Seo, J.E.; Kim, S.; Kim, B.H. In vitro skin absorption tests of three types of parabens using a Franz diffusion cell. *J. Expo. Sci. Environ. Epidemiol.* **2017**, *27*, 320–325. [CrossRef]
48. El Hussein, S.; Muret, P.; Berard, M.; Makki, S.; Humbert, P. Assessment of principal parabens used in cosmetics after their passage through human epidermis-dermis layers (ex-vivo study). *Exp. Dermatol.* **2007**, *16*, 830–836. [CrossRef]
49. Hatami, A.; Carr, K.; Whiteley, P.; Wilkinson, S.; Dodou, K. Ex vivo studies for the passive transdermal permeation and extent of metabolism of methyl and butyl paraben from a cream. *Br. J. Pharm.* **2017**, *2*, 2.
50. Padula, C.; Pappani, A.; Santi, P. In vitro permeation of levothyroxine across the skin. *Int. J. Pharm.* **2008**, *349*, 161–165. [CrossRef]

51. Suchý, T.; Bartoš, M.; Sedláček, R.; Šupová, M.; Žaloudková, M.; Martynková, G.S.; Foltán, R. Various Simulated Body Fluids Lead to Significant Differences in Collagen Tissue Engineering Scaffolds. *Materials* **2021**, *14*, 4388. [CrossRef]
52. Braumann, T.; Weber, G.; Grimme, L.H. Quantitative structure—activity relationships for herbicides: Reversed-phase liquid chromatographic retention parameter, log kw, versus liquid-liquid partition coefficient as a model of the hydrophobicity of phenylureas, s-triazines and phenoxycarbonic acid derivatives. *J. Chromatogr. A* **1983**, *261*, 329–343.

molecules

MDPI

Article

Monitoring of Pollutants Content in Bottled and Tap Drinking Water in Italy

Giacomo Russo [1], Sonia Laneri [2], Ritamaria Di Lorenzo [2], Ilaria Neri [2], Irene Dini [2], Roberto Ciampaglia [2] and Lucia Grumetto [2,3,*]

1 School of Applied Sciences, Sighthill Campus, Edinburgh Napier University, 9 Sighthill Ct, Edinburgh EH11 4BN, UK; g.russo@napier.ac.uk

2 Department of Pharmacy, School of Medicine and Surgery, University of Naples Federico II, Via D. Montesano, 49, I-80131 Naples, Italy; slaneri@unina.it (S.L.); ritamaria.dilorenzo@unina.it (R.D.L.); ilaria.neri@unina.it (I.N.); irdini@unina.it (I.D.); roberto.ciampaglia@unina.it (R.C.)

3 Consorzio Interuniversitario IIstituto Nazionale di Biostrutture e Biosistemi, Viale Medaglie d'Oro, 305, I-00136 Rome, Italy

* Correspondence: grumetto@unina.it; Tel.: +39-(081)-678628

Abstract: The concentration levels of thirteen organic pollutants and selected heavy metals were investigated in 40 plastics bottled and tap water samples. Some of the selected contaminants have an ascertained or suspected endocrine disrupting activity, such as Bisphenol A (BPA) and its analogs, and Bis 2-ethylhexyl phthalate (DEHP), which are used by industries as plasticizers. The most frequently detected pollutants were Bisphenol AF (BPAF) (detection frequency (DF) = 67.5%, mean 387.21 ng L^{-1}), DEHP (DF = 62.5%, mean 46.19 μg L^{-1}) and BPA (DF = 60.0%, mean 458.57 ng L^{-1}), with higher concentration levels found in tap waters. Furthermore, a possible level of exposure to thirteen pollutants via drinking water intake was calculated. Our findings show that, even though the occurrence of contaminants and heavy metals in drinking waters does not pose an immediate, acute health risk for the population, their levels should be constantly monitored and "hard-wired" into everyday practice. Indeed, the health impact to the continuous and simultaneous intake of a huge variety of xenobiotics from various sources by humans is complex and still not fully understood.

Keywords: drinking water; endocrine disruptors; bisphenols; DEHP; water monitoring; heavy metals

Citation: Russo, G.; Laneri, S.; Di Lorenzo, R.; Neri, I.; Dini, I.; Ciampaglia, R.; Grumetto, L. Monitoring of Pollutants Content in Bottled and Tap Drinking Water in Italy. *Molecules* **2022**, *27*, 3990. https://doi.org/10.3390/molecules27133990

Academic Editors: Victoria Samanidou, Eleni Deliyanni and Dimitra Voutsa

Received: 31 May 2022
Accepted: 17 June 2022
Published: 21 June 2022

1. Introduction

Human water demands depend upon various factors such as climate, physical activity, and diet. However, the average daily water requirement for a sedentary adult's daily intake, ranges from 1.5 to 2.0 L, besides the water already contained in foodstuff [1]. Therefore, due to the very high quantities needed by the human body, water is considered a macronutrient. Unfortunately, one of the significant sources of human exposure to pollutants is paradoxically drinking water, because of its universal solvent properties of many environmental contaminants. European legislation [2] refers to the natural mineral waters as waters "originated from an aquifer or underground reservoir, spring from one or more natural or bore sources and having specific hygienic features and, eventually, healthy properties" [3]. Natural mineral waters differ from drinking water for their spring purity, constant mineral composition, and the specific effects that these can determine. Unfortunately, chemicals released from industrial discharges, runoff, landfill leachate, and wastewater effluents can spread into the groundwater due to their incomplete removal after wastewater treatment and, therefore, pollute water already from water springs. Among the many anthropogenic pollutants that humans discharge in the environment, endocrine-disrupting chemicals (EDCs) have gained the world's scientific attention because of their being hormonally active agents, potentially interfering with endocrine systems [4]. EDCs can be both natural and synthetic, and can mimic hormones, causing developmental, reproductive, brain, and immune diseases, as well as other disorders, including cancer [5–7]. People can be simultaneously exposed to multiple EDCs, with a synergistic action affecting metabolic processes from early development and organ functions [8–10]. People are mostly exposed to EDCs via the intake of foodstuffs and beverages. Indeed, plastic bottles and metal food cans lined with epoxy resins can release free monomers, which have EDC properties, migrating into the food and getting therefore absorbed by the human body [11,12]. This phenomenon may be magnified for foodstuff consumed in high quantities, such as drinking water. Among the huge number of EDCs, Bisphenol A (BPA) and Bis 2-ethylhexyl phthalate (DEHP) are among the most well-characterized EDCs utilized by industries managing plastics material to store beverages, and components of water supply lines. BPA is commonly used in the production of polycarbonates and epoxy resins intended as food contact materials, including tanks for drinking water [13–16], and DEHP is a plasticizer modulating the flexibility and hardness of plastic materials [17]. DEHP, BPA and/or other monomers can be added to the polymer during its synthesis to tune its mechanical properties, such as flame resistance, color, plasticity, viscosity, and lubricity. Free monomers can migrate into the food, especially if the packaging is stored under conditions that promote this, such as long times, high temperature, or sunlight and UV rays exposition [18]. In addition, every pollutant may already be present in the water at its source due to groundwater contamination. Even if the EDCs presence in water and other beverages is assessed as the lowest among other food categories, the tap or bottled water occurrence can be justified by the possible migration from the coating materials of cans, bottle tops, and water supply pipes [19]. European Plastics Regulation No. 10/2011 [20] authorizes BPA as a monomer in the production of plastic with a specific migration limit of 0.05 mg Kg^{-1} of food and sets DEHP migration limit to a maximum of 1.5 mg Kg^{-1}. Despite the sufficiently high limits that make acute intoxication unlikely, the US Environmental Protection Agency (EPA) and US National Toxicology Program (NTP) committees assessed "low dose" effects for several EDCs, defining them "as any biological change occurring in the range of typical human exposures or at doses lower than those tested in traditional toxicology assessments" [21]. Furthermore, due to restrictions set by these laws, industries replaced BPA with compounds having the same structural scaffold, such as bisphenol F (BPF), bisphenol S (BPS), bisphenol A diglycidyl ether (BADGE), bisphenol E (BPE), bisphenol B (BPB), and Bisphenol AF (BPAF), which are currently covered by less stringent or no restrictions at all. These are erroneously considered less toxic than BPA, but they are conversely suspected to be associated with cell toxicity and adverse health effect equal to or even higher than that of the parental compound [22–25]. The present study aims at determining thirteen organic and seven inorganic contaminants in plastic bottled natural mineral water and in tap water, marketed or collected from running taps in Italy, respectively. The organic chemicals investigated included BPA and several of its analogues mentioned above, but also organic pollutants from other classes such as:

1. 2-chlorophenol (2-CP), as s representative of chlorophenols (CPs), banned or controlled by many countries due to its high toxicity and slow biodegradation, but still used as a wood preservative and frequently detected in aquatic and human samples [26,27];
2. 4-nonylphenol (4-NP) is a more toxic [28] biodegradation product than its parental compound, nonylphenol ethoxylate (NPEO), a surfactant used in various consumer products as detergents and wetting products to lower interfacial surface tension [29];
3. 1,4-dichlorobenzene (DCB), and
4. 1,2,4,5-tetrachlorobenzene (TCB) volatile organic compounds mainly used as solvents [30];
5. Triclosan (TCS), an antibacterial agent in consumer products [31–33] and well known environmental pollutants, having suspected or assessed estrogenic activity [34].

Furthermore, the monitoring was extended to several inorganic pollutants to include an array of selected heavy metals, i.e., As, Hg, Pb, Cr, Co, Ni, and Cd. Indeed, heavy metal exposure to living organisms may increase the oxidation stress of cells, possibly inducing DNA damage, protein modification, lipid peroxidation, and, for some of them as Cd, there is rising concern about their action on the endocrine system [35]. Due to their high degree of toxicity, Cd, Cr, Pb, and Hg rank among the priority metals that are of public health significance. These metallic contaminants persist in the environment and can accumulate in the body, leading to multiple organ damage and development of cancer [36]. The metals evaluated in this study shall be those operated by the DM 10-02-2015 (Available online: https://www.gazzettaufficiale.it/eli/id/2015/03/02/15A01419/sg (accessed on 20 May 2022)).

2. Results

2.1. Method Validation

Calibration curves were performed in the range 15.63–250.00 ng mL^{-1} for bisphenol analogs and 4 NP, and in the range 1.25–100 µg mL^{-1} for the different UV detected analytes and a linear R-squared values (r^2) of all the calibration curves ranged from 0.986 to 1. The sensitivity of the method was supported by LOD parameters, with values ranging from 0.68 to 5.65 ng mL^{-1} for FLD detected analytes and from 0.16 to 7.24 µg mL^{-1} for UV detected analytes. LOQ parameters, ranged from 2.26 to 18.83 ng mL^{-1} for analytes FLD detected and from 0.53 to 24.15 µg mL^{-1} for analytes UV detected. Selectivity, i.e., the ability to discriminate the analyte under study from the occurrence of other chemicals possible interfering with its signal, was assessed by utilizing blank processed samples using Milli-Q® water and any interfering signal was observed. Matrix effect values were found ranging from the lowest value of 1.025 to 1.140. The accuracy of the method using the water matrix was assessed via recovery at low and high spiking concentrations (UV detected chemicals 20–60 µg mL^{-1}, and for FLD detected chemicals 1.0–2.0 µg mL^{-1}) of all the investigated chemicals resulting in keeping to what previous reported ranging from 75.2 to 112.5%. Method validation was performed according to 2002/657/EC guidelines [37]. These results indicate that this analytical method provides a reliable response regardless of the concentrations utilized. This is straightforward given that the biological matrices studied in [38] are much more complex than the waters assayed in this work. After running of each sample in triplicate, methanol injections did not show carryover.

2.2. Real Samples

The water samples, both tap and bottled waters, were checked for chemical and physical parameters and found to be compliant to the current Italian legislation [3,39]. The selected EDCs were detected in 37 out of 40 investigated water samples, while only three samples (i.e., water samples, 9, 10 and 13) did not contain any of the thirteen screened compounds to any detectable extent (Tables 1 and 2). Among the FLD molecules, BPF was found in 14 water samples (seven tap water and seven bottled waters), BPE in 11 waters (four tap water and seven bottled waters), BPA in 23 waters (10 tap water and 13 bottled waters), BPB in only seven waters (three tap water and four bottled waters), BADGE was detected in 20 waters (11 tap water and nine bottled waters). BPAF was detected in 26 waters (11 tap waters and 15 bottled waters) (Table 1). The mean concentrations were BPF 55.26 ng L^{-1}, BPE 130.01 ng L^{-1}, BPA 458.57 ng L^{-1}, BPB 43.54 ng L^{-1}, BPAF 387.21 ng L^{-1}, BADGE 353.71 ng L^{-1}, and 4-NP 89.91 ng L^{-1}, respectively. Regarding UV detected investigated chemicals: BPS was found in six water samples (four tap water and two bottled waters), 2-CP in only two water samples and both at <LOQ level (one tap water and one bottled waters), DCB in 18 waters (six tap waters and 12 bottled waters), TCB in two water samples (one tap water and one bottled water), TCS was detected in five waters (one tap water and four bottled waters), and DEHP in 25 waters (eight tap waters and 17 bottled waters) (Table 2). The mean concentrations were BPS 30.74 µg L^{-1}, DCB 61.84 µg L^{-1}, TCB 64.07 µg L^{-1}, TCS 151.08 µg L^{-1}, DEHP 46.19 µg L^{-1}. The most commonly chemical detected by FLD was BPAF with a DF of 67.50% along with BPA (DF = 60%), while DEHP was the most frequently screened contaminant among those UV detected with a DF of 62.50%. Mean and median concentrations, detection frequency (%), and concentration range of all the EDCs under our survey are listed in Table 3 (UV and FLD).

Table 1. FLD detected chemicals under investigation.

Sample	*Origin*	*BPF* ng L^{-1}	*BPE* ng L^{-1}	*BPA* ng L^{-1}	*BPB* ng L^{-1}	*BPAF* ng L^{-1}	*BADGE* ng L^{-1}	*4-NP* ng L^{-1}
Water 1	South Italy spring 1	*	88.37	*	*	811.61	*	117.95
Water 2	South Italy spring 2	*	11.65	*	192.36	512.39	*	81.55
Water 3	Central Italy	*	43.11	*	*	1544.68	*	156.71
Water 4	South Italy	30.29	*	48.94	*	57.35	23.04	*
Water 5	South Italy	9.13	5.63	4.34	*	639.51	29.74	67.81
Water 6	North Italy	*	*	*	*	322.37	*	188.09
Water 7	North Italy	*	*	*	*	179.95	*	*
Water 8	Central Italy	*	*	*	21.78	*	21.15	*
Water 9	Central Italy	*	*	*	*	*	*	*
Water 10	Central Italy	*	*	*	*	*	*	*
Water 11	South Italy	*	*	*	*	110.05	*	*
Water 12	North Italy	*	*	50.57	*	*	375.49	*
Water 13	Central Italy	*	*	*	*	*	*	*
Water 14	South Italy spring 1	*	16.03	*	*	54.57	52.10	41.80
Water 15	South Italy spring 2	14.07	9.24	4.34	2.56	62.53	30.96	67.39
Water 16	South Italy spring 3	7.96	67.91	28.85	*	863.45	*	97.34
Water 17	South Italy spring 1	*	*	*	*	165.17	19.04	33.64
Water 18	South Italy spring 2	*	*	*	*	51.58	*	*
Water 19	South Italy spring 3	42.02	*	18.97	*	*	*	136.22
Water 20	Central Italy spring 1	*	*	*	*	485.90	214.64	128.09
Water 21 #	South Italy	*	23.20	4.34	*	582.82	49.01	66.80
Water 22 #	South Italy	63.05	*	27.47	*	866.37	177.72	34.53
Water 23 #	South Italy	*	*	4.34	3.90	37.71	17.70	4.39
Water 24 #	South Italy pickup site1	*	*	*	*	942.61	101.17	133.47
Water 25 #	South Italy pickup site 2	*	*	*	1.16	1.02	*	*
Water 26 #	South Italy pickup site 3	327.42	164.01	204.54	*	373.44	75.85	120.18
Water 27 #	South Italy	*	*	26.83	*	345.55	*	*
Water 28 #	South Italy	*	*	*	*	35.15	*	57.56
Water 29 #	South Italy	*	*	46.17	*	109.59	1246.91	*
Water 30 #	South Italy	18.18	*	12.82	*	871.92	*	84.87
Water 31	Central Italy	*	*	85.44	*	19.82	*	*
Water 32	South Italy	9.60	*	30.64	*	*	*	*
Water 33	France	6.24	*	25.44	*	*	*	*
Water 34 #	South Italy	77.29	779.53	68.19	67.75	20.30	67.97	*
Water 35 #	South Italy	127.42	*	329.19	*	*	2363.46	*
Water 36 #	South Italy	11.60	*	56.69	*	*	17.09	*
Water 37	Central Italy	*	*	52.13	*	*	*	*
Water 38	North Italy	*	*	716.00	15.24	*	175.15	*
Water 39	South Italy	*	*	1050.00	*	*	173.45	*
Water 40	Central Italy	29.37	223.3	900	*	*	1842.64	*

Values < LOQ = LOD values # Tap waters. * not detected.

Table 2. UV detected chemicals under investigation.

Sample	*Origin*	*BPS* μg L^{-1}	*2-CP* μg L^{-1}	*DCB* μg L^{-1}	*TCS* μg L^{-1}	*TCB* μg L^{-1}	*DEHP* μg L^{-1}
Water 1	South Italy spring 1	*	*	26.28	*	*	21.51
Water 2	South Italy spring 2	6.58	*	*	19.14	*	21.59
Water 3	Central Italy	*	*	*	*	*	22.53
Water 4	South Italy	0.24	*	1.49	*	*	5.89
Water 5	South Italy	*	*	6.76	*	*	5.33
Water 6	North Italy	*	*	*	*	*	20.0
Water 7	North Italy	*	*	*	*	*	*
Water 8	Central Italy	*	*	*	*	*	*

Table 2. *Cont.*

Sample	*Origin*	*BPS* µg L^{-1}	*2-CP* µg L^{-1}	*DCB* µg L^{-1}	*TCS* µg L^{-1}	*TCB* µg L^{-1}	*DEHP* µg L^{-1}
Water 9	Central Italy	*	*	*	*	*	*
Water 10	Central Italy	*	*	*	*	*	*
Water 11	South Italy	*	*	*	*	*	*
Water 12	North Italy	*	*	*	*	*	*
Water 13	Central Italy	*	*	*	*	*	*
Water 14	South Italy spring 1	*	*	15.30	*	*	21.55
Water 15	South Italy spring 2	*	*	*	*	*	6.15
Water 16	South Italy spring 3	*	0.16	27.22	0.26	*	32.46
Water 17	South Italy spring 1	*	*	5.62	*	*	6.20
Water 18	South Italy spring 2	*	*	*	*	*	2.10
Water 19	South Italy spring 3	*	*	16.29	*	*	*
Water 20	Central Italy spring 1	*	*	*	*	*	24.65
Water 21 #	South Italy	*	*	20.18	*	*	3.57
Water 22 #	South Italy	1.83	*	11.03	*	*	5.06
Water 23 #	South Italy	*	*	*	*	*	2.10
Water 24 #	South Italy pickup site1	*	*	24.26	*	*	6.95
Water 25 #	South Italy pickup site 2	*	*	*	*	*	*
Water 26 #	South Italy pickup site3	11.84	0.16	22.18	*	*	35.28
Water 27 #	South Italy	*	*	*	*	*	*
Water 28 #	South Italy	*	*	*	*	31.03	39.24
Water 29 #	South Italy	51.45	*	*	*	*	*
Water 30 #	South Italy	*	*	19.05	*	*	5.04
Water 31	Central Italy	*	*	*	*	*	*
Water 32	South Ital	*	*	207.98	*	*	*
Water 33	France	10.04	*	16.51	*	*	34.90
Water 34 #	South Italy	*	*	*	162.29	*	383.48
Water 35 #	South Italy	*	*	233.40	*	*	*
Water 36 #	South Italy	*	*	*	*	*	*
Water 37	Central Italy	*	*	112.80	*	*	121.26
Water 38	North Italy	*	*	127.47	67.16	97.12	134.52
Water 39	South Italy	*	*	*	355.73	*	75.00
Water 40	Central Italy	*	*	219.36	*	*	118.46

Values < LOQ = LOD values. # Tap waters. * not detected.

Table 3. Mean, median, range of concentration, and detection frequency (DF%) of the investigated chemicals. Values are expressed as ng L^{-1} for FLD detected chemicals and as µg L^{-1} for UV detected chemicals. NA = not applicable.

Compound	*Mean*	*Median*	*Range*	*DF%*
		FLD detected		
BPF	55.26	23.78	6.24–327.42	37.5
BPE	130.01	45.56	2.92–779.53	30.0
BPA	458.57	46.17	4.34–1050.00	60.0
BPB	43.54	9.57	1.16–192.36	20.0
BPAF	387.21	165.17	1.02–1544.68	67.5
BADGE	353.71	71.91	17.7–2363.46	52.5
4-NP	89.91	67.81	4.39–188.09	47.5
		UV detected		
2-CP	NA	NA	NA	5.0
BPS	30.74	8.31	0.24–51.45	15.0
DCB	61.84	13.77	1.49–233.4	45.0
TCB	64.07	NA	31.03–97.12	5.0
TCS	151.08	19.14	0.26–355.73	12.5
DEHP	46.19	21.59	2.10–383.48	62.5

2.3. Heavy Metals

All heavy metal concentrations in the samples studied were found to be below the national standard limits (DM 10-02-2015 Available online https://www.gazzettaufficiale.it/eli/id/2015/03/02/15A01419/sg (accessed on 20 May 2022) (Tables 4 and 5). However, the control and monitoring of heavy metals are important, as their presence could cause serious harm to the population, considering the presence of other substances creating a mixture harmful to human health.

Table 4. Analytical parameters for the heavy metals screened in the present study.

Element	*Wavelength* (nm)	*Slit Width* (nm)	*LOD* (mg/L)	*LMA* (mg/L) *	*Calibration Curve*	*LR*	R^2
Hg	253.7	0.2	0.0001	0.0010	y = 0.0005 x + 0.0001	0–20	0.9810
Pb	283.3	0.2	0.001	0.010	y = 0.0315 x − 0.01454	0–20	0.9990
Cr	357.9	0.5	0.0001	0.050	y = 0.0008 x + 0.0089	0–10	0.9972
Co	240.7	0.2	0.0001	NA.	y = 0.0994 x + 0.0279	0–20	0.9609
Ni	232.0	0.2	0.0001	0.020	y = 0.0754 x + 0.0143	0–20	0.9983
Cd	228.8	0.5	0.0003	0.003	y = 0.0127 x + 0.00175	0–10	0.9983
As	193.7	0.2	0.001	0.010	y = 0.0016 x + 0.0006	0–20	0.9797

LMA: Limit maximum admissible. LR: Linear range. NA = not applicable. * DM 10-02-2015.

Table 5. Heavy metals analysis in forty investigated water samples.

Sample	*Hg*	*Pb*	*Cr*	*Co*	*Ni*	*Cd*	*As*
Water 1	<0.0001	0.000223	0.00056	<0.0001	0.0005	0.00056	0.00081
Water 2	<0.0001	0.000223	0.00056	<0.0001	0.0005	0.00056	0.00081
Water 3	<0.0001	0.000162	0.00015	<0.0001	0.00016	0.00018	0.00018
Water 4	<0.0001	0.000555	0.00034	<0.0001	0.00063	0.00084	0.00088
Water 5	<0.0001	0.000398	0.00022	<0.0001	0.00068	0.00088	0.00238
Water 6	<0.0001	0.00022	0.00088	<0.0001	0.00045	0.00059	0.00281
Water 7	<0.0001	0.000162	0.00015	<0.0001	0.00016	0.00018	0.00018
Water 8	<0.0001	0.000162	0.00015	<0.0001	0.00016	0.00018	0.00018
Water 9	<0.0001	0.000224	0.00056	<0.0001	0.00022	0.00021	0.0033
Water 10	<0.0001	0.000433	0.00023	<0.0001	0.00027	0.00034	0.00029
Water 11	<0.0001	0.000432	0.00022	<0.0001	0.00033	0.00033	0.00344
Water 12	<0.0001	0.000162	0.00015	<0.0001	0.00016	0.00018	0.00018
Water 13	<0.0001	0.000311	0.00039	<0.0001	0.00044	0.00022	0.00026
Water 14	<0.0001	0.000331	0.00025	<0.0001	0.00361	0.00079	0.00178
Water 15	<0.0001	0.000313	0.00061	<0.0001	0.00264	0.00079	0.00105
Water 16	<0.0001	0.000422	0.00038	<0.0001	0.00033	0.00085	0.00109
Water 17	<0.0001	0.00038	0.00024	<0.0001	0.00048	0.00092	0.00077
Water 18	<0.0001	0.000231	0.00034	<0.0001	0.00019	0.00018	0.00055
Water 19	<0.0001	0.000231	0.00034	<0.0001	0.00019	0.00018	0.00055
Water 20	<0.0001	0.000299	0.00021	<0.0001	0.00056	0.00036	0.00018
Water 21	<0.0001	0.000561	0.00027	<0.0001	0.00244	0.00085	0.00126
Water 22	<0.0001	0.000259	0.00015	<0.0001	0.00054	0.00081	0.00091

Table 5. *Cont.*

Sample	*Hg*	*Pb*	*Cr*	*Co*	*Ni*	*Cd*	*As*
Water 23	<0.0001	0.00047	0.00044	<0.0001	0.00066	0.00077	0.00158
Water 24	<0.0001	0.000404	0.00017	<0.0001	0.00033	0.00085	0.00084
Water 25	<0.0001	0.000404	0.00017	<0.0001	0.00033	0.00085	0.00084
Water 26	<0.0001	0.000404	0.00017	<0.0001	0.00033	0.00085	0.00084
Water 27	<0.0001	0.000368	0.00023	<0.0001	0.00079	0.00083	0.00109
Water 28	<0.0001	0.00047	0.00044	<0.0001	0.00066	0.00077	0.00158
Water 29	<0.0001	0.000563	0.00055	<0.0001	0.00019	0.00026	0.00054
Water 30	<0.0001	0.000579	0.00046	<0.0001	0.00175	0.00088	0.00103
Water 31	<0.0001	0.000299	0.00021	<0.0001	0.00056	0.00036	0.00018
Water 32	<0.0001	0.000311	0.00034	<0.0001	0.00041	0.00045	0.0019
Water 33	<0.0001	0.000162	0.00015	<0.0001	0.00016	0.00018	0.00018
Water 34	<0.0001	0.000579	0.00046	<0.0001	0.00175	0.00088	0.00103
Water 35	<0.0001	0.000579	0.00046	<0.0001	0.00175	0.00088	0.00103
Water 36	<0.0001	0.000579	0.00046	<0.0001	0.00175	0.00088	0.00103
Water 37	<0.0001	0.000341	0.00067	<0.0001	0.00016	0.00032	0.00034
Water 38	<0.0001	0.000162	0.00015	<0.0001	0.00016	0.00018	0.00018
Water 39	<0.0001	0.000162	0.00025	<0.0001	0.0002	0.00083	0.00077
Water 40	<0.0001	0.000456	0.00031	<0.0001	0.00071	0.00067	0.00067

LMA: Limit maximum admissible; LMD: Limit minimum of detection. LMA Hg = 0.001 for all water samples; LMA Pb, As = 0.01 for all water samples; LMA Cr = 0.05 for all water samples; LMA Ni = 0.02 for all water samples; LMA Cd = 0.003 for all water samples; LMA Co = ND for all water samples. LMD Hg, Cr, Co, Ni = 0.0001 for all water samples; LMD Pb, As = 0.001; LMD Cd = 0.0003 for all water samples.

2.4. Estimated Daily Intake

Regarding human safety, the estimated daily intakes of individual EDCs considered in our study for male and female adults and young people were plotted and shown in Figure 1. The results indicate that the highest values of EDI concern BPA, BPAF, BADGE and TCS for all the gender and age groups, even if adult female group EDI values are slightly lower than the others, due to, probably, their IR lower if compared to IR of male adult, and due to the higher body weight if compared to the youngest categories. Instead, EDI values were found not considerably affected by the age. Furthermore, we also calculated EDI/TDI ratio (Table 6), of BPA, BPAF, BADGE and TCS, which are the chemicals found at the highest concentration values, based on TDI EFSA values (BPA 4 μg/kg BW/day), (BADGE 0.15 mg/Kg BW/day) and (BPAF and TCS 5 mg/Kg BW/day).

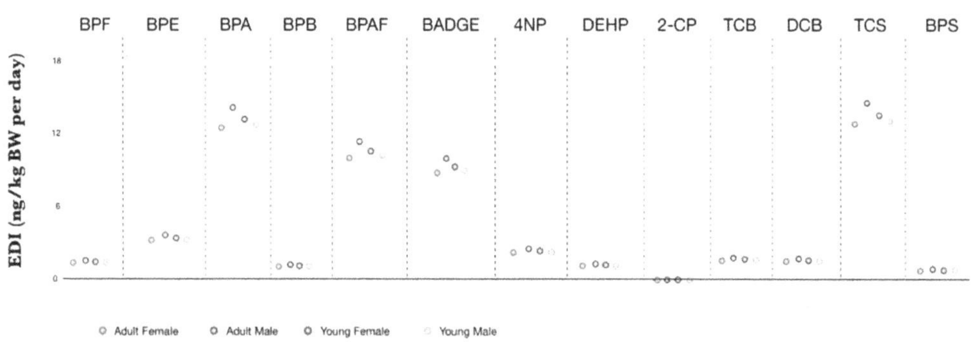

Figure 1. Estimated daily intake (ng/kg BW per day) of detected EDCs via drinking water consumption for age and gender-specific groups.

Table 6. Calculated EDI/TDI ratio for adult and young, male, and female, based on TDI EFSA values (BPA 4 µg/kg BW/day), (BADGE 0.15 mg/kg BW/day) and BPAF and TCS (5 mg/kg BW/day).

Analyte	*Adult Male*	*Adult Female*	*Young Male*	*Young Female*
BPA	0.003568	0.003147	0.003216	0.003319
BPAF	2.29×10^{-6}	2.02×10^{-6}	2.06×10^{-6}	2.13×10^{-6}
BADGE	6.71×10^{-5}	5.91×10^{-5}	6.04×10^{-5}	6.24×10^{-5}
TCS	2.94×10^{-6}	2.59×10^{-6}	2.65×10^{-6}	2.73×10^{-6}

3. Discussion

BPA, BPAF, and DEHP (Figure 2a,b), were observed in waters samples with a detection frequency of more than 60%, confirming that the observed contamination by these chemicals is due to the anthropogenic activities impacting on the environment. Indeed, packaging industries widely exploit monomers as plasticizers to manufacture food or beverage contact materials [40]. However, these, and other monomers, often leach from packaging. The highest DF and average concentration values of BPAF and its parent compound BPA found in the water samples (Figure 2a), indicate that industries still use the parent chemical, but at the same time increased the exploitation of structurally related bisphenols, probably due to legal restriction set for BPA.

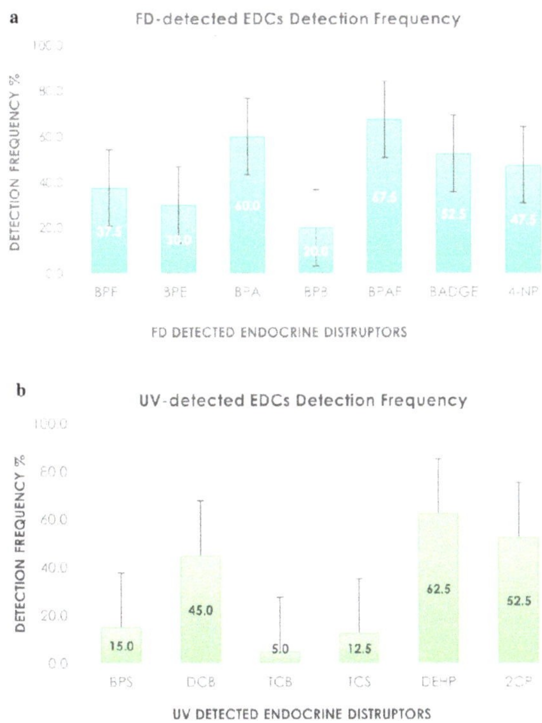

Figure 2. Detection frequency of FD (**a**) and UV (**b**) detected EDCs.

Literature reports aimed at studying the occurrence of EDCs in drinking waters in various countries around the world were used to compare the data achieved in this study [41–52]. For most pollutants, we realized that no straightforward comparison on pollutant contamination can be drawn as current data on the drinking waters retailed in most countries is either missing or outdated. Nevertheless, we still believe it is worth comparing our beverage surveillance data with the most representative papers discussing the contamination levels of the main chemicals covered in the present article, in Europe and overseas. BPA mean concentration (458.57 ng L^{-1}) is almost ten times higher than those reported by Valcarcel et al. [42] on 30 chemicals in drinking water of Central Spain (51.23 ng L^{-1}), and five times higher than that determined by Maggioni et al. [50] performed only on drinking water from public fountains in 35 Italian cities (up to 102 ng L^{-1}). However, the concentration values we retrieved are rather similar to those observed in China by Li et al. [43], who detected BPA concentrations up to 317 ng L^{-1}. Furthermore, bisphenol analogs in drinking waters were investigated by Zhang and coworkers in source and drinking waters in China [44], with lower concentration values than those reported in our current study.

The 4-NP concentration values we retrieved (89.91 ng L^{-1}) were close to those by Maggioni et al. (84 ng L^{-1}) [50], but higher than those by Li et al. [43] in China (1.987 ng L^{-1}). However, 4-NP concentration in drinking water retailed in Spain by Esteban et al. in Spain (15.0 ng L^{-1}) [41], and by Kuch and Ballschmiter [45] in Germany (16.0 ng L^{-1}) were about one five times lower than those recorded in the present study.

Triclosan is an antimicrobial agent added to many household and personal care products such as bar soaps, detergents, body, and hair wash. Several studies have reported its occurrence in water samples in concentrations up to 9.74 ng L^{-1}, which are much lower than those found in our study [51].

Our results concerning DEHP concentrations showed values in the range 2.10–383.48 μg L^{-1}, which is higher than those reported by Jeddi et al. [48] in Iran (0.217 μg L^{-1}) and in a literature review by Luo et al. [49], which reporting average levels of DEHP in bottled water retailed in Thailand, Croatia, and the Czech Republic, as high as 61.1, 8.8, and 6.3 μg L^{-1}, respectively.

We point out that the daily dietary intakes estimated in our study may underestimate the actual exposures because we considered only water consumption, while many foodstuffs other than drinking water can be contaminated with EDCs and or other chemicals [15,52]. With regard to the comparison between tap and bottled waters, our data support higher concentration levels for each compound (Figure 3) in tap water than bottled water, except for TCS, probably due to the higher purity of source water and the absence of any water treatments.

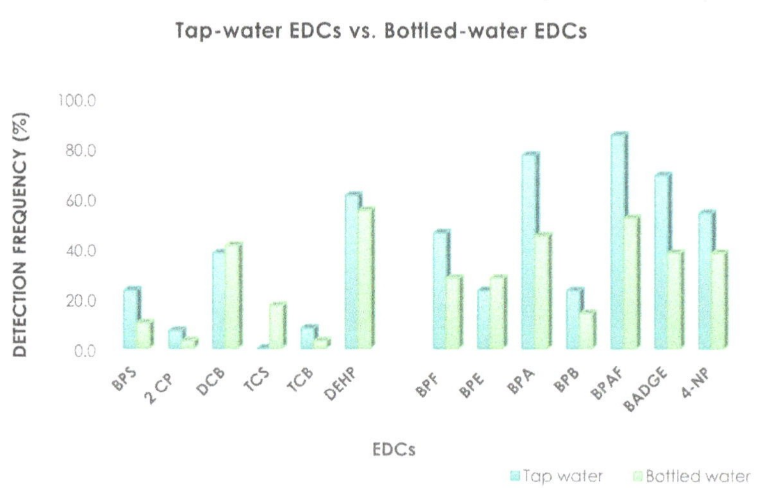

Figure 3. Detection frequency of each investigated chemicals in tap water vs. bottled waters.

4. Materials and Methods

4.1. Analysis of Organic Pollutants

4.1.1. Reagents and Chemicals

For the analysis of organic pollutants, analytical standards, BPF (CAS No.620-92-8), BPS (CAS No.80-09-1), BPA (CAS No.80-05-7), BADGE (CAS No.1675-54-3),2-CP (CAS No.95-57-8), 4-NP (CAS No.104-40-5), DCB (CAS No.106-46-7), TCB (CAS No.95-94-3), DEHP (CAS No.117-81-7), and TCS (CAS No.3380-34-5) were purchased from Sigma-Aldrich (Dorset, United Kingdom). BPE (CAS No.2081-08-5), BPB (CAS No.: 77-40-7), and BPAF (CAS No.1478-61-1) were purchased from TCI Europe (Zwijndrecht, Belgium). Methanol (HPLC analytical grade) and formic acid (minimum purity $\geq$ 95%) were both purchased from Sigma-Aldrich (Milan, Italy). Milli Q water (from here onward named "distilled water") was produced in-house, and its conductivity was 0.055 μS cm^{-1} at 25 °C (resistivity equals 18.2 MΩ cm).

4.1.2. Samples and Standard Solutions

All the standard stock solutions (2.0 mg mL^{-1}) of each compound were prepared in methanol, as well as those of the mixed standard solution of all the investigated chemicals; the solutions were stored at −24 °C in the dark until the use. The calibration solutions (15.63–250.00 ng mL^{-1} for bisphenols and 4-NP, which were detected by fluorescence detection and 1.25–100.00 μg mL^{-1}, for all the other chemicals, which were detected by UV) were freshly prepared for instrumental analysis. Plasticware was treated with a solution of 50:50 n-hexane: tetrahydrofuran [53], water and solvents used for sample preparation were previously analytically [38] verified as chemicals free to avoid any possible background contamination.

From June until December 2019, we retailed 40 water samples, of which 27 drinking water from various Italian wellsprings (North, Central and South of Italy, and one marketed in Italy but from French spring), packaged in bottles of polyethylene terephthalate (PET), one of the most used types of plastic for beverages for its convenience and low cost, and 13 tap water samples from the water distribution Italian system. Figure 4 shows the sampling sites of tap waters and the spring sites of marketed bottled water. For each brand, we took two samples from two different bottles with the same batch number. Tap waters were directly collected from running taps after 2 min of flushing. Each sampling was taken into two pre-rinsed glass bottles, stored at 4 °C, and analyzed within one week.

Figure 4. Map of tap water sampling and marketed bottled water springs. Waters 1, 2, 14–16, and 17–19 were of different brand and different spring in the same area; tap waters, were sampled in different cities (Waters 21, 22, 27, 29, 30, 34–36) or in the same city, but served by different aqueducts or pipelines (city 1: Waters 23, 28; city 2 Waters 24–26).

4.1.3. Sample Preparation

Figure 5 synthetizes samples preparation and analyses steps applied in this study. In brief, the excessive residual chlorine in tap water was removed using 0.2 g L^{-1} of ascorbic acid to fix the concentration of chlorination by products in time [54]. After that, a 200.0 mL volume of each water sample was ultrasonicated (Elmasonic S 30/H, frequency 40 kHz) for 15 min to remove CO_2 and then filtered through a 0.45 μm nylon membrane. SPE (200-mg Strata X Polymeric Sorbent from Phenomenex, Torrance, CA, USA) was performed on water samples (175 mL) added of 75 mL of methanol to quantitatively solubilize the investigated chemicals. The SPE reversed-phase cartridge were conditioned with 4.0 mL of methanol and activated by 4.0 mL of distilled water according to the SPE manufacturer guidelines. Afterward the sample solutions were loaded onto the cartridge, washed with a total volume of 4.0 mL of a 95/5 (*v*/*v*) water/methanol solution, and subsequently with 12.0 mL of 80/20 (*v*/*v*) water/methanol. Finally, the target contaminants were eluted using 8.0 mL of methanol. All the procedures were vacuum driven. The eluate was evaporated to dryness (Thermo Scientific Savant DNA SpeedVac Concentrator Kits), and the residue was reconstituted with 1.0 mL of acetonitrile for the analysis.

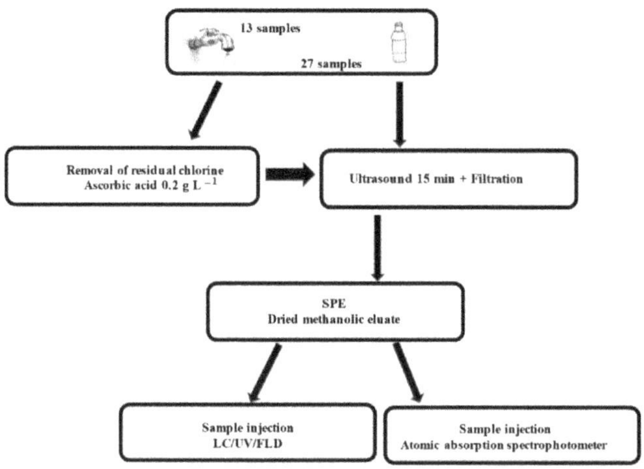

Figure 5. Samples preparation and analysis procedure.

4.1.4. Analysis

A separation chromatographic method, already reported in the literature [38], was fully applied. In brief, each sample was injected three times. Analyses were performed at room temperature (22 ± 2 °C). A high-performance liquid chromatography (LC-20 AD VP; Shimadzu Corp., Kyoto, Japan), equipped with an ultraviolet (UV)–visible detector (Shimadzu Model SPD10 AV) set at λ 220 nm and Fluorescence detector (FLD) set at the excitation wavelength of 263 nm and at the emission wavelength of 305 nm was conducted. The analytical column was a reversed-phase LC column Kinetex phenyl-hexyl (100 Å, 150 × 4.6 mm, 5.0 μm particle size), equipped with a precolumn (4 × 3.0 mm) (Phenomenex, Torrance, CA, USA). All mobile phases were vacuum filtered through 0.45 μm nylon membranes (Millipore, Burlington, MA, USA). BPF, BPE, BPA, BPB, BPAF, BADGE and 4-NP were determined by FLD, while BPS, 2-CP, DCB, TCB, TCS and DEHP were Ultraviolet detected (UV). Data acquisition and integration were accomplished by Cromatoplus 2011 software. Methanol injections were made randomly to assess that no carryover occurred.

4.2. Heavy Metals Analysis

4.2.1. Sample Preparation

For the analysis of heavy metals, multi-element standard solution IV 1000 mg L^{-1}: Ag, Al, Ba, Bi, Ca, Cd, Co, Cr, Cu, Fe, Ga, In, K, Mg, Mn, Na, Ni, Pb, Sr, Tl, Zn and a Hg and As standard solution of 1000 mg L^{-1},—were purchased from Merck KGaA (Darmstadt, Germany). Water with same specifications as discussed in 4.1.1 was used.

4.2.2. Heavy Metal Analysis

Calibration blank (Cal Blk) and calibration standard (Cal Std) solutions were prepared at concentrations of 0 μg L^{-1} (Cal Blk), 4 μg L^{-1} (Cal Std 1), 12 μg L^{-1} (Cal Std 2), and 20 μg L^{-1} (Cal Std 3), using the selenium standard stock solution, or multi-element standard stock solution, into separate 50 mL DigiTUBE® tubes (SCP SCIENCE, Baie-D'Urfe, QC, Canada) with the addition of 0.5 mL internal standard stock solution, 0.5 mL methanol and 5 mL concentrated nitric acid, and then diluted to the final volume with water. A Shimadzu GFA-EX7i graphite furnace mounted on a Shimadzu model 6300 atomic absorption spectrophotometer was used with a Shimadzu ASC-6100 autosampler. Pyrolytically coated graphite tubes and background correction (D2 lamp) were utilized. Working solutions were prepared by diluting immediately before use standard solutions for absorption a spectroscopy.

4.3. Exposure Assessment

The estimated daily intake (EDI ng/kg BW per day) of each chemical under our research, was calculated considering the water consumption, gender, age, and body weight (BW) according to the following Equation reported by Shi et al. [55]:

$$EDI = (C \times IR \times AP)/BW$$

where C is the average concentration for each detected chemicals in water sample (ng L^{-1}), IR is the ingestion rate of water, AP is the absorption percent of intake, which is assumed to be 100% for drinking water, and BW is body weight (kg). Concentrations below the LOQs were assumed as LODs, to calculate the average values. Age and gender-specific BW along with the intake of drinking water were considered, therefore IR for water and BW were calculated as follows: 2.23 L/day for a 78.4 kg adult male (≥18 years old), 1.65 L/day for a 62.2 kg adult female (≥18 years old), 0.81 L/day for a 31.6 kg boy (0–14 years old), and 0.76 L/day for a 28.73 kg girl (0–14 years old), respectively, and according to literature [55]. For young people (0–14 years old) BW used for EDI calculation was based on the mean value of three different subcategories (0–2.9, 3–9.9 and 10–14 years).

4.4. Data Analysis

All statistical analysis was done using a commercially available statistical package for personal computers (Microsoft Excel® 2016, Washington, DC, USA). Data are expressed as the mean of three values.

5. Conclusions

The occurrence of organic and inorganic contaminants still raises many questions about risk they may pose to human health. In this pilot monitoring study, several pollutants with suspected or ascertained EDC properties were detected in drinking water at different concentration ranges spanning from ng to μg per L. All forty waters had a total content not exceeding 10 milligrams of total constituents released per dm^2 of food contact surface (mg/dm^2) according to the Regulation (EU) No 10/2011 and its amendment Regulation (EU) 2018/213, on plastic materials and articles intended to come into contact with foodstuffs [20].

These findings evidence the need of more frequent surveillance practices aimed at detecting and quality control freshwaters intended for human consumption. However, it is essential to emphasize that the co-occurrence of other contaminants, as well as "natural" elements such as heavy metals, despite their low concentration values, can cause additive/synergistic interactions with possible unexpected effects on human health, especially if the exposure is over a long period of time. Consistently, a proposal to revise the Drinking Water Directive, aimed at updating the water quality standards, including for the first-time EDCs, has been recently approved [56]. This study highlights the importance of assessing their occurrence, regardless of its being low doses, even more in drinking water than other foods. This is because the consumption of water is high and frequent, and assessing the human exposure to contaminants is also highly relevant to inform the legislative work of the main government bodies.

Author Contributions: Conceptualization: G.R. and L.G.; data curation: R.C. and G.R.; formal analysis: R.D.L., I.N., R.C.; methodology: G.R.; software: I.N., R.D.L. and R.C.; supervision: L.G.; validation: I.N.; visualization: I.D.; writing—original draft preparation: L.G.; writing—review and editing: S.L. All authors have read and agreed to the published version of the manuscript.

Funding: This research received no external funding.

Institutional Review Board Statement: Not applicable.

Informed Consent Statement: Not applicable.

Data Availability Statement: Not applicable.

Conflicts of Interest: The authors declare no conflict of interest.

Sample Availability: Samples of the waters are not available from the authors.

References

1. Jéquier, E.; Constant, F. Water as an essential nutrient: The physiological basis of hydration. *Eur. J. Clin. Nutr.* **2010**, *64*, 115–123. [CrossRef] [PubMed]
2. European Commission. *Directive 2009/54/EC of the European Parliament and of the Council of 18 June 2009 on the Exploitation and Marketing of Natural Mineral Waters (Recast) (Text with EEA Relevance)*; European Commission: Brussels, Belgium, 2009.
3. Gazzetta Ufficiale Repubblica Italiana DECRETO LEGISLATIVO 8 Ottobre 2011, n. 176 Attuazione della Direttiva 2009/54/CE, Sull'utilizzazione e la Commercializzazione delle Acque Minerali Naturali.(11G0218). 2011. Available online: https://www.gazzettaufficiale.it/eli/id/2011/11/05/011G0218/sg (accessed on 20 January 2022).
4. Kabir, E.R.; Rahman, M.S.; Rahman, I. A review on endocrine disruptors and their possible impacts on human health. *Environ Toxicol. Pharmacol.* **2015**, *40*, 241–258. [CrossRef] [PubMed]
5. Küblbeck, J.; Vuorio, T.; Niskanen, J.; Fortino, V.; Braeuning, A.; Abass, K.; Rautio, A.; Hakkola, J.; Honkakoski, P.; Levonen, A.L. The EDCMET Project: Metabolic Effects of Endocrine Disruptors. *Int. J. Mol. Sci.* **2020**, *21*, 3021. [CrossRef] [PubMed]
6. Marotta, V.; Russo, G.; Gambardella, C.; Grasso, M.; La Sala, D.; Chiofalo, M.G.; D'Anna, R.; Puzziello, A.; Docimo, G.; Masone, S.; et al. Human exposure to bisphenol AF and diethylhexylphthalate increases susceptibility to develop differentiated thyroid cancer in patients with thyroid nodules. *Chemosphere* **2019**, *218*, 885–894. [CrossRef]
7. Shafei, A.; Ramzy, M.M.; Hegazy, A.I.; Husseny, A.K.; El-Hadary, U.G.; Taha, M.M.; Mosa, A.A. The molecular mechanisms of action of the endocrine disrupting chemical bisphenol A in the development of cancer. *Gene* **2018**, *647*, 235–243. [CrossRef]
8. Brenker, C.; Rehfeld, A.; Schiffer, C.; Kierzek, M.; Kaupp, U.B.; Skakkebæk, N.E.; Strünker, T. Synergistic activation of CatSper Ca^{2+} channels in human sperm by oviductal ligands and endocrine disrupting chemicals. *Hum. Reprod.* **2018**, *33*, 1915–1923. [CrossRef]
9. Hamid, N.; Junaid, M.; Pei, D.S. Combined toxicity of endocrine-disrupting chemicals: A review. *Ecotoxicol. Environ. Saf.* **2021**, *215*, 112136. [CrossRef]
10. Mao, W.; Song, Y.; Sui, H.; Cao, P.; Liu, Z. Analysis of individual and combined estrogenic effects of bisphenol, nonylphenol and diethylstilbestrol in immature rats with mathematical models. *Environ. Health Prev. Med.* **2019**, *24*, 32. [CrossRef]
11. Fattore, M.; Russo, G.; Barbato, F.; Grumetto, L.; Albrizio, S. Monitoring of bisphenols in canned tuna from Italian markets. *Food Chem. Toxicol.* **2015**, *83*, 68–75. [CrossRef]
12. Russo, G.; Barbato, F.; Grumetto, L. Development and Validation of a LC-FD Method for the Simultaneous Determination of Eight Bisphenols in Soft Drinks. *Food Anal. Methods* **2016**, *9*, 2732–2740. [CrossRef]
13. Bae, B.; Jeong, J.H.; Lee, S.J. The quantification and characterization of endocrine disruptor bisphenol-A leaching from epoxy resin. *Water Sci. Technol.* **2002**, *46*, 381–387. [CrossRef] [PubMed]
14. Lane, R.F.; Adams, C.D.; Randtke, S.J.; Carter, R.E., Jr. Bisphenol diglycidyl ethers and bisphenol A and their hydrolysis in drinking water. *Water Res.* **2015**, *72*, 331–339. [CrossRef] [PubMed]

15. Russo, G.; Barbato, F.; Mita, D.G.; Grumetto, L. Occurrence of Bisphenol A and its analogues in some foodstuff marketed in Europe. *Food Chem. Toxicol.* **2019**, *131*, 110575. [CrossRef] [PubMed]
16. Vandenberg, L.N.; Hauser, R.; Marcus, M.; Olea, N.; Welshons, W.V. Human exposure to bisphenol A (BPA). *Reprod. Toxicol.* **2007**, *24*, 139–177. [CrossRef] [PubMed]
17. Jayaweera, M.; Perera, H.; Bandara, N.; Danushika, G.; Gunawardana, B.; Somaratne, C.; Manatunge, J.; Zoysa, K.; Thathsara, T. Migration of phthalates from PET water bottle in events of repeated uses and associated risk assessment. *Environ. Sci. Pollut. Res. Int.* **2020**, *27*, 39149–39163. [CrossRef] [PubMed]
18. Kako, S.; Isobe, A.; Kataoka, T.; Hinata, H. A decadal prediction of the quantity of plastic marine debris littered on beaches of the East Asian marginal seas. *Mar. Pollut. Bull.* **2014**, *81*, 174–184. [CrossRef]
19. Amiridou, D.; Voutsa, D. Alkylphenols and phthalates in bottled waters. *J. Hazard. Mater.* **2011**, *185*, 281–286. [CrossRef]
20. European Commission. *Commission Regulation (EU) No 10/2011 of 14 January 2011 on Plastic Materials and Articles Intended to Come into Contact with Food Law*; E.U., Ed.; European Commission: Brussels, Belgium, 2011.
21. Melnick, R.; Lucier, G.; Wolfe, M.; Hall, R.; Stancel, G.; Prins, G.; Gallo, M.; Reuhl, K.; Ho, S.M.; Brown, T.; et al. Summary of the National Toxicology Program's report of the endocrine disruptors low-dose peer review. *Environ. Health Perspect.* **2002**, *110*, 427–431. [CrossRef]
22. Chen, D.; Kannan, K.; Tan, H.; Zheng, Z.; Feng, Y.L.; Wu, Y.; Widelka, M. Bisphenol Analogues Other Than BPA: Environmental Occurrence, Human Exposure, and Toxicity-A Review. *Environ. Sci. Technol.* **2016**, *50*, 5438–5453. [CrossRef]
23. Huang, M.; Liu, S.; Fu, L.; Jiang, X.; Yang, M. Bisphenol A and its analogues bisphenol S, bisphenol F and bisphenol AF induce oxidative stress and biomacromolecular damage in human granulosa KGN cells. *Chemosphere* **2020**, *253*, 126707. [CrossRef]
24. Mustieles, V.; D'Cruz, S.C.; Couderq, S.; Rodríguez-Carrillo, A.; Fini, J.B.; Hofer, T.; Steffensen, I.L.; Dirven, H.; Barouki, R.; Olea, N.; et al. Bisphenol A and its analogues: A comprehensive review to identify and prioritize effect biomarkers for human biomonitoring. *Environ. Int.* **2020**, *144*, 105811. [CrossRef] [PubMed]
25. Russo, G.; Capuozzo, A.; Barbato, F.; Irace, C.; Santamaria, R.; Grumetto, L. Cytotoxicity of seven bisphenol analogues compared to bisphenol A and relationships with membrane affinity data. *Chemosphere* **2018**, *201*, 432–440. [CrossRef] [PubMed]
26. Khalil, F.; Amir, A.M.; Hatam, A.; Pourya, B.; Hossein, T.; Amir, A.F.; Hossein, P. A novel dispersive micro solid phase extraction using zein nanoparticles as the sorbent combined with headspace solid phase micro-extraction to determine chlorophenols in water and honey samples by GC–ECD. *Talanta* **2014**, *128*, 493–499.
27. Zheng, D.; Jiao, H.; Zhong, H.; Qiu, J.; Yan, X.; Duan, Q.; Chai, L. Chlorophenols in marine organisms from the southern coast of Hangzhou Bay, China, and an assessment of risks posed to human health. *J. Oceanol. Limnol.* **2018**, *36*, 726–737. [CrossRef]
28. Ying, G.; Williams, B.; Kookana, R. Environmental fate of alkylphenols and alkylphenol ethoxylates—A review. *Environ. Int.* **2002**, *28*, 215–226. [CrossRef]
29. Ji, X.; Li, N.; Yuan, S.; Zhou, X.; Ding, F.; Rao, K.; Ma, M.; Wang, Z. A comparison of endocrine disruption potential of nonylphenol ethoxylate, vanillin ethoxylate, 4-n-nonylphenol and vanillin in vitro. *Ecotoxicol. Environ. Saf.* **2019**, *175*, 208–214. [CrossRef]
30. Lehmann, I.; Roder-Stolinski, C.; Nieber, K.; Fischader, G. In vitro models for the assessment of inflammatory and immuno-modulatory effects of the volatile organic compound chlorobenzene. *Exp. Toxicol. Pathol.* **2008**, *60*, 185–193. [CrossRef]
31. Montagnini, B.G.; Pernoncine, K.V.; Borges, L.I.; Costa, N.O.; Moreira, E.G.; Anselmo-Franci, J.A.; Kiss, A.C.I.; Gerardin, D.C.C. Investigation of the potential effects of triclosan as an endocrine disruptor in female rats: Uterotrophic assay and two-generation study. *Toxicology* **2018**, *410*, 152–165. [CrossRef]
32. Sharma, P.; Tseng, H.H.; Lee, J.L.; Tsai, E.M.; Suen, J.L. A prominent environmental endocrine disruptor, 4-nonylphenol, promotes endometriosis development via plasmacytoid dendritic cells. *Mol. Hum. Reprod.* **2020**, *26*, 601–614. [CrossRef]
33. Makita, Y. Effects of perinatal, combined exposure to 1,4-dichlorobenzene and 1,1-dichloro-2,2-bis(p-chlorophenyl)ethylene on rat female reproductive system. *Basic Clin. Pharmacol. Toxicol.* **2008**, *102*, 360–364. [CrossRef]
34. Stoker, T.E.; Gibson, E.K.; Zorrilla, L.M. Triclosan exposure modulates estrogen-dependent responses in the female wistar rat. *Toxicol. Sci.* **2010**, *117*, 45–53. [CrossRef] [PubMed]
35. Tchounwou, P.B.; Yedjou, C.G.; Patlolla, A.K.; Sutton, D.J. Heavy metal toxicity and the environment. *Exp. Suppl.* **2012**, *101*, 133–164. [PubMed]
36. Wu, X.; Cobbina, S.J.; Mao, G.; Xu, H.; Zhang, Z.; Yang, L. A review of toxicity and mechanisms of individual and mixtures of heavy metals in the environment. *Environ. Sci. Pollut. Res. Int.* **2016**, *23*, 8244–8259. [CrossRef] [PubMed]
37. European Commission. EC implementing Council Directive 96/23/EC, Concerning the Performance of Analytical Methods and the Interpretation of Results. 2002/657/EC. *Off. J. Eur. Commun.* **2002**, *221*, 8–36.
38. Russo, G.; Barbato, F.; Mita, D.G.; Grumetto, L. Simultaneous determination of fifteen multiclass organic pollutants in human saliva and serum by liquid chromatography-tandem ultraviolet/fluorescence detection: A validated method. *Biomed. Chromatogr.* **2019**, *33*, e4427. [CrossRef] [PubMed]
39. Gazzetta Ufficiale Repubblica Italiana. Attuazione della Direttiva 98/83/CE Relativa alla Qualita' delle Acque Destinate al Consumo Umano. Gazzetta Ufficiale n. 52 del 3 Marzo 2001—Supplemento Ordinario n. 41. 2001. Available online: https://www.gazzettaufficiale.it/eli/id/2001/03/03/001G0074/sg (accessed on 20 January 2022).
40. Alamri, M.S.; Qasem, A.A.A.; Mohamed, A.A.; Hussain, S.; Ibraheem, M.A.; Shamlan, G.; Alqah, H.A.; Qasha, A.S. Food packaging's materials: A food safety perspective. *Saudi J. Biol. Sci.* **2021**, *28*, 4490–4499. [CrossRef]

41. Esteban, S.; Gorga, M.; Petrovic, M.; Gonzalez-Alonso, S.; Barcelo, D.; Valcarcel, Y. Analysis and occurrence of endocrine-disrupting compounds and estrogenic activity in the surface waters of Central Spain. *Sci. Total Environ.* **2014**, *466–467*, 939–951. [CrossRef]
42. Valcarcel, Y.; Valdehita, A.; Becerra, E.; Lopez de Alda, M.; Gil, A.; Gorga, M.; Petrovic, M.; Barcelo, D.; Navas, J.M. Determining the presence of chemicals with suspected endocrine activity in drinking water from the Madrid region (Spain) and assessment of their estrogenic, androgenic and thyroidal activities. *Chemosphere* **2018**, *201*, 388–398. [CrossRef]
43. Li, X.; Ying, G.G.; Su, H.C.; Yang, X.B.; Wang, L. Simultaneous determination and assessment of 4-nonylphenol, bisphenol A and triclosan in tap water, bottled water and baby bottles. *Environ. Int.* **2010**, *36*, 557–562. [CrossRef]
44. Zhang, H.; Zhang, Y.; Li, J.; Yang, M. Occurrence and exposure assessment of bisphenol analogues in source water and drinking water in China. *Sci. Total Environ.* **2019**, *655*, 607–613. [CrossRef]
45. Kuch, H.M.; Ballschmiter, K. Determination of endocrine-disrupting phenolic compounds and estrogens in surface and drinking water by HRGC-(NCI)-MS in the picogram per liter range. *Environ. Sci. Technol.* **2001**, *35*, 3201–3206. [CrossRef] [PubMed]
46. Coogan, M.A.; Edziyie, R.E.; La Point, T.W.; Venables, B.J. Algal bioaccumulation of triclocarban, triclosan, and methyl-triclosan in a North Texas wastewater treatment plant receiving stream. *Chemosphere* **2007**, *67*, 1911–1918. [CrossRef]
47. Halden, R.U.; Paull, D.H. Co-occurrence of triclocarban and triclosan in U.S. water resources. *Environ. Sci. Technol.* **2005**, *39*, 1420–1426. [CrossRef] [PubMed]
48. Jeddi, M.Z.; Rastkari, N.; Ahmadkhaniha, R.; Yunesian, M. Endocrine disruptor phthalates in bottled water: Daily exposure and health risk assessment in pregnant and lactating women. *Environ. Monit. Assess.* **2016**, *188*, 534. [CrossRef]
49. Luo, Q.; Liu, Z.H.; Yin, H.; Dang, Z.; Wu, P.X.; Zhu, N.W.; Lin, Z.; Liu, Y. Migration and potential risk of trace phthalates in bottled water: A global situation. *Water Res.* **2018**, *147*, 362–372. [CrossRef] [PubMed]
50. Maggioni, S.; Balaguer, P.; Chiozzotto, C.; Benfenati, E. Screening of endocrine-disrupting phenols, herbicides, steroid estrogens, and estrogenicity in drinking water from the waterworks of 35 Italian cities and from PET-bottled mineral water. *Environ. Sci. Pollut. Res. Int.* **2013**, *20*, 1649–1660. [CrossRef] [PubMed]
51. Wee, S.Y.; Aris, A.Z.; Yusoff, F.M.; Praveena, S.M. Tap water contamination: Multiclass endocrine disrupting compounds in different housing types in an urban settlement. *Chemosphere* **2021**, *264 Pt 1*, 128488. [CrossRef]
52. Carnevali, O.; Notarstefano, V.; Olivotto, I.; Graziano, M.; Gallo, P.; Di Marco Pisciottano, I.; Vaccari, L.; Mandich, A.; Giorgini, E.; Maradonna, F. Dietary administration of EDC mixtures: A focus on fish lipid metabolism. *Aquat. Toxicol.* **2017**, *185*, 95–104. [CrossRef]
53. Olivieri, A.; Degenhardt, O.S.; McDonald, G.R.; Narang, D.; Paulsen, I.M.; Kozuska, J.L.; Holt, A. On the disruption of biochemical and biological assays by chemicals leaching from disposable laboratory plasticware. *Can. J. Physiol. Pharmacol.* **2012**, *90*, 697–703. [CrossRef]
54. Urbansky, E.T.; Freeman, D.M.; Rubio, F.J. Ascorbic acid reduction of residual active chlorine in potable water prior to halocarboxylate determination. *J. Environ. Monit.* **2000**, *2*, 253–256. [CrossRef]
55. Shi, W.; Zhang, F.; Zhang, X.; Su, G.; Wei, S.; Liu, H.; Cheng, S.; Yu, H. Identification of trace organic pollutants in freshwater sources in Eastern China and estimation of their associated human health risks. *Ecotoxicology* **2011**, *20*, 1099–1106. [CrossRef] [PubMed]
56. DIRECTIVE (EU) 2020/2184 on the Quality of Water Intended for Human Consumption. *Off. J. Eur. Union* **2020**, *435*, 1–62. Available online: https://eur-lex.europa.eu/legal-content/EN/TXT/PDF/?uri=CELEX:32020L2184&from=EN (accessed on 20 January 2022).

molecules MDPI

Article

Selective Removal of the Emerging Dye Basic Blue 3 via Molecularly Imprinting Technique

Maria Sadia [1,*], Izaz Ahmad [1], Faiz Ali [1], Muhammad Zahoor [2,*], Riaz Ullah [3], Farhat Ali Khan [4], Essam A. Ali [5] and Amir Sohail [6]

1 Department of Chemistry, University of Malakand, Chakdara, Lower Dir 18800, Pakistan; izazahmadij@gmail.com (I.A.); faizy186@gmail.com (F.A.)
2 Department of Biochemistry, University of Malakand, Chakdara, Lower Dir 18800, Pakistan
3 Department of Pharmacognosy, College of Pharmacy, King Saud University, Riyadh 11451, Saudi Arabia; rullah@ksu.edu.sa
4 Department of Pharmacy, Shaheed Benazir Bhutto University, Sheringal, Dir Upper 18000, Pakistan; farhatkhan2k9@yahoo.com
5 Department of Pharmaceutical Chemistry, College of Pharmacy, King Saud University, Riyadh 11451, Saudi Arabia; esali@ksu.edu.sa
6 MSC Construction Project Management, University of Bolton, Bolton BL3 5AB, UK; syedaamirsohail89@gmail.com
* Correspondence: mariasadia@gmail.com (M.S.); mohammadzahoorus@yahoo.com (M.Z.)

Abstract: A molecularly imprinting polymer (MIP) was synthesized for Basic Blue 3 dye and applied to wastewater for the adsorption of a target template. The MIPs were synthesized by bulk polymerization using methacrylic acid (MAA) and ethylene glycol dimethacrylate (EGDMA). Basic Blue 3 dye (BB-3), 2,2′-azobisisobutyronitrile (AIBN) and methanol were used as a functional monomer, cross linker, template, initiator and porogenic solvent, respectively, while non-imprinting polymers (NIP) were synthesized by the same procedure but without template molecules. The contact time was 25 min for the adsorption of BB-3 dye from 10 mL of spiked solution using 25 mg polymer. The adsorption of dye (BB-3) on the MIP followed the pseudo-second order kinetic (k_2 = 0.0079 $mg{\cdot}g^{-1}{\cdot}min^{-1}$), and it was according to the Langmuir isotherm, with maximum adsorption capacities of 78.13, 85.4 and 99.0 $mg{\cdot}g^{-1}$ of the MIP at 283 K, 298 K and 313 K, respectively and 7 $mg{\cdot}g^{-1}$ for the NIP. The negative values of ΔG° indicate that the removal of dye by the molecularly imprinting polymer and non-imprinting polymer is spontaneous, and the positive values of ΔH° and ΔS° indicate that the process is endothermic and occurred with the increase of randomness. The selectivity of the MIP for BB-3 dye was investigated in the presence of structurally similar as well as different dyes, but the MIP showed higher selectivity than the NIP. The imprinted polymer showed 96% rebinding capacity at 313 K towards the template, and the calculated imprinted factor and Kd value were 10.73 and 2.62, respectively. In this work, the MIP showed a greater potential of selectivity for the template from wastewater relative to the closely similar compounds.

Keywords: molecularly imprinting polymer; adsorption; Basic Blue; environmental pollution

Citation: Sadia, M.; Ahmad, I.; Ali, F.; Zahoor, M.; Ullah, R.; Khan, F.A.; Ali, E.A.; Sohail, A. Selective Removal of the Emerging Dye Basic Blue 3 via Molecularly Imprinting Technique. *Molecules* **2022**, *27*, 3276. https://doi.org/10.3390/molecules27103276

Academic Editors: Victoria Samanidou, Eleni Deliyanni and Dimitra Voutsa

Received: 26 April 2022
Accepted: 16 May 2022
Published: 19 May 2022

Publisher's Note: MDPI stays neutral with regard to jurisdictional claims in published maps and institutional affiliations.

1. Introduction

Environmental pollution caused by wastewater discharging from industries containing dyes is a worldwide problem. About 7×10^5–1×10^6 tons of synthetic dyes are produced annually around the world [1]. In these dyes, nearly 50% are discarded during the coloring process, and approximately 10–15% are lost as effluent into the environment. Synthetic and natural dyes are mostly used in industries such as food, pharmaceuticals, leathers, pulps, plastics, cosmetics, textiles, etc. [2]. When the effluents of these industries release into the river; they cause the degradation of that environment, such as changing the natural color and the formation of foam on the water surface. Synthetic dyes are generally toxic, mutagenic, carcinogenic and lethal for the health of different organisms (algae and phytoplankton) [3]. Dye concentration of even less than 1 mg/L causes intense water coloration. It can reduce the penetration of oxygen and light to the aquatic environment and result in a decrease in photosynthetic activity.

The common wastewater plant treatment is not efficient in removing toxic compounds such as dyes from water [4]. For the removal and determination of these toxic dyes from industrial effluent, different physical and chemical techniques are used such as chemical oxidation, chemical precipitation, adsorption, photocatalysis, microbial or enzymatic treatment, natural sorbent materials [5], mineral composite, chemical oxidation, coagulation–flocculation, irradiation, adsorption, precipitation, membrane technologies, a combination of an aerobic and anaerobic and ion exchange processes [6]. Therefore, all of these techniques have some drawbacks and problems such as high cost, large solvent consumption, large analysis time and tough sample preparation, etc. Among these techniques, adsorptions using solid adsorbent are effective because of the simplicity and ease of their operation. However, simple solid sorbent also has some limitations such as high cost and lack of selectivity [7]. The biological degradation method is also used but cannot eliminate the color of many synthetic dyes completely because of their biodegradable difficulty. To overcome the above limitation, more specific and selective methods are needed to remove pollutants from environmental water. So, MIP is the best choice because of its specificity and selectivity [8]. Among the published work, MIPs for dyes are used as a sorbent for the solid phase extraction because of their high selectivity, even in complex samples [9].

Molecularly imprinting polymers (MIPs) are the synthetic polymer used for the recognition of specific substances and are able to bind with target molecules. MIPs can be synthesized by mixing template (analyte) and functional monomer to form pure complex. Then, a cross linker is added to stabilize the MIP's 3-dimensional structure [10]. After the polymerization, the template is removed by washing, creating cavities which are complementary to the template's shape, size and functional group position.

Most commonly, MIPs interact with target molecules through ionic interaction, hydrogen interaction or non-covalent bonding [11]. The successful synthesis of MIP depends on the cross linker, monomer and the appropriate condition of polymerization for the target analyte. There are different methods for MIP preparation such as suspension polymerization, bulk polymerization, precipitation polymerization and two-step polymerization. The preparation of MIP through Bulk polymerization is the simplest method to form a large-sized monolith, which can be sieved and ground by mortar and pestle into a smaller size of 5–50 μm of irregular shape particles [12]. These synthetic, tailor-made receptors have some advantages over the recognition of biological substances such as antibodies and enzymes as they are cheaper, easily prepared and resistant to harsh conditions such as base, acid, high temperature, organic solvents and high pressure. In contrast to biological molecules, they are less expensive as well [13].

MIPs have been applied successfully to remove pesticides, pharmaceutical products and products of personal care as sensors and in solid phase extraction (SPE). MIPs and non-imprinted polymers (NIPs) can also be used in many areas such as control drug delivery, biomedical or analytical diagnosis, bio sensor SPE and chromatography, as well as for viruses and toxins [14].

The main goal of this research work is to design a highly selective MIP sorbent for Basic Blue 3 dye and its selectivity, rebinding and application in different effluents. For the fast determination of Basic Blue 3 dye in water samples, MIPs and NIPs were synthesized through bulk polymerization, in which Basic Blue 3 dye shows more selectively toward MIPs as compared to NIPs because of the recognition property of the MIP network.

2. Materials and Methods

2.1. Materials

All solvents and dyes were of analytical grade and were supplied from Sigma–Aldrich. For MIP synthesis, methacrylic acid (MAA) (Merck; Darmstadt, Germany) was used as a monomer, ethylene glycol dimethacrylate (EGDMA) (Xiya Reagent (Chengdu, China) was used as a cross linker, methanol (J.T. Baker, Phillipsburg, NJ, USA) was used as a porogenic solvent, glacial acetic acid (Brazil) was used for the removal of the template from MIP, 2,2′-azobis(isobutyronitrile) (AIBN) (Shanghai, China) was used as an initiator and Basic Blue 3 (Sigma Aldrich, Darmstadt, Germany) was used as a template molecule. A selectivity study of MIP for BB3 was carried out with MB, BB41, safranin and Thioflavin T. The water was deionized by the Milli-Q system. The chemical structures of the dyes used as given in Figure 1.

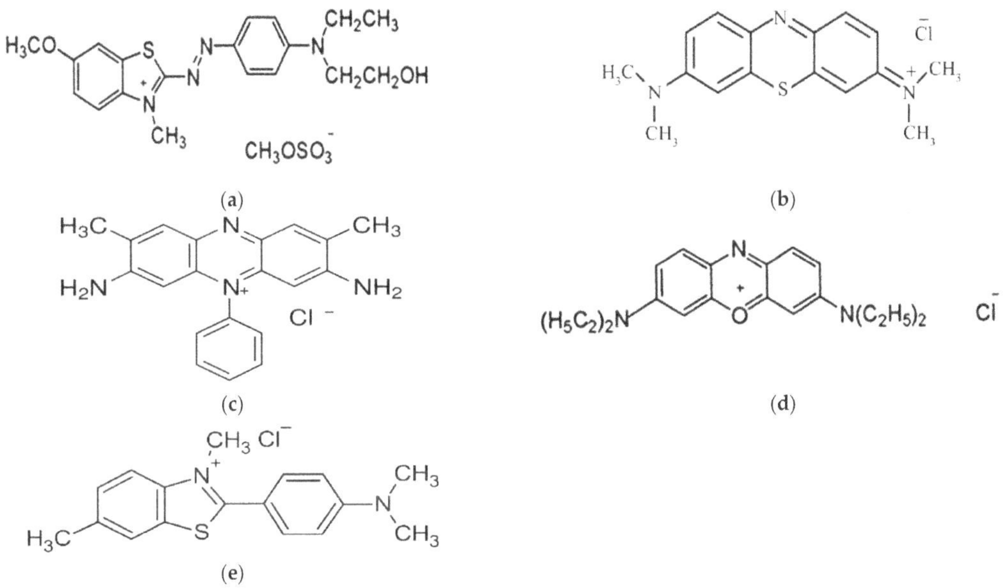

Figure 1. Structure of the different dyes: (**a**) Basic blue 41 (BB41), (**b**) Methylene blue (MB), (**c**) Safranin, (**d**) Basic blue 3, (**e**) Thioflavin T.

2.2. Characterization of Molecularly Imprinting Polymer and Non Imprinting Polymer

The structure of the polymer was analyzed using the Fourier-transform infrared spectroscopy (FTIR) Vertex 70 (Bruker, Billerica, MA USA) technique with a DLaTGS detector and a He Ne laser in the range of 4000–500 cm^{-1}. The polymer size and morphology were determined using scanning electron microscopy (JSM-IT500, Zaventem, Belgium), and the surface area was determined by multipoint Brunauer–Emmett–Teller from the isotherm of nitrogen adsorption (ASAP 2010, Norcross, GA, USA). A total of 25 mg of dry sample was inserted in a sample holder before the measurement with the removal of N_2 at 75 °C for 5 h using micrometrics at liquid nitrogen temperature. The absorbance measurements were carried out by a UV-Visible 1800 spectrophotometer using a quartz cuvette (Shimadzu, Kyoto, Japan).

2.3. Preparation of MIP and NIP

Mostly, the processes for MIP chemical synthesis involve noncovalent preassembly of the template functional monomer in solution followed by bulk polymerization [15]. In the current work, 2 mmol of the functional monomer (MAA) and 0.2 mmol of the template (BB3) were initially dissolved in 10 mL of methanol in a glass tube and then stirred for 4 min. After mixing, the solution was allowed to rest for 2 h. Then, 5 mmol of the cross liker (EGDMA) and 0.025 g of the initiator (AIBN) were added. The oxygen gas was removed from the mixture by nitrogen purging for 10 min. The sealed test tube was then placed in a water bath at 65 °C for 14 h. The polymer was grounded by mortar and pestle and then added into soxlet for washing with 3 rounds of methanol/acetic acid (7:3, *v*/*v*) and 3 rounds of methanol for the complete extraction of analyte. The same procedure as that described above was applied for non-molecularly imprinting polymer (NIP) synthesis, but without the template addition.

2.4. Binding Adsorption Analysis

The binding study of MIP was achieved using 20 mL vials containing 25 mg of MIP/NIP and 10 mL of 100 mg/L dye and adjusting parameters such as mass, concentration, pH and time. The mixture was then kept for 30 min on a shaker followed by centrifugation at 14,000 rpm and then filtration of the supernatant over 0.45 μm membrane (PTFE) before UV/V is spectrophotometric analysis. For the binding adsorption capacity, the following equation was used.

$$Q = \frac{(C_0 - C_e)}{V} m \tag{1}$$

C_0 (mg·L^{-1}) is the initial dye concentration, C_e (mg·L^{-1}) is the equilibrium dye concentration, Q (mg·g^{-1}) is the experimental adsorption quantity, V (L) is the volume of solution and m (g) is the mass of MIP and NIP.

2.5. Selectivity Study

In order to check the formation of selective cavities in polymers, competitive adsorption analysis was carried out in the presence of those molecules that are similar as well as different in structure to BB3. In this regard different dyes such as Methylene blue (MB), Safranin, Basic blue 41 (BB41) and Thioflavin T were used. Each compound's selectivity recognition experiments were carried out by dissolving 25 mg of MIP/NIP in a 10 mL solution (pH 11) containing 100 mg/L of each dye and equilibrated for 25 min. The concentration of these dyes in the supernatant was measured with a UV-Vis spectrophotometer.

2.6. Application in Real Water

To determine the application of MIPs for the preconcentration of reactive dyes in water, different samples were spiked with known amounts of dye concentration (100 mg/L) and subjected to the optimized extraction process. The extraction power of MIP seems to be affected in tap water and river water for BB3 dye, because dissolved organic and in-organic matter in the water samples played a role in slightly reducing the selective efficiency of MIP.

3. Results and Discussion

3.1. Choice of the Materials

Typically, stable and high affinity MIP synthesis requires the following reagents:

A high nominal level of cross linker is used for the template site preservation. A porogenic solvent and one or more functional monomer is needed to form a stable complex with the target molecule [16].

3.2. Characterization

3.2.1. Characterization of Synthesized Polymer by SEM

With the help of scanning electron microscopy (SEM), the size information and geometry of the MIP can be obtained. Different research papers report the use of SEM to investigate MIP particles [17]. Therefore, SEM micrographs of the synthesized MIP and NIP were taken at different resolutions, as shown in Figure 2. The SEM images of both the MIP and NIP are not much different from each other. An overall assumption could be made from the general morphological architecture of both the MIP and NIP images for better general adsorption. The surface morphological appearance of all the SEM images is in favor of good adsorption ability in general. The SEM images are not very informative regarding the selective nature of MIPs, but they could be a good source for the general adsorption point of view [18].

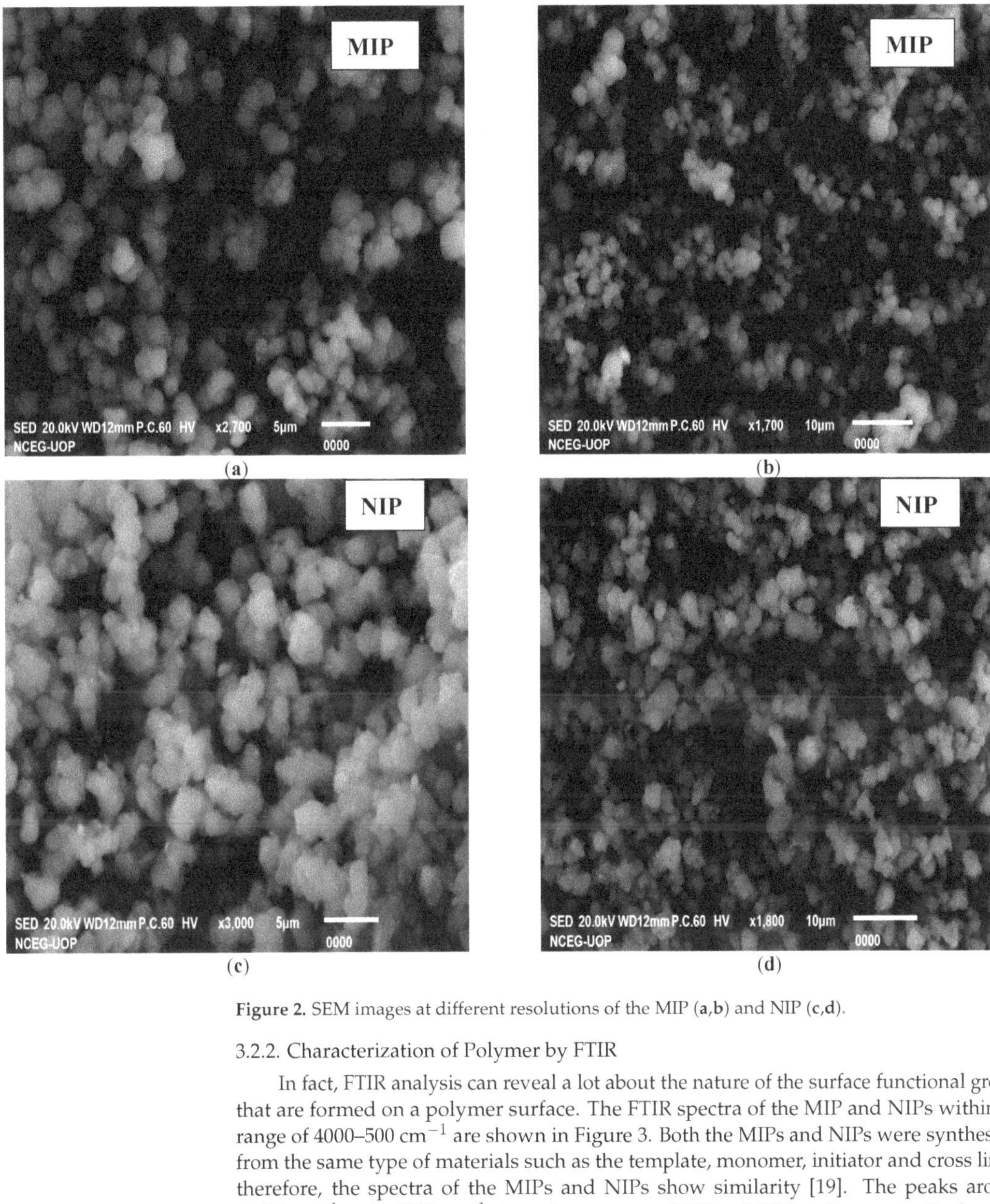

Figure 2. SEM images at different resolutions of the MIP (**a**,**b**) and NIP (**c**,**d**).

3.2.2. Characterization of Polymer by FTIR

In fact, FTIR analysis can reveal a lot about the nature of the surface functional groups that are formed on a polymer surface. The FTIR spectra of the MIP and NIPs within the range of 4000–500 cm^{-1} are shown in Figure 3. Both the MIPs and NIPs were synthesized from the same type of materials such as the template, monomer, initiator and cross linker; therefore, the spectra of the MIPs and NIPs show similarity [19]. The peaks around ~3400 cm^{-1} and ~2900 cm^{-1} show the presence of OH and CH bending, respectively, while the stretching at around~1700 cm^{-1} is due to the presence of C=O [20]. Therefore, the MIPs show a clear bend of the ester group at 1720 cm^{-1} and 1256 cm^{-1} due to the carbonyl C=O and C-O stretching. Additionally, the peaks at ~1400 cm^{-1} and ~1300 cm^{-1} are due to the presence of -CH_2, and -CH_3, respectively, while the strong peaks at 1159 cm^{-1} and 1046 cm^{-1} are due to the presence of the C-O functional group [21]. The FT-IR analysis of the MIPs and NIPs show that the MAA and EGDMA C=C double bond signal at 1637 cm^{-1} was absent, indicating that the polymer is successfully synthesized.

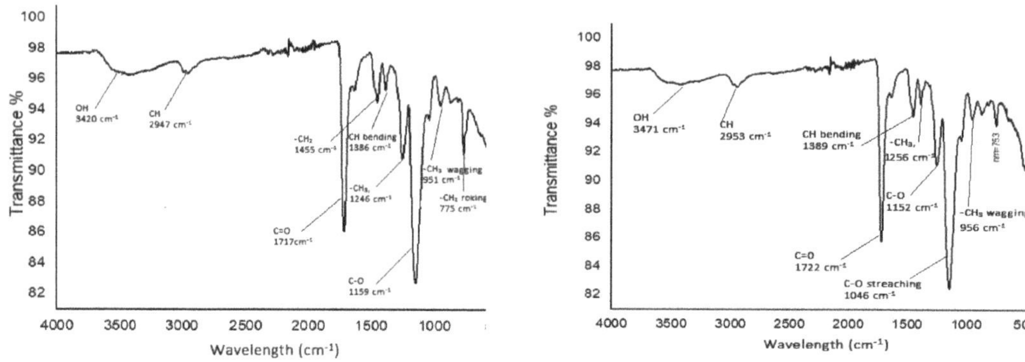

(a) (b)

Figure 3. FTIR spectra of (**a**) MIP (**b**) NIP.

3.2.3. Analysis by Brunauer–Emmett–Teller

The BET study determines the porosity and specific area of the selected BB3 molecularly imprinted polymers. The MIPs showed a specific surface area of 245.321 (m^2/g), a pore volume of 0.078 (cc/g) and a pore radius of 14.492 (Å), while the NIPs have a 34.072 (m^2/g), 0.0098 (cc/g) and 2.145 (Å) surface area, pore volume and pore radius, respectively (Table 1). The surface area of the NIP is almost seven times smaller than that of the MIP. The larger surface area of the MIP compared to the NIP indicates that specific cavities are formed for the recognition of BB3. The porosity of a particle refers to the volume ratio of the open pore to the total volume [22]. Previous research works show that the greater adsorption capacity of MIPs compared to NIPs is due to the higher surface area of MIPs as compared to NIPs. Additionally, MIPs have a higher total volume than NIPs due to their greater load capacity [23]. The larger pores of the MIP show that the structure of the NIP is more compact. The surface area and porosity of the MIP is mostly affected by the template in polymerization, and the average diameter of the pores for the MIP and NIP falls in the range of 2–50 nm, showing that both polymers are mesoporous [24]. A BET plot of the MIP and BJH plots for the MIP and NIP is shown in Figure 4.

Table 1. BET analysis of MIP and NIP.

Polymers	Specific Surface Area (m^2/g)	Pore Volume (cc/g)	Pore Radius (Å)
MIP	245.321	0.078	14.492
NIP	34.072	0.0098	2.145

3.3. *Effect of Adsorbent Dose and pH*

For the optimization of BB3 dye adsorption, the dose of the MIP was used in the range of (0.005–0.050 g) and 10 mL of 100 mg/L dye concentration. Figure 5 shows a linear increase in the removal of dye with an increase in the weight of the MIP and NIP from 0.0025 to 0.025 g due to the fact that the number of active sites of MIP/NIP increased. Further increase in the dose of MIP/NIP showed no effect on the uptake of the dye. The mechanism of dye adsorption depends on the degree of interaction with the adsorbent (MIP and NIP) surface and the protonation or deprotonation of dye, which may change adsorption efficiency [25]. So, for further studies, 0.025 g of polymer was used, and the pH of the test samples was varied in a range of 1–13, keeping the other parameters constant. It was observed that, at low pH, the adsorption is small. The adsorption capacity increases with increasing pH, and maximum adsorption was achieved at pH 11. So, this pH was selected as an optimum pH and used for further studies. With the change in pH, the surface charge varies. At lower pH, the surface charge is positive, whereas at higher pH, it becomes negative. Since the subject dye BB3 is a cationic dye, at higher pH, there will be strong interaction between the negative surface charge and the cationic BB3 dye. Due to this strong interaction, the optimum adsorption took place at pH 11. At higher pH, all sites became covered with negative charge; therefore, no further adsorption of BB3 dye took place. The results are shown in Figure 5.

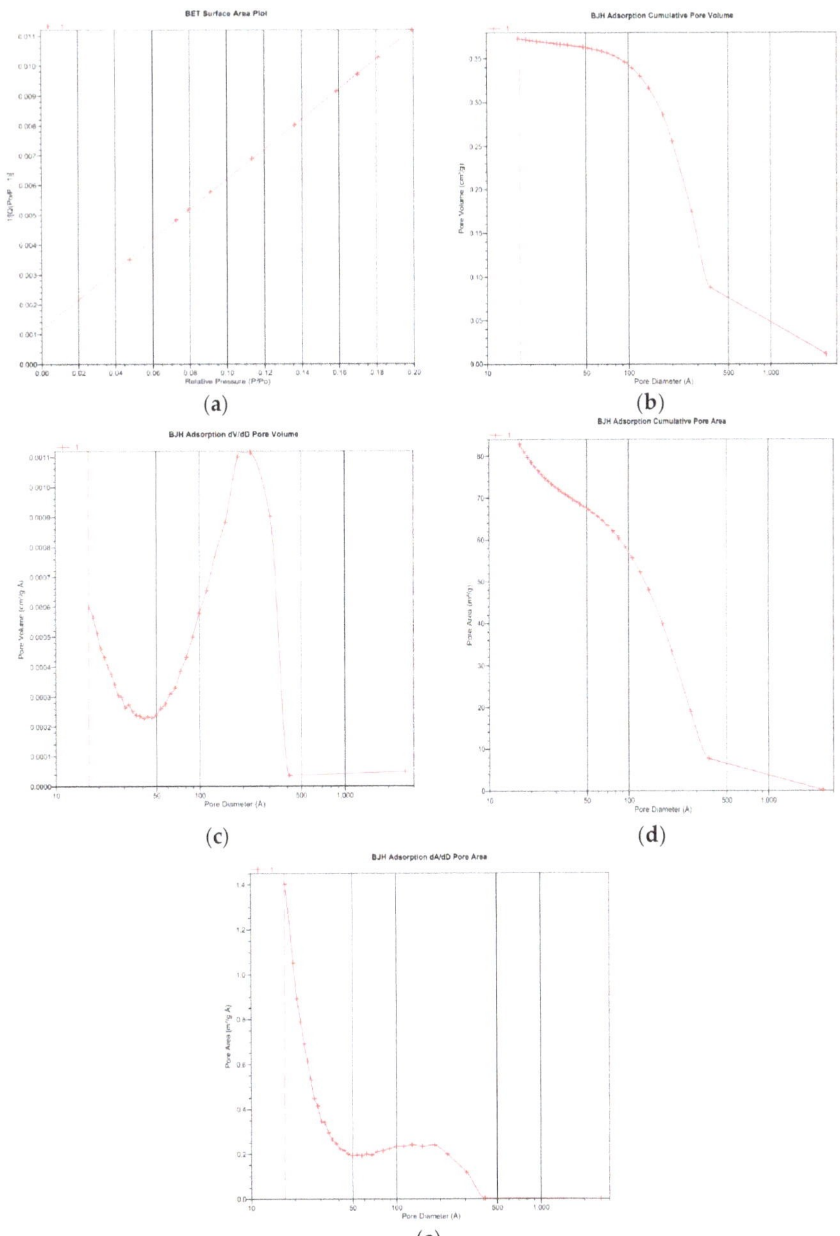

Figure 4. BET and BJH adsorption curves for MIP (**a**–**c**) and NIP (**d**,**e**).

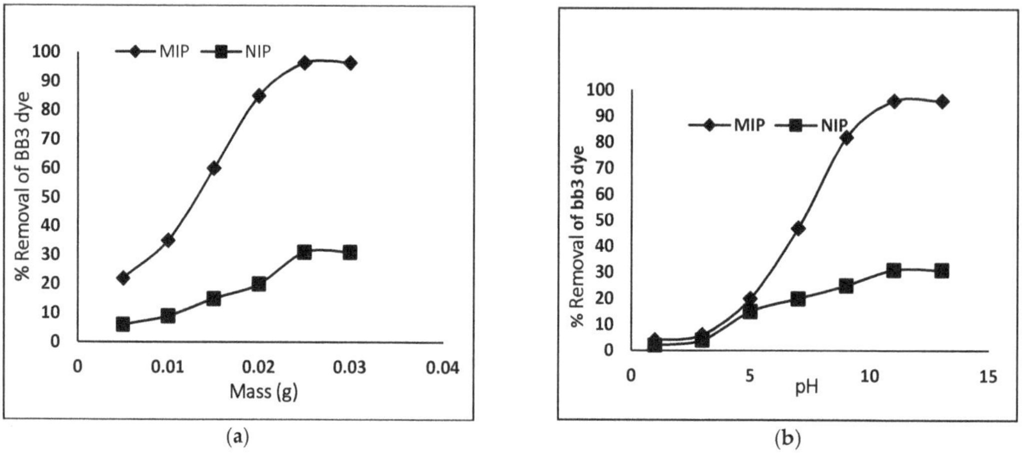

Figure 5. Removal efficiency of MIP/NIP for BB3 dye; (**a**) adsorbent dose (**b**) pH.

3.4. Contact Time Studies as a Function of Temperature

Chemical kinetics is necessary to obtain and to find information about the binding mechanism and rate controlling process. The effect of contact time was examined as a function of temperature. This experiment was achieved using a fixed amount of polymer, 25 mg, with 10 mL of 100 mg/L dye solution at the optimized condition of pH, while varying the time from 1–35 min and the temperature from 283–313 K. The maximum dye adsorption was observed at 25 min, and then it remained constant. With the increase in temperature, a small change in the uptake of dye was observed, as shown in Figure 6.

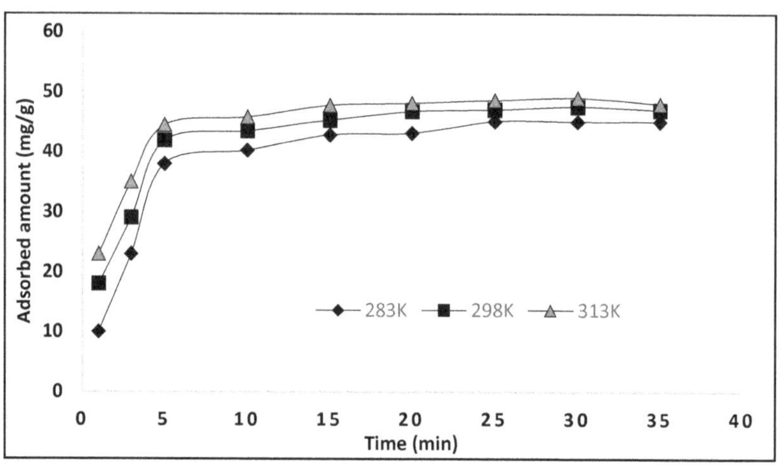

Figure 6. Adsorption kinetics for BB3 dye adsorption on MIP. Mass of polymer 25 mg; V = 10 mL; pH; 11; [BB3] = 100 mg/L.

3.5. Kinetic Models

To study the mechanism of dye adsorption on the adsorbent (MIP/NIP), we used three kinetics models: pseudo-first order, pseudo-second order and the Weber and Morris intraparticle diffusion model.

3.5.1. Pseudo 1st t Order Kinetics

The pseudo-first kinetic model gives information about the rate of occupation of adsorption sites, which is proportional to unoccupied sites and is represented in Equation (2) [26].

$$\log\left(q_e - q_t\right) = \log q_e - K_1 \frac{t}{2.303} \quad (2)$$

The values of q_t and q_e are the amount of BB3 dye (mg g^{-1}) adsorbed at time t (min) and at equilibrium, respectively. K_1 (min^{-1}) is the pseudo-first order constant and is calculated by plotting log ($q_e - q_t$) against "t". The pseudo-first order kinetic model at different temperatures (283 K, 298 K, 313 K) is shown in Figure 7, and different parameters are given in Table 2. As the Q_e (cal) and Q_e (exp) do not match with each other, and the values of correlation coefficient are less than 0.99, we came to the conclusion that the adsorption of BB3 on MIP did not follow pseudo-first order kinetics.

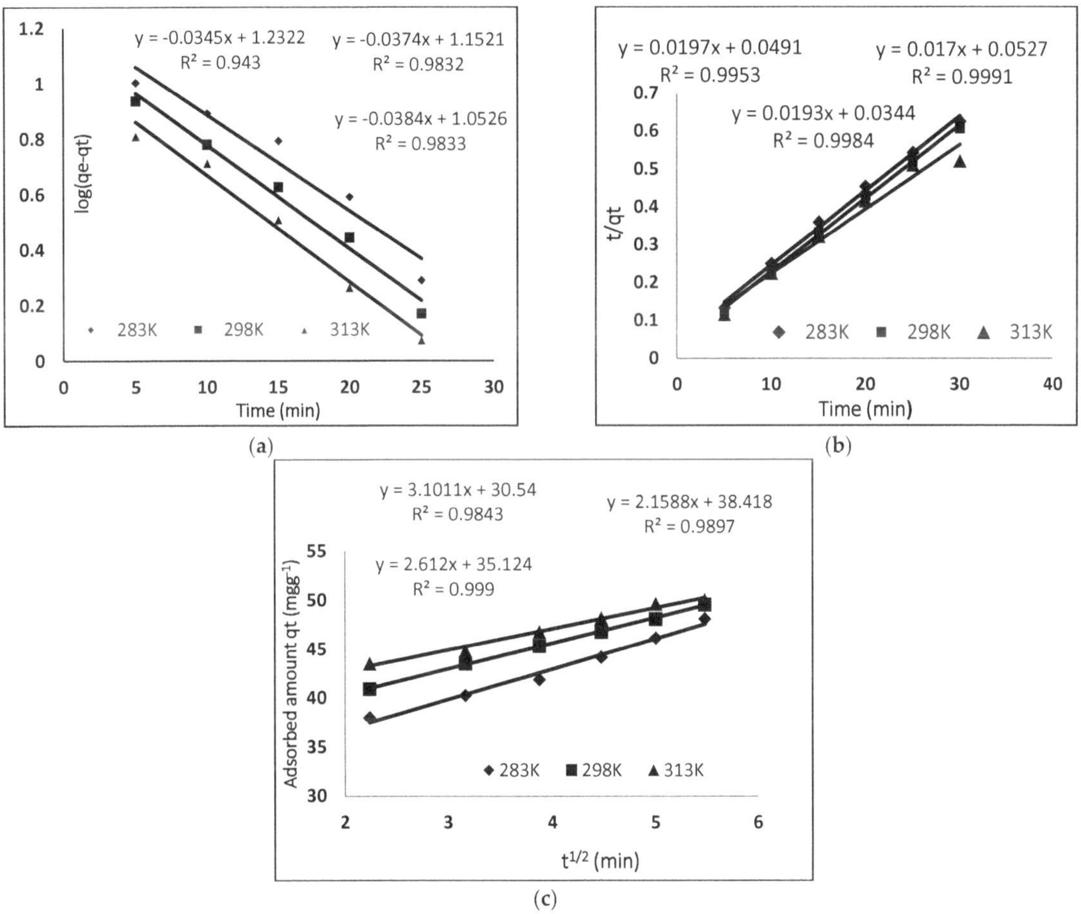

Figure 7. Different kinetic models for the adsorption of BB-3 on MIP; (**a**) pseudo-first order kinetic (**b**) pseudo-second order kinetic (**c**) intra particle diffusion model.

Table 2. Different kinetic models' parameters for the adsorption of BB3 on MIP.

Parameter	283 K	298 K	313 K
	Pseudo 1st order kinetic		
K_1	0.0794	0.0861	0.0951
Q_e (cal)	17.068	14.197	11.287
Q_e (exp)	48.049	49.528	49.996
R^2	0.943	0.9832	0.9833
	Pseudo 2nd second kinetic		
K_2	0.0079	0.0108	0.0146
Q_e (cal)	50.761	51.813	51.54
Q_e (exp)	48.049	49.528	49.996
R^2	0.9953	0.9984	0.9991
	Intraparticle diffusion		
K_{id} (mg/g min $^{-1/2}$)	3.09	2.61	2.05
C	30.54	35.12	38.72
R^2	0.9843	0.999	0.9897

3.5.2. Pseudo 2nd Order Kinetic

The adsorption process can also be described by the pseudo-second order kinetic using the following Equation (3) [27].

$$\frac{t}{q_t} = \frac{1}{k_2\, q_e^{\,2}} + \frac{t}{q_e} \tag{3}$$

In the above equation, k_2 (mg g^{-1} min^{-1}) is the constant of the pseudo-second order, which is calculated from the plot of t/q_t versus "t". For the adsorption of BB3 dye on MIP, the pseudo-second order model was applied at different temperatures, as shown in Figure 7, while the different parameters calculated from this plot are given in Table 2. The R^2 value for second-order kinetics is higher as compared to pseudo-first order kinetics, and a close resemblance between the Qe (cal) and Qe (exp) was found. So, we can conclude that the adsorption data of the pseudo second-order model are the best fit. These kinetic results show that the adsorption of BB3 dye depends on adsorbents as well as adsorbates.

3.5.3. Intraparticle Diffusion Model

The kinetics data were also evaluated with the help of the intraparticle diffusion model proposed by Morris and Weber, which is given in Equation (4) [28].

$$q_t = k_{id}\, t^{1/2} + C \tag{4}$$

where C is the intercept which indicates the thickness of the boundary and k_{id} is the rate constant of intraparticle diffusion. According to this equation, a plot of qt versus $t^{1/2}$ would be linear if the adsorption follows the process of the intraparticle diffusion model. Figure 7 shows a linear plot of the intraparticle diffusion for the adsorption of BB-3 on MIP. Multi linearity shows that the process of adsorption occurs in three steps. The first displays the diffusion of dye to the external surface of the adsorbent from the solution. The diffusion of the adsorbate from the external surface into the adsorbent pore is described by the second step. The final step is the adsorption of the adsorbate (BB-3) on the internal surface of the pore. However, at each temperature, the line fails to pass from the origin, which is due to the variance in the rate of mass transfer of initial and final. Furthermore, such a straight-line deviation from the origin reveals that the pore diffusion is not a sole rate control step [29]. Comparing the value of "C" and the rate constant indicates that the intraparticle diffusion is not the only rate limiting step. The adsorption of BB-3 on MIP is a complex phenomenon and is controlled by the surface sorption and intraparticle diffusion.

3.6. Adsorption Isotherms

The equilibrium adsorption isotherm study was carried out with varying dye concentrations (25, 50, 100, 150, 200 mg/L) to determine the efficiency of the prepared polymer, as shown in Figure 8. The figure shows that the isotherm rises abruptly at first, indicating that there are plenty of readily available sites for adsorption. When the adsorbent (MIP) becomes saturated after equilibration, a plateau is reached, indicating that no more sites are accessible for further adsorption. The experiments were conducted at 283 K, 298 K and 313 K to better realize the effect of temperature on dye adsorption. When the adsorption isotherms are compared, it is clear that adsorption increases as temperature rises, indicating that the process is endothermic.

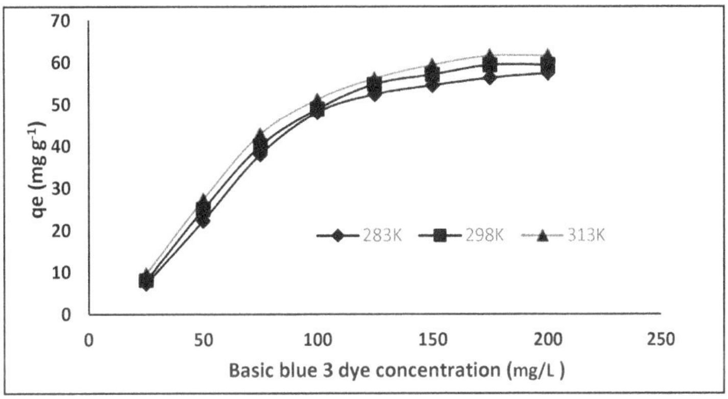

Figure 8. Adsorption isotherms of Basic Blue 3 dye adsorption on MIP adsorbent at different temperatures.

The adsorption isotherm provides mechanism information of the adsorption process. Adsorption isotherms are used to assign the adsorption system, as they reveal the mechanism of adsorption. Generally, the Langmuir isotherm defines the monolayer formation and the non-covalent behavior of MIP, as described by the Freundlich model [30]. Therefore, the adsorption data were evaluated using the Langmuir and Freundlich isotherms.

3.6.1. Langmuir Model

The Langmuir model defines a uniform and homogenous surface of the adsorbate that forms a monolayer. This model is based on a hypothesis: adsorption occurs at specific homogeneous sites within the body of the adsorbent. There is no contact between the adsorbed species and the adsorbent. The Langmuir model is described by Equation (5) [31].

$$\frac{C_e}{Q_e} = \frac{1}{K_L Q_m} + \frac{C_e}{Q_m} \tag{5}$$

where C_e ($mg{\cdot}L^{-1}$) is the liquid phase equilibrium concentration of the dye, Q_m ($mg{\cdot}g^{-1}$) is the maximum adsorption capacity of the adsorbent, K_L ($L{\cdot}mg^{-1}$) is the amount of dye adsorbed, the energy or the net enthalpy of adsorption and Q_e ($mg{\cdot}g^{-1}$) is the quantity of dye adsorbed. The relationship between C_e/q_e and C_e must be linear, with a slope of $1/q_m$ and an intercept of $1/(q_m K_L)$. The values of K_L, Q_m and R^2 obtained from the curve are given in Table 3. The maximum adsorption capacities (Qmax) for MIP and NIP were 75.13 $mg{\cdot}g^{-1}$ and 7 $mg{\cdot}g^{-1}$, respectively. In Table 3, the Langmuir model provides an R^2 value that best fits to the experimental value.

Table 3. Different parameters of the Langmuir and Freundlich models for the adsorption of BB3 on MIP.

313 K	298 K	283 K	Parameters
	Langmuir isotherm model		
91.743	83.33	78.125	Q_m (mg·g^{-1})
0.1372	0.1872	0.1626	K_L
0.9737	0.9837	0.9792	R^2
	Freundlich		
15.33	13.63	12.96	K_f (mg·g^{-1}) (L mg·g^{-1})
0.4152	0.3439	0.5362	1/n
0.8211	0.8703	0.8801	R^2

3.6.2. Freundlich Model

The Freundlich isotherm is used to determine the adsorption properties of multilayer and heterogeneous surfaces with unequal adsorption sites and unusually available adsorption energies. The Freundlich isotherm model equation is given below [32].

$$\ln q_e = \ln K_f + \frac{1}{n} \ln C_e \tag{6}$$

where C_e (mg·L^{-1}) is the liquid phase concentration at equilibrium, q_e mg·g^{-1} is the adsorb amount of dye, K_f (mg·g^{-1}) is taken as a relative indicator of the adsorption capacity and $1/n$ is the heterogeneity factor of the surface, which shows the adsorption nature. The $1/n$ value should be smaller than 1 for favorable adsorption, while for unfavorable adsorption, the value is greater than 1 and shows weak bond adsorption [33]. The Freundlich model is shown in Figure 9, and different parameters are given in Table 3.

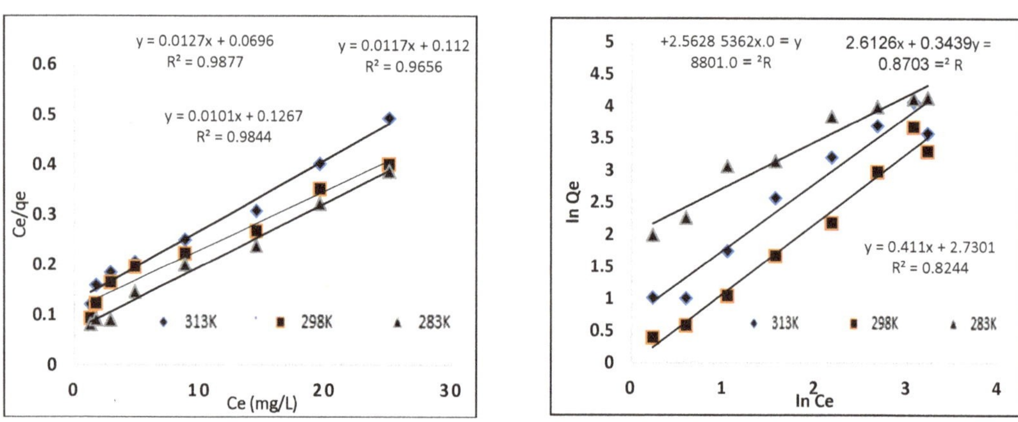

(**a**) (**b**)

Figure 9. Isotherm models (**a**) Langmuir (**b**) Freundlich.

3.7. Thermodynamic Study

Thermodynamic studies were undertaken to explore the adsorption mechanism feasibility of dye (BB-3) onto the MIP. In this study, a very significant decision has to be ended whether the process is spontaneous or not. Many thermodynamic parameters such as standard free energy (ΔG°), enthalpy (ΔH°) and entropy (ΔS°) were calculated using the following equation [34]:

$$\log K_c = \frac{\Delta S^\circ}{2.303R} - \frac{\Delta H^\circ}{2.303RT} \quad (7)$$

$$\Delta G^\circ = \Delta H^\circ - T\Delta S^\circ \quad (8)$$

where T (K) is the absolute temperature, R (8.314 J·mol^{-1} K^{-1}) is the universal gas constant and K_c (L·g^{-1}) is thermodynamic equilibrium constant described by q_e/C_e. The values of ΔS° and ΔH° were calculated from the intercept and the slope of a plot of log K_c versus 1/T. Different thermodynamic parameters were studied at the different temperatures given in Table 4, and the Van't Hoff plot is shown in Figure 10. At all temperatures, the value of ΔG° was found to be negative, which shows that the adsorption of dye (BB-3) onto the MIP is spontaneous [35]. By increasing the temperature, the value of ΔG° decreased, which indicates that high temperature facilitates the adsorption of dye (BB3) on MIP. The positive value of ΔH° specifies that this adsorption is endothermic, because increasing the temperature leads the rate of adsorbate diffusion to also increase on the adsorbent (external and internal surface). The positive value of ΔS° indicates that disorderedness increased during adsorption (solution–solid interface) [36,37].

Table 4. Thermodynamic parameters for the removal of BB3 on MIP.

Temperature	ΔG° KJ mol^{-1}	ΔH° KJ mol^{-1}	ΔS° J mol^{-1} K^{-1}
283 K	−2057	10.38	44.82
298 K	−2735		
313 K	−3595		

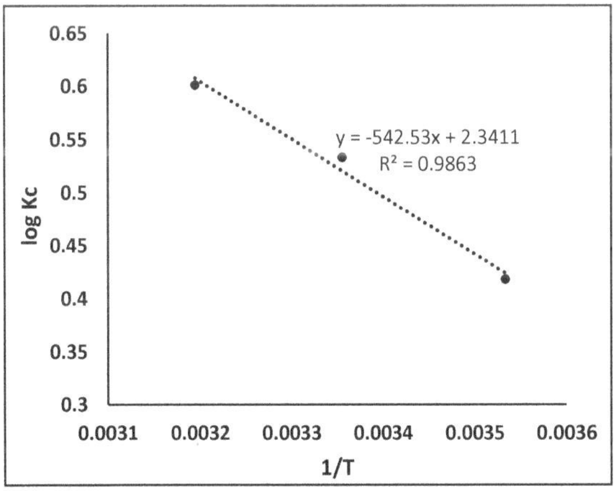

Figure 10. Van't Hoff plotting for MIP.

3.8. Selectivity Study

For the confirmation of selective cavities formation in a polymer, competitive adsorption analysis was carried out in the presence of those dyes which are similar as well different in structure to the template Basic Blue 3 (BB-3), such as Basic Blue 41 (BB-41), Thioflavin T, Methylene blue (MB) and Safranin, whose structures are shown in Figure 11. The MIP was selective for the specific template (BB-3) as compared to the NIP. This study was performed by adding 100 mg/L of BB-3 dye and the interfering dyes to each Erlenmeyer flask at optimum conditions, and the mixture was stirred in a thermostat shaker at 125 rpm. At predetermined times, the samples were removed and the absorbance was measured by the UV-vis spectrophotometer. The amount of dye adsorbed on the polymer was calculated by substracting the final dye conentration from the initial dye added to a mixture. For BB-3, the MIP is more selective at 96% as compared to other dyes whose adsorption was between 5% and 41%. So, it confirms that the MIP cavities are selective only for the template (BB-3) molecule.

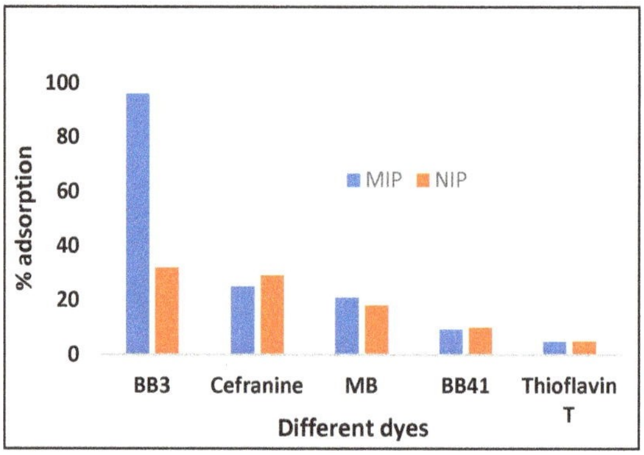

Figure 11. Effect of different dyes, showing the selectivity behavior of the MIP towards the adsorption of BB3 as compared to the NIP.

3.9. Imprinting Factor and Distribution Ratio

The imprinting factor describes the interaction and strength of the template molecule (BB3) toward the polymer (MIP/NIP). It indicates the recognition properties of the MIP and NIP toward specific analyte. The imprinting factor (IF) was calculated for the molecularly imprinting polymer using Equation (9).

$$IF = \frac{Q_{MIP}}{Q_{NIP}} \tag{9}$$

where Q_{MIPs} is the adsorption capacity of the MIPs for the dye (BB3) and Q_{NIPs} is the adsorption capacity of the NIPs for BB3. While, for the determination of the distribution ratio, the following equation was used.

$$K_d = \frac{(C_i - C_f)}{C_f\, m} V \tag{10}$$

where K_d (L/g) is the distribution coeffecient, Ci is the initial dye concentration, C_f is the final dye concentarion, V is the volume used and "m" is the mass of the polymer (MIP/NIP). The different adsorption parameters estimated from various models are summurized in Table 5.

Table 5. Different adsorption parameters for dyes by the polymer (MIP/NIP) at 283 K.

Dyes	% Removal		Adsorption Capacity Q (mg·g^{-1})		Distribution Coefficient Kd (L·g^{-1})		Imprinting Factor IF $\propto = \frac{Q_{MIP}}{Q_{NIP}}$	Selectivity $S = \frac{I^{BB3}}{I^{interfering}}$
	MIP	NIP	MIP	NIP	MIP	NIP		
BB3	96	31	78.4	7.0	0.48	0.07	10.73	-
Safranin	25	29	5.34	6.4	0.3	0.27	0.83	6.18
MB	21	18	4.03	3.26	0.02	0.03	1.23	9.66
BB41	9.3	10	1.75	1.87	0.19	0.05	0.93	7.37
Thioflavin T	4.76	5.0	2.0	2.11	0.20	0.21	0.94	7.29

3.10. Real Sample Application

This study was undertaken to determine the application of MIPs for the preconcentration of reactive dyes in water. Molecularly imprinting polymers (MIP) were taken as a solid phase extracting material to evaluate their efficiency for the extraction of BB3 dye from environmental samples. The samples were spiked with known amounts of dye concentration (100 mg/L) and subjected to the optimized extraction process. It was found that the removal efficiency of dye (BB3) from the environmental samples was between 60–80%. The extraction power of MIP seems to be affected in tap water and river water for the extraction of BB3 dye. The dissolved organic and inorganic matter in the water samples played a role in slightly reducing the selective efficiency of the MIPs. Moreover, MIPs offer high selectivity towards reactive dyes, which is not present in expensive adsorbents such as activated carbon, biological treatment, etc. The results of BB3 dye extraction from different water samples are summarized in Table 6 whereas Table 7 summarizes the adsorption capacities of different adsorbents for the mentioned dye.

Table 6. MIP efficiency for the removal of BB3 from different environmental samples.

Samples	Amount of BB3 Added (mg/L)	Amount of BB3 Found (mg/L)	%Recovery ±SD
Distilled water	100	96.2	96.2 ± 0.9
River water	100	61.1	31.1 ± 0.1
Tap water	100	82.09	32.09 ± 1.2

Table 7. Comparison of the adsorption capacities of different adsorbents to BB3 dye.

Adsorbents	Adsorption Capacity ($mg \cdot g^{-1}$)	Ref
Chitosan-based	166.5	[38]
Amberlite XAD 1180	66.5	[39]
Aleppo pine-tree sawdust	65.4	[40]
Pineapple stem	58.9	[41]
Durian husk	49.5	[42]
Acrylic resin	46.95	[43]
Peat	41.00	[44]
Risk Hull	13.41	[45]
MIP	75.125	Present study

4. Conclusions

The current study focuses on the use of synthesized polymers (MIP) for the removal of a specific analyte (BB-3) from different water samples under optimized conditions. About 96.6% removal of BB-3 was achieved at 313 K, which indicates high selectivity toward the specific analyte as compared to other dyes such as safranin, Basic Blue 41 (BB-41), Thioflavin T (TT) and Methylene blue (MB), with 25%, 21%, 9.3% and 4.7%, respectively. Kinetic and isotherm studies showed that the adsorption of BB-3 on the polymer followed second order kinetics and the Langmuir model. The linear plot of qt vs. $t^{1/2}$ in the intraparticle model revealed the surface sorption and intraparticle diffusion. In a thermodynamic study, the negative values of ΔG°, ΔH° and ΔS° demonstrate that the spontaneity, endothermic nature and disorder increased during the adsorption of BB-3 on the MIP, respectively. The MIP proved to be a workable material for the extraction and concentration of BB3 from effluent, as MIP can be easily removed from the media and recovered using centrifugation methods, allowing it to be reused without considerable activity loss.

Author Contributions: Conceptualization, M.S., M.Z. and F.A.; methodology, F.A. and I.A.; formal analysis, R.U., E.A.A., F.A.K. and A.S.; investigation, F.A., M.S. and I.A.; resources, R.U., E.A.A., F.A.K. and A.S.; data curation, R.U., E.A.A., F.A.K. and A.S.; writing—original draft preparation, M.S., M.Z. and F.A.; writing—review and editing, M.S., M.Z. and F.A.; supervision, M.S., M.Z. and F.A.; project administration, M.S., M.Z. and F.A. All authors have read and agreed to the published version of the manuscript.

Funding: The study was supported by Researchers Supporting Project Number RSP/2021/45, King Saud University Riyadh, Saudi Arabia.

Institutional Review Board Statement: Not applicable.

Informed Consent Statement: Not applicable.

Acknowledgments: Authors are thankful to Researchers Supporting Project Number RSP/2021/45, King Saud Univer-sity Riyadh, Saudi Arabia.

Conflicts of Interest: The authors declare no conflict of interest.

References

1. Foguel, M.V.; Pedro, N.T.B.; Wong, A.; Khan, S.; Zanoni, M.V.B.; Sotomayor, M.D.P.T. Synthesis and evaluation of a molecularly imprinted polymer for selective adsorption and quantification of Acid Green 16 textile dye in water samples. *Talanta* **2017**, *170*, 244–251. [CrossRef] [PubMed]
2. Yang, X.Q.; Zhao, X.X.; Liu, C.Y.; Zheng, Y.; Qian, S.J. Decolorization of Azo, Triphenylmethane and Anthraquinone Dyes by a Newly Isolated Trametes sp. SQ01 and Its Laccase. *Process Biochem.* **2009**, *44*, 1185–1189. [CrossRef]
3. Bisgin, A.T.; Surme, Y.; Ucan, M.; Narin, I. Simultaneous spectrophotometric determination and column solid-phase extraction of two Lanaset textile dyes in environmental water samples. *J. Ind. Eng. Chem.* **2016**, *38*, 186–192. [CrossRef]
4. Ton, X.A.; Acha, V.; Bonomi, P.; Bui, B.T.S.; Haupt, K. A disposable evanes-cent wave fiber optic sensor coated with a molecularly imprinted polymer as a selective fluorescence probe. *Biosens. Bioelectron.* **2015**, *64*, 359–366. [CrossRef] [PubMed]
5. Munagapati, V.S.; Yarramuthib, V.; Kim, Y.; Lee, K.M.; Kim, D.S. Removal of anionic dyes (Reactive black 5 and Congo red) from aqueous solutions using banana peel powder as an adsorbent. *Ecotoxicol. Environ. Saf.* **2018**, *148*, 601–607. [CrossRef]
6. Salahi, S.; Parvini, M.; Ghorbani, M. Equilibrium studies in adsorption of Hg (II) from aqueous solutions using biocompatible polymeric polypyrrole-chitosan nanocomposite. *Polycycl. Aromat. Compd.* **2014**, *34*, 225–236. [CrossRef]
7. Zhang, Y.; Xie, Z.; Teng, X.; Fan, J. Synthesis of molecularly imprinted polymer nanoparticles for the fast and highly selective adsorption of sunset yellow. *J. Sep. Sci.* **2016**, *39*, 1559–1566. [CrossRef]
8. Shafqat, S.R.; Bhawani, S.A.; Bakhtiar, S.; Ibrahim, M.N.M. Synthesis of molecularly imprinted polymer for removal of Congo red. *BMC Chem.* **2020**, *14*, 27. [CrossRef]
9. Qiujin, Z.; Liping, W.; Shengfang, W.; Wasswa, J.; Xiaohong, G.; Jian, T. Selectivity of molecularly imprinted solid phase extraction for sterol compounds. *Food Chem.* **2009**, *113*, 608–615.
10. Regal, P.; Bao, M.D.; Barreiro, R.; Cepeda, A.; Fente, C. Application of molecularly imprinted polymers in food analysis: Clean-up and chromatographic improvements. *Cent. Eur. J. Chem.* **2012**, *10*, 766–784. [CrossRef]
11. Feng, Q.; Zhao, L.; Lin, J.M. Molecularly imprinted polymer as micro-solid phase extraction combined with high performance liquid chromatography to determine phenolic compounds in environmental water samples. *Anal. Chim. Acta* **2009**, *650*, 70–76. [CrossRef] [PubMed]
12. Ashley, J.; Shahbazi, M.A.; Kant, K.; Chidambara, V.A.; Wolff, A.; Bang, D.D.; Sun, Y. Molecularly imprinted polymers for sample preparation and biosensing in food analysis: Progress and perspectives. *Biosens. Bioelectron.* **2017**, *91*, 606–615. [CrossRef] [PubMed]
13. Foguel, M.V.; Ton, X.A.; Zanoni, M.V.B.; Sotomayor, M.D.P.T.; Haupt, K.; Bui, B.T.S. A molecularly imprinted polymer-based evanescent wave fiber optic sensor for the detection of basic red 9 dye. *Sens. Actuators B* **2015**, *218*, 222–228. [CrossRef]
14. Moczko, E.; Poma, A.; Guerreiro, A.; Sansalvador, I.P.D.V.; Caygill, S.; Canfarotta, F.; Whitcombe, M.J.; Piletsky, S. Surface-modified multifunctional MIP nanoparticles. *Nanoscale* **2013**, *5*, 3733–3741. [CrossRef] [PubMed]
15. Dirion, B.; Cobb, Z.; Schillinger, E.; Andersson, L.I.; Sellergren, B. Water-compatible molecularly imprinted polymers obtained via high-throughput synthesis and experimental design. *J. Am. Chem. Soc.* **2003**, *125*, 15101–15109. [CrossRef]
16. Piyush, S.; Agnieszka, P.L.S.; Francis, D.S.; Wlodzimierz, K. Electrochemically synthesized polymers in molecular imprinting for chemical sensing. *Anal. Bioanal. Chem.* **2012**, *402*, 3177–3204.
17. Zhou, T.C.; Shen, X.T.; Chaudhary, S.; Ye, L. Molecularly imprinted polymer beads prepared by pickering emulsion polymerization for steroid recognition. *J. Appl. Polym. Sci.* **2014**, *131*, 39606. [CrossRef]
18. Yang, J.J.; Li, Y.; Wang, J.C.; Sun, X.L.; Cao, R.; Sun, H. Molecularly imprinted polymer microspheres prepared by Pickering emulsion polymerization for selective solid-phase extraction of eight bisphenols fromhuman urine samples. *Anal. Chim. Acta* **2015**, *872*, 35–45. [CrossRef]

19. Abdallah, N.A.; Ibhrahim, H.F.; Hegabe, N.H. Comparative Study of Molecularly Imprinted Polymer and Magnetic Molecularly Imprinted Nanoparticles as Recognition Sites for the Potentiometric Determination of Gemifloxacin Mesylate. *Int. J. Electrochem. Sci.* **2017**, *12*, 10894–10910. [CrossRef]
20. Zhang, W.; She, X.; Wang, L.; Fan, H.; Zhou, Q.; Huang, X.; Tang, J.Z. Preparation, Characterization and pplication of Molecularly Imprinted Polymer for Selective Recognition of Sulpiride. *Materials* **2017**, *10*, 475. [CrossRef]
21. Awokoya, K.N.; Oninla, V.O.; Babalola, J.O.; Mbaeyi, N.N.; Folorunso, F.T.; Ndukwe, N.A. Adsorption of malachite green onto styrene-methacrylate based molecularly imprinted polymer. *J. Life Sci.* **2019**, *21*, 67–80. [CrossRef]
22. Asadi, E.; Deilami, S.A.; Abdouss, M.; Kordestani, D.; Rahimi, A.; Asadi, S. Synthesis, recognition and evaluation of molecularly imprinted polymer nanoparticle using miniemulsion polymerization for controlled release and analysis of risperidone in human plasma samples. *Korean J. Chem. Eng.* **2014**, *31*, 1028–1035. [CrossRef]
23. Farrington, K.; Regan, F. Investigation of the nature of MIP recognition development and characterisation of a MIP for Ibuprofen. *Biosens. Bioelectron.* **2007**, *22*, 1138–1146. [CrossRef] [PubMed]
24. Cormack, P.A.G.; Elorza, A.Z. Molecularly imprinted polymers: Synthesis and characterization. *J. Chromatogr. B* **2004**, *804*, 173–182. [CrossRef]
25. Farooq, S.; Nie, J.; Cheng, Y.; Yan, Z.; Li, J.; Bacha, S.A.S.; Mushtaqa, A.; Zhang, H. Molecularly imprinted polymers' application in pesticide residue detection. *Analyst* **2018**, *143*, 3971–3989. [CrossRef]
26. Maryam, F.; Mohammad, A.T.; Daryoush, A.; Ali, M. Preparation of molecularly imprinted polymer coated magnetic multi-walled carbon nanotubes for selective removal of dibenzothiophene. *Mater. Sci. Semicond. Process.* **2015**, *40*, 501–507.
27. Ho, Y.S.; McKay, C. Pseudo second order model for sorption processes. *Proc. Biochem.* **1999**, *34*, 451–465. [CrossRef]
28. Weber, J.; Morris, J.C. Kinetics of adsorption on carbon from solution. *J. Sanit. Eng. Div. Asce.* **1963**, *89*, 31–59. [CrossRef]
29. Poots, V.J.P.; McKay, G.; Healy, J.J. Removal of basic dye from effluent using wood as an adsorbent. *J. Water Pollut. Control Fed.* **1978**, *50*, 926–935.
30. Zhang, Y.L.; Zhang, J.; Dai, C.M.; Zhou, X.F.; Liu, S.G. Sorption of carbamazepine from water by magnetic molecularly imprinted polymers based on chitosan-Fe_3O_4. *Carbohydr. Polym.* **2013**, *97*, 809–816. [CrossRef]
31. Gautam, R.K.; Rawat, V.; Banerjee, S.; Sanroman, M.A.; Soni, S.; Singh, S.K.; Chattopadhyaya, M.C. Synthesis of bimetallic Fe–Zn nanoparticles and its application towards adsorptive removal of carcinogenic dye malachite green and Congo red in water. *J. Mol. Liq.* **2015**, *212*, 227–236. [CrossRef]
32. Pandian, C.J.; Palanivel, R.; Dhananasekaran, S. Green synthesis of nickel nanoparticles using Ocimum sanctum and their application in dye and pollutant adsorption. *Sep. Sci. Technol.* **2015**, *23*, 1307–1315. [CrossRef]
33. Hameed, H.; Ahmad, A.A. Batch adsorption of methylene blue from aqueous solution by garlic peel, an agricultural waste biomass. *J. Hazard. Mater.* **2009**, *164*, 870–875. [CrossRef] [PubMed]
34. Gautam, R.K.; Gautam, P.K.; Banerjee, S.; Soni, S.; Singh, S.K.; Chattopadhyaya, M.C. Removal of Ni (II) by magnetic nanoparticles. *J. Mol. Liq.* **2015**, *204*, 60–69. [CrossRef]
35. Jain, R.; Gupta, V.K.; Sikarwar, S. Adsorption and desorption studies on hazardous dye Naphthol Yellow S. *J. Hazard. Mater.* **2010**, *182*, 749–756. [CrossRef]
36. Zahoor, M.; Nazir, N.; Iftikhar, M.; Naz, S.; Zekker, I.; Burlakovs, J.; Uddin, F.; Kamran, A.W.; Kallistova, A.; Pimenov, N.; et al. A Review on Silver Nanoparticles: Classification, Various Methods of Synthesis, and Their Potential Roles in Biomedical Applications and Water Treatment. *Water.* **2021**, *13*, 2216. [CrossRef]
37. Harsini, N.N.; Ansari, M.; Kazemipour, M. Synthesis of molecularly imprinted polymer on magnetic core-shell silica nanoparticles for recognition of congo red. *Eurasian J. Anal. Chem.* **2018**, *13*, 1–13. [CrossRef]
38. Crini, G.; Gimbert, F.; Robert, C.; Martel, B.; Adama, O.; Crini, N.M.; Giorgi, F.D.; Badot, P.M. The removal of Basic Blue 3 from aqueous solutions by chitosan-based adsorbent: Batch studies. *J. Hazard. Mater.* **2008**, *153*, 96–106. [CrossRef]
39. Wawrzkiewicz, M. Removal of C.I. Basic Blue 3 dye by sorption onto cation exchange resin, functionalized and non-functionalized polymeric sorbents from aqueous solutions and wastewaters. *Chem. Eng. J.* **2013**, *217*, 414–425. [CrossRef]
40. Ouazene, N.; Lounis, A. Adsorption characteristics of CI Basic Blue 3 from aqueous solution onto Aleppo pine-tree sawdust. *Color Technol.* **2012**, *128*, 21–27. [CrossRef]
41. Chana, S.L.; Tana, Y.P.; Abdullaha, A.H.; Ong, S.T. Equilibrium, kinetic and thermodynamic studies of a new potential biosorbent for the removal of Basic Blue 3 and Congo Red dyes: Pineapple (Ananas comosus) plant stem. *J. Taiwan Inst. Chem. Eng.* **2016**, *61*, 306–315. [CrossRef]
42. Ong, T.; Tan, S.Y.; Khoo, E.C.; Lee, S.L.; Ha, S.T. Equilibrium studies for Basic blue 3 adsorption onto durian peel (Durio zibethinus Murray). *Desalin Water Treat.* **2012**, *45*, 161–169. [CrossRef]
43. Barsanescu, A.; Buhaceanu, R.; Dulman, V. Removal of Basic Blue 3 by sorption onto a weak acid acrylic resin. *J. Appl. Polym. Sci.* **2009**, *113*, 607–614. [CrossRef]
44. Contreras, E.; Martinez, B.; Sepúlveda, L.; Palma, C. Kinetics of basic dye adsorption onto Sphagnum Magellanicum peat. *Adsorpt. Sci. Technol.* **2007**, *25*, 637–646. [CrossRef]
45. Ong, S.T.; Lee, C.K.; Zainal, Z. Removal of basic and reactive dyes using Ethylenediamine modified rice hull. *Bioresour. Technol.* **2007**, *98*, 2792–2799. [CrossRef] [PubMed]

molecules

Article

Total Release of 21 Indicator Pharmaceuticals Listed by the Swedish Medical Products Agency from Wastewater Treatment Plants to Surface Water Bodies in the 1.3 Million Populated County Skåne (Scania), Sweden

Erland Björklund * and **Ola Svahn ***

Department of Environmental Science and Bioscience, Kristianstad University, Elmetorpsvägen 15, SE-291 88 Kristianstad, Sweden
* Correspondence: erland.bjorklund@hkr.se (E.B.); ola.svahn@hkr.se (O.S.)

Abstract: In 2017, the Swedish Environmental Protection Agency published a report on advanced wastewater treatment for the removal of pharmaceutical residues and stated that advanced treatment should be implemented where it will make the largest difference from an environmental perspective. However, the report also concluded that this need cannot be specified with existing data, but consideration must be made of local conditions. Two considerations are (1) the discharged amount of pharmaceutical into receiving water bodies and (2) the turnover of water in the recipient, where the highest risks are related to recipients with a low water turnover and low dilution. The current project comprised eight different WWTPs distributed throughout the entire County Skåne (Scania) in Sweden, with a population of ca. 1,300,000 persons. In total, 21 of 22 pharmaceuticals were analyzed according to the list proposed by the Swedish Medical Products Agency 2015. The results show that large amounts of pharmaceuticals are released from the WWTPs yearly to Scanian recipients. The total discharge of pharmaceuticals from the eight treatment plants adds up to 71 kg of these 21 substances alone, mainly comprising metoprolol, which is a drug that lowers blood pressure, and the analgesic drug diclofenac. Additionally, carbamazepine, losartan, naproxen and oxazepam were present in significant concentrations. These represented three illnesses that are very common: high blood pressure, inflammation/pain and depression/anxiety. The concentrations were generally in line with previous national Swedish screenings. It was estimated that, when one million cubic meters (1,000,000 m^3) of wastewater is discharged, almost 4 kg of the 21 pharmaceuticals is released. The total volume wastewater release by the >90 WWTPs in Scania was estimated to 152,887,000 m^3, which corresponded to 590 kg/year. The investigated 21 drugs cover only a small part of many hundred pharmaceuticals that are in use in Sweden. Thus, most likely, one or several tons of pharmaceuticals leak out to the Scanian recipients annually. The analysis of river samples shows that the dilution of wastewater is a key parameter in reducing concentrations. However, some locations have remarkably high concentrations, which occur when the volume wastewater is large in relation to the flow in the river. These kinds of regional results are of importance when selecting where advanced treatment should be prioritized in a first instance, as requested by the Swedish EPA.

Keywords: indicator pharmaceuticals; wastewater; surface water

Citation: Björklund, E.; Svahn, O. Total Release of 21 Indicator Pharmaceuticals Listed by the Swedish Medical Products Agency from Wastewater Treatment Plants to Surface Water Bodies in the 1.3 Million Populated County Skåne (Scania), Sweden. *Molecules* **2022**, *27*, 77. https://doi.org/10.3390/molecules27010077

Academic Editors: Victoria Samanidou, Eleni Deliyanni and Dimitra Voutsa

Received: 17 November 2021
Accepted: 20 December 2021
Published: 23 December 2021

1. Introduction

Pharmaceuticals in the environment (PIE) is an environmental topic that no longer can be considered new within the field of environmental contamination and toxicology. A well-known research paper with the title "Occurrence, fate and effects of pharmaceutical substances in the environment-A review" was published by Halling-Sørensen and coworkers more than 20 years ago and has today been cited over 4000 times [1]. Since then, a massive number of scientific articles have been released on basically all imaginable aspects of PIE [2–14]. Despite the large research funding spent on PIE, there are still many unknowns, as the topic is very complex. Nevertheless, today there is an increasing consensus among scientists that the present leakage of pharmaceuticals into the aquatic environment from our existing wastewater treatment plants (WWTPs) is unsustainable. It has been shown that some individual pharmaceuticals might be present in concentrations downstream WWTPs that could have adverse effects on wildlife, such as fish and birds [15–17]. Additionally, there is a concern that the overall effect of a large number of compounds released from WWTPs, in what often is referred to as a cocktail of micropollutants, may cause unwanted changes to receiving aquatic ecosystems [18,19]. Apart from these negative reports on the release of micropollutants from WWTPs, there is a parallel trend that this partly treated water could (and should) be seen as a future water resource, but more research and development is required to assure a safe use [20–22]. This is driven by a scarcity of water in many areas, as a consequence of increased water consumption by societies, combined with higher temperatures and draught due to a changing climate.

From a Swedish perspective, the awareness of the constant leakage of micropollutants, such as pharmaceuticals, to our recipients has been a topic of high public or political concern only in the past decade. Unlimited access to high-quality water has also always been nearly indisputable in Sweden, but after the summer of 2018, with extremely dry and hot weather conditions [23], the need for water increased. Especially agriculture was highlighted after historically low harvests. This justifies a better reuse of our limited water resources. In Sweden, several projects related to PIE have been performed in the past 15 years. Summarizing all of these is not the scope of this work; however, looking at some of the major funding made available during this period may be warranted, as it puts this and coming Swedish investigations into a broader context. It also points towards a change in how funding of PIE might (and possibly should) change from 2021 and onwards in order to make an actual impact by reducing the regional pollutant load to our recipients. Between 200 and 2009, a large Swedish project named "Pharmaceuticals-Presence and effects in the aquatic environment, preventive measures and possible treatment methods" was funded by Miljömiljarden, City of Stockholm, with a budget of 1.4 million Euro (Figure 1).

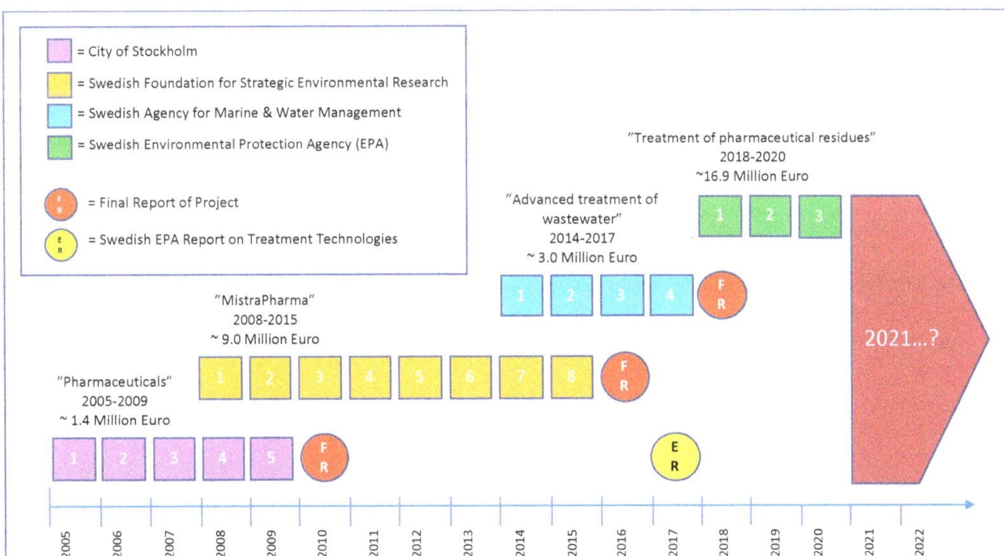

Figure 1. Example of major research funding made available in Sweden for 2005–2020 in relation to pharmaceuticals in the environment (PIE) and their removal from wastewater in WWTPs.

The final report was published in 2010 and concluded that there is a need to remove pharmaceuticals in Stockholm and that activated carbon or ozonation might be suitable technologies [24]. In 2008, a very general project called "MistraPharma" was launched, a project that, in the end, would receive a total of ~9.0 million Euros between 2008 and 2015 from the Swedish Foundation for Strategic Environmental Research (Figure 1). In its time, "MistraPharma" was one of largest research programs in the world within the PIE field. For 8 years, several research groups in Sweden tried to identify human pharmaceuticals that might be of concern to various aquatic ecosystems. "MistraPharma" also proposed risk management strategies, including better regulatory test requirements, as well as improved technologies for wastewater treatment. In the end, this resulted in 130 scientific papers, and, in 2016, the final report was published [25]. One of the main outcomes was 10 recommendations for improved environmental risk assessment [26]; however, some research was also dedicated to wastewater treatment technologies. Mainly two technologies showed to be promising: activated carbon and ozonation. As a logical continuation of these projects, the Swedish Agency for Marine and Water Management (on behalf of the Swedish Government) allocated ~3.0 million Euros to targeted projects that intended to develop novel techniques for "Advanced treatment of wastewater" from drug residues and other harmful compounds. The projects were operative between 2014 and 2017, mainly investigating ozonation and, to some extent, activated carbon (Figure 1). Most projects were headed by scientists, but the work was performed in collaboration with municipalities. Consequently, focus was partly moved from being primarily scientific to being more broadly connected to society by involving personnel at selected WWTPs. The final report was published in 2018 [27] and (once again) showed that ozonation and activated carbon were suitable to remove micropollutants and gave estimates of costs in doing so.

In parallel, the Swedish Government in December 2015 appointed the Swedish Environmental Protection Agency to find out if it would be conceivable to implement advanced treatment of wastewater to protect the aquatic environment from pharmaceuticals. This included analyzing which technical solutions were available and the pros and cons of these, as well as other effects that might occur if advanced treatment was introduced. The results were presented in a final report in 2017 [28] and concluded that the emission of pharmaceuticals can be reduced if Swedish WWTPs are equipped with more advanced technologies (Figure 1). However, it was notable that the Swedish EPA clearly stated that the Swedish society now must investigate where the technology should be introduced first, but concluded that, with existing occurrence data, this is not possible to specify. Several factors are important to make adequate prioritizations on where the needs are greatest, and consideration must be taken of local conditions, such as the following:

- The amount of pharmaceutical residues that are discharged into the recipients;
- The recipient's water turnover;
- The number of WWTPs that discharge to the same recipient;
- The recipient's sensitivity;
- Variations over the year;
- Variations in discharged amounts from the WWTP.

This is especially important as the Swedish Government, via the Swedish EPA, recently allocated a total of ~16.9 million Euros to fund different technological projects for "Treatment of pharmaceutical residues" at various WWTPs. This time, scientists at universities could be part of the projects but could not apply for the funding. Instead, the funding was now intended for Swedish municipalities who wanted to investigate advanced treatment technologies in a small or large scale on-site.

In order to harmonize the Swedish investigations, 22 pharmaceuticals were proposed as indicators by the Swedish Medical Products Agency in 2015 [29]. These pharmaceuticals should be analyzed and monitored in Swedish recipients. In our study, we utilized a method capable of analyzing all of these compounds simultaneously [30,31]. The year before, in 2014, the County Administrative Board of Scania released a document where guidelines concerning pharmaceuticals residues in wastewater were presented [32]. The document states that samples should be taken in outlet wastewater from WWTPs that are dimensioned for 200 PE or more, along with samples upstream and downstream the WWTP in question. Up to now, no Swedish or county studies have been performed at this large geographical resolution, as suggested by the County Administrative board of Scania [32], and, at the same time, employing the Swedish Medical Products Agency's analytical protocol [29]. This paper attempts to start investigating where actions should be taken at a regional level as requested by the Swedish Environmental Protection Agency [28], as well as meeting the request from the Swedish Medical Products Agency [29] to estimate the burden of these indicator pharmaceuticals. This is important, since actual figures on burden will aid in the dialogue with politicians and stakeholder at local WWTPs as to why and where action should be taken to reduce the pharmaceutical release from Scanian WWTPs to the aquatic environment.

2. Results and Discussion

2.1. Wastewater Treatment Plants (WWTPs)

The eight WWTPs were distributed over the entire County Scania (Figure 2a), and the annual volumes of treated wastewater differed largely, as seen in Supplementary Materials, Table S1a. The extent of variation ranged from 77,000 m^3 in the tiniest WWTP in Gärds Köpinge to 8,000,000 m^3 at Kristianstad WWTP. The relative size of the eight WWTPs, using the annual volume of treated wastewater as a base and thereafter attributing Gärds Köpinge WWTP a value of 1, showed that the WWTPs varied by a factor more than 100, as seen in Figure 2b.

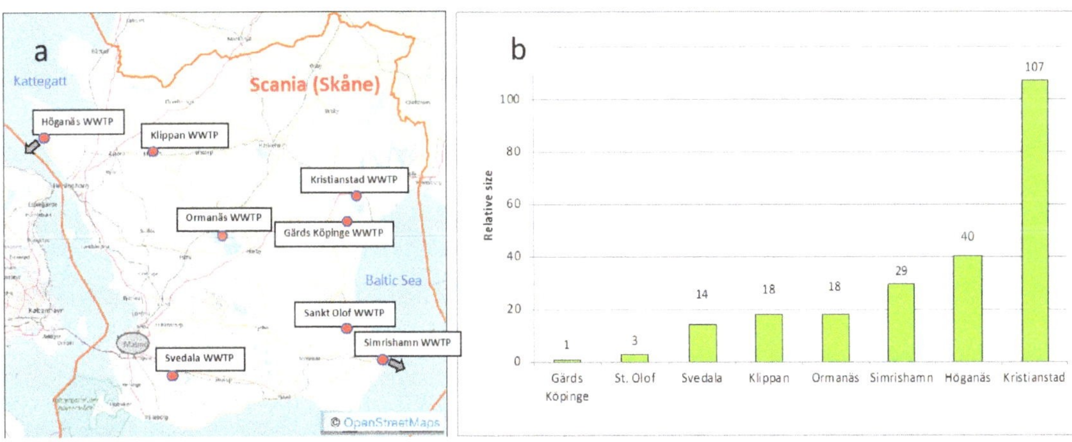

Figure 2. Eight investigated WWTPs. (**a**) Geographical spread of the WWTPs in Scania, Sweden. Gray arrows indicate that the WWTP discharges directly into the sea. (**b**) Relative size of the WWTPs based on annual volumes of treated wastewater attributing Gärds Köpinge WWTP a value of 1 (corresponding to approximately 77,000 m^3 treated water/year).

The water flow expressed as an average in m^3/h varied by a factor of 109 from Gärds Köpinge WWTP, at 8.75 m^3/h, to Kristianstad WWTP, at 958 m^3/h. The PE differed from 425 to 118,300 PE, corresponding to a factor of 278. Kristianstad WWTP and Simrishamn WWTP stood out, as they contain a large proportion industrial water. Consequently, the included WWTPs represented very different types of wastewater treatment plants. Complementary information about the treatment techniques applied in the different WWTPs can be found in Supplementary Materials, Table S1b, which is based on the information provided by the participants in the questionnaires.

2.2. *Results of Pharmaceutical Analyses*

2.2.1. Chemical Emissions of Pharmaceuticals as Concentrations (ng/L)

The recipients' chemical loads, expressed as outlet concentrations of pharmaceuticals, from all eight WWTPs are presented in Supplementary Materials, Table S2. From these data, the average emission concentration from the eight Scanian WWTPs of each specific pharmaceutical was calculated in order to get an overview of which substances that had the highest concentrations in Scanian wastewater in general, as seen in Figure 3.

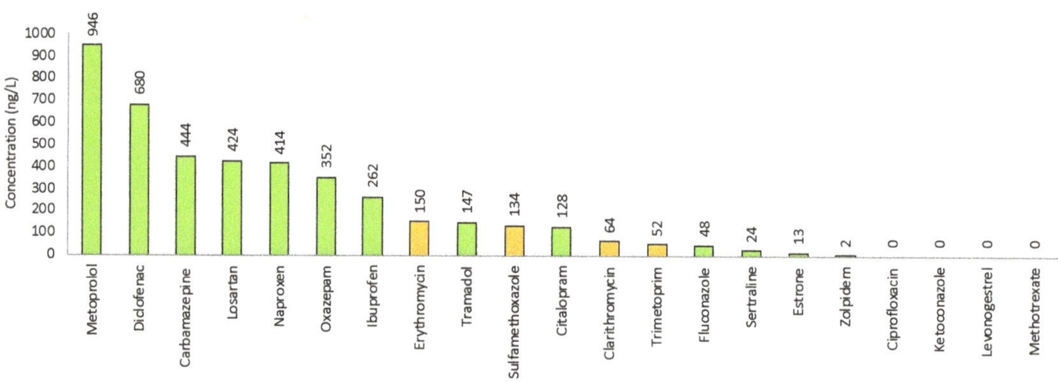

Figure 3. Average concentrations (ng/L) of the 21 analyzed pharmaceuticals in outlet water from 8 Scanian WWTPs (see Supplementary Materials Table S2). Antibiotics are marked in orange, except for ciprofloxacin, which was not found in measurable concentrations.

Based on this information, the compounds are discussed below in groups based in outlet concentrations.

Metoprolol, Diclofenac, Carbamazepine, Losartan, Naproxen, Oxazepam and Ibuprofen

These compounds comprise several types of pharmaceuticals, representing three general and common illnesses: (1) high blood pressure, (2) inflammation/pain and (3) depression/anxiety. The two top pharmaceuticals were metoprolol and diclofenac, occurring at average concentrations between 500 and 1000 ng/L, followed by five pharmaceuticals (carbamazepine, losartan, naproxen, oxazepam and ibuprofen) occurring between 250 and 499 ng/L.

Metoprolol is a ß-blocker used against high blood pressure and is well-known to be recalcitrant during conventional wastewater treatment [33]. Metoprolol had the highest average concentration of all substances at 946 ng/L (1 µg/L). The variation among the WWTPs was small. The lowest concentration was observed in Gärds Köpinge, at 692 ng/L, while Svedala had the highest value, at 1430 ng/L. The concentrations are close to those measured in earlier Swedish studies. In 2010, Stockholm Vatten showed that Henriksdal WWTP had an outlet concentration of 1161 ng/L (RSD = 53%), whereas Bromma WWTP had 1320 ng/L (RSD = 68%) [24]. Several WWTPs were also tested in a Swedish National Screening Programme in 2011 (National Screening), where outlet concentrations from WWTPs at the four cities, i.e., Skövde, Stockholm, Uppsala and Umeå, were reported [34]. A comparison of metoprolol between the eight Scanian WWTPs and theses four cities is shown in Figure 4a.

Figure 4. Measured concentrations (ng/L) of (**a**) metoprolol and (**b**) diclofenac in outlet water from the 8 Scanian WWTPs and in the Swedish National Screening Programme (Report 2011) for the cities Skövde, Stockholm, Uppsala and Umeå. The WWTPs in these 4 cities were analyzed three times each, as indicated by 1, 2 and 3. The 8 Scanian WWTPs are shown in green bars.

The concentrations in the Scanian WWTPs are on a par or somewhat below the results from the National Screening. It should be noted that no Scanian WWTPs were included in either Stockholm Vatten [24] or in National Screening [34], thus highlighting the need for a regional assessment of the pharmaceutical load in this highly populated Swedish area. Metoprolol was also recently identified at up to 10 ng/L in the large Swedish lake Mälaren, showing that this compound is present in the Swedish aquatic environment [35].

Diclofenac is a NSAID belonging to a group of anti-inflammatory drugs and has shown to be semi-persistent in the Swedish environment [36]. Swedish researchers recently demonstrated that, when three-spine stickleback (*Gasterosteus aculeatus*) was exposed to even low μg/L concentrations of diclofenac, it caused histological changes, and the researchers expressed concern about fish populations exposed to treated sewage effluents [37]. Diclofenac was recently also highlighted by the Swedish Medical Products Agency for being so heavily used and easily accessible in Sweden [38]. The total Swedish sales amounted to 4.5 tons, with 2.6 tons OTC and 1.9 tons on prescription. In Scania, more than 600 kg were sold, with 50% OTC. In our study, diclofenac presented the second highest concentration with an average concentration of 680 ng/L (0.7 μg/L) and a small variation between WWTPs.

The lowest concentration was observed at Ormanäs, at 442 ng/L, while the highest concentration was observed at Svedala, at 1117 ng/L. Stockholm Vatten reported that the outlet concentration in the Henriksdal WWTP was 288 ng/L (RSD = 36%), while, in the Bromma WWTP, it was 257 ng/L (RSD = 32%) [24]. When comparing with the National Screening [34] (Figure 4b) pronounced similarities in concentrations are observed. In an EU-wide monitoring covering 90 WWTPs (EU Monitoring) [39], the detection frequency was 89%, the maximum concentration 174 ng/L and the average concentration 49 ng/L. These are somewhat lower concentrations; however, the authors commented that their diclofenac concentrations might be underestimated. The concentrations of metoprolol and diclofenac in the eight Scanian WWTPs are compared in Supplementary Materials, Figure S1: Figure S1a shows the similarities in measured concentrations in outlet wastewater, while Figure S1b shows that the size of the WWTP has no influence on the outlet concentrations. Diclofenac was also recently quantified in Swedish lake Mälaren at 3.5 ng/L [35].

Carbamazepine is administered to treat epilepsy and alcohol abstinence and has a well-known persistence in water [40]. Carbamazepine occurred in all WWTPs between 139 ng/L (Gärds Köpinge) and 699 ng/L (Svedala), with an average concentration of 444 ng/L. Stockholm Vatten reported a concentration of 373 ng/L (RSD = 32%) at Henriksdal and 305 ng/L at Bromma (RSD = 35%) [24]. The concentrations in the four cities in National Screening [34] varied between 460 and 1100 ng/L, while the EU Monitoring had a detection frequency of 90%, a maximum concentration 4609 ng/L and a somewhat higher average concentration of 832 ng/L [39]. Concentrations in Scania are in line with previous Swedish screenings, but somewhat lower than the average concentration in Europe. In a recent Swedish study covering the entire Baltic Sea, carbamazepine was identified with a frequency of >90% in 43 samples, as a consequence of extensive usage, low removal in WWTPs and environmental persistence [41]. The low detection limits, in combination with its persistence, have made carbamazepine a good indicator of wastewater intrusion in natural waters [42]. The environmental concentrations in Swedish lake Mälaren reached as high as 20 ng/L [35].

Losartan is a blood-pressure-lowering medicine. Recently, researchers observed effects on the brown mussel (*Perna perna*), even at ng/L levels, and suggested it as a suitable marine model in environmental assessments [43]. The average concentration in Scanian WWTPs was 424 ng/L, with a relatively high variation; Ormanäs had the lowest concentration, at 83 ng/L; and Svedala had the highest, at 921 ng/L. The concentration at Henriksdal was 204 ng/L (RSD = 48%), and at Bromma, it was 187 ng/L (RSD = 48%), according to Stockholm Vatten [24]. Losartan was neither analyzed in the National Screening [34] nor in EU Monitoring [39], but the concentrations in Scania are within a factor of two compared to Stockholm Vatten [24]. Losartan was quantified at 5 ng/L in lake Mälaren revealing its environmental occurrence [35].

Naproxen is an anti-inflammatory NSAID pharmaceutical. Swedish researchers recently pointed out that naproxen and diclofenac produce highly similar toxic effects in fish, but based on new experiments, they showed that the environmental hazards and risks are lower for naproxen [44]. Hence, they stated that substitution would be advisable when naproxen presents an adequate alternative from a clinical point-of-view. The Scanian WWTPs had an average concentration of 414 ng/L. The concentrations vary in a way that resembles losartan with the lowest concentration of 119 ng/L at Kristianstad and the highest of 1430 ng/L at Sankt Olof. The concentration in Sankt Olof is much higher than in the other WWTPs, considering that the Simrishamn has the second highest value of only 379 ng/L. Excluding Sankt Olof reduces the average concentration to 269 ng/L. Sankt Olof is one of the smallest WWTPs, and an explanation for the high value could be a high consumption of naproxen in this small population during this explicit sampling event, causing extraordinary concentrations this day. The concentration at Henriksdal was 476 ng/L (RSD = 84%), and at Bromma, it was 565 ng/L (RSD = 46%), as shown by Stockholm Vatten [24]. The concentrations of naproxen in the four cities investigated in National Screening [34] ranged between 26 and 490 ng/L. The EU Monitoring reported a detection frequency of 66%, while maximum and average concentration were 958 and 27 ng/L, respectively [39]. Just as for diclofenac, this might be an underestimate. In general, the concentrations in Scania were close to previous Swedish studies.

Oxazepam is a benzodiazepine administered to treat anxiety and depression. It is classified as a narcotic and is well-known for its observed effects on perch (*Perca fluviatilis*) [16] and common roach (*Rutilus rutilus*) [45], as Swedish researchers demonstrated behavior changes in these fish species when exposed to low µg/L levels. The average concentration was 352 ng/L, with a low in Gärds Köpinge of 95 ng/L and a high in Kristianstad of 475 ng/L. Gärds Köpinge stands out with a 3–4 times lower concentration. Stockholm Vatten reported concentrations at Henriksdal of 324 ng/L (RSD = 49%), and at Bromma of 190 ng/L (RSD = 31%) [24]. A comparison between the four cities in National Screening [34] and the eight Scanian WWTPs is shown in Supplementary Materials, Figure S2. The average concentration of these four cities was 463 ng/L, which agrees well with Scania. The detection frequency in the EU Monitoring was 90% and a maximum concentration of 1766 ng/L [39]. The average concentration was lower (162 ng/L), but within a factor of two compared to Scania. Thirteen benzodiazepines were monitored in a study including 30 European rivers [46]. In the study, oxazepam had the highest detection frequency (85%), with a maximum concentration of 61 ng/L, showing widespread environmental occurrence. The Swedish lake Mälaren also contained oxazepam up to 5 ng/L [35].

Ibuprofen is a third NSAID drug (similar to diclofenac and naproxen) and is available as an OTC pharmaceutical with a widespread use. The average concentration was 262 ng/L; however, the differences in observed concentrations were very large. At Gärds Köpinge, Höganäs and Kristianstad WWTPs ibuprofen could not be detected. In contrast, at Ormanäs the observed concentration exceeded 1 µg/L (1158 ng/L). Stockholm Vatten found a concentration of 42 ng/L (RSD = 136%) at Henriksdal and 80 ng/L (RSD = 108%) at Bromma [24]. The ibuprofen concentrations varied largely in National Screening [34] just as they did in the Scanian WWTPs, with concentrations from 42 to 990 ng/L in the four cities. The EU Monitoring showed a detection frequency of 57%, with maximum and average concentrations of 2129 and 81 ng/L, respectively [39]. This is comparable to Scanian observations, where both frequency and concentration vary largely between WWTPs.

Seasonal Variation in Concentrations of the Top Six Pharmaceuticals at Kristianstad WWTP

To investigate how pharmaceutical concentrations vary over time and disclose the representativeness of the selected sampling period (April), a study was conducted at Kristianstad WWTP. Samples were taken at 13 occasion to cover all four seasons. The top six pharmaceuticals, which all occurred in all WWTPs (Supplementary Materials, Table S2), were analyzed in these samples. The results are seen in Supplementary Materials, Table S3, while the seasonal variation is shown in Figure 5 for metoprolol and diclofenac. This shows a tendency towards highest concentrations in December/January and a slight drop in June/July to then increase again. Similar results were observed for all six pharmaceuticals (Supplementary Materials, Figure S3). Notably, metoprolol had an average annual concentration of 638 ng/L (n = 13; Supplementary Materials, Table S3), which is close to the concentration of metoprolol at Kristianstad WWTP, with a value of 714 ng/L (n = 1; see Supplementary Materials, Table S2). The month of April thereby seems to mirror the annual average concentration for Kristianstad WWTP reasonably well. The similarities seen for metoprolol between the month of April (n = 1) and the entire season 2016/2017 (n = 13), were also observed for diclofenac (746 vs. 655 ng/L), carbamazepine (470 vs. 441 ng/L), losartan (217 vs. 219 ng/L), naproxen (119 vs. 173 ng/L) and oxazepam (475 vs. 499 ng/L).

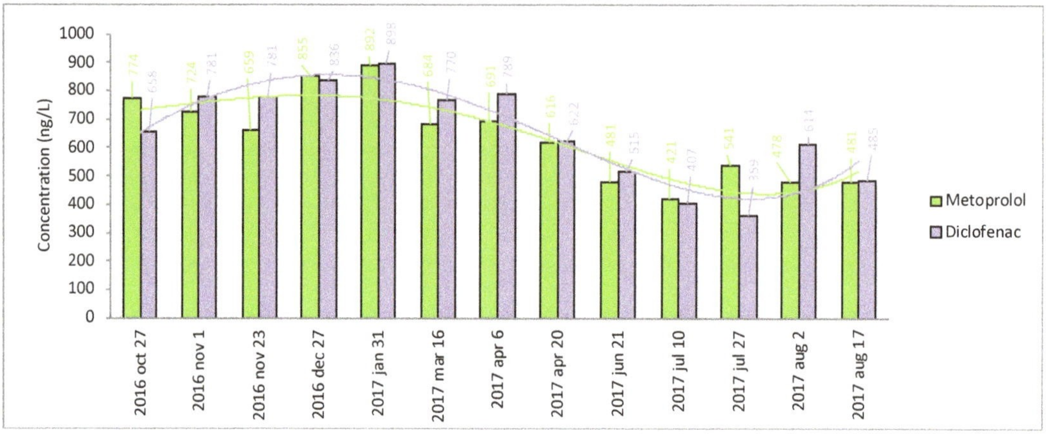

Figure 5. Measured concentrations (ng/L) of metoprolol and diclofenac in outlet water from Kristianstad WWTP during the season 2016/2017 (n = 13).

Antibiotics

Five antibiotics were analyzed: ciprofloxacin, clarithromycin, erythromycin, sulfamethoxazole and trimethoprim (Supplementary Materials, Table S2 and Figure 3).

Ciprofloxacin is an antibiotic (broad-spectrum) used to treat bacterial infections of various types. Ciprofloxacin is known to adsorb to the sludge phase [47], and this probably explains why it was not detected in any wastewater outlets. In previous Swedish studies, ciprofloxacin has been quantified in sludge up to a few thousand μg/kg [47,48]. Stockholm Vatten [24] reported low concentrations of 20 ng/L (RSD = 50%) at Henriksdal and 40 ng/L (RSD = 44%) at Bromma. The concentrations in the for the four cities in National Screening [34] ranged from 0 up to 65 ng/L. In the EU Monitoring [39], the detection frequency was 90% and the maximum concentration found was 264 ng/L, while the average concentration was 96 ng/L. In the most recent Swedish study [47] covering 11 WWTPs, the detection frequency in treated effluent water was 27% with maximum and average concentrations of 62 and 38 ng/L, respectively. This is similar to a previous Swedish study including five WWTPs, where the outlet concentration varied between <LOQ and 60 ng/L [49]. Overall, the reported ciprofloxacin outlet concentrations in Swedish WWTPs are below 70 ng/L.

Clarithromycin and erythromycin are both macrolides with a chemical structure that is very similar. Both are administered to treat bacterial infections. Clarithromycin concentrations ranged from <LOQ to 213 ng/L and erythromycin from 1 to 640 ng/L. Stockholm Vatten [24] did not analyze clarithromycin, while the concentration of erythromycin was 236 ng/L (RSD = 67%) which can be compared to the Scanian average concentration of 151 ng/L. Noteworthy is that in the two smallest WWTPs Gärds Köpinge and Sankt Olof, erythromycin was only found in low concentrations of <3 ng/L which is discussed in detail below. By removing Gärds Köpinge and Sankt Olof from the average calculation, a new average of 200 ng/L is obtained for Scania, which is close to Stockholm Vatten [24]. In National Screening [34], clarithromycin varied in the range of <LOQ–780 ng/L, and erythromycin varied between 53 and 530 ng/L. Both were close to the Scanian WWTPs, but the range was broader. In the Swedish study including 11 WWTPs, the detection frequencies were 61% and 88% for clarithromycin and erythromycin, respectively [47]. Clarithromycin had maximum and average concentrations of 86 and 36 ng/L, respectively, while corresponding concentrations for erythromycin were 350 and 129 ng/L, which is in line with this study. None of the macrolides was analyzed in the EU Monitoring [39].

Sulfamethoxazole and Trimethoprim are used in combination. The ratio of SMX/TMP in raw wastewaters has been suggested as a marker of wastewater origin where hospital effluents and WWTP influents show similar SMX/TMP footprints, while livestock effluents exhibit higher SMX/TMP ratios, due to usage of SMX alone in farming [50]. Recently researchers investigated the ecotoxic effects of sulfamethoxazole–diclofenac in a binary mixture to bacteria (*A. fischeri*), crustaceans (*D. magna*) and vascular plants (*L. minor*) [51]. Sulfamethoxazole alone showed low toxicity with low environmental risk, while diclofenac revealed moderate toxicity with a high risk to aquatic organisms. However, the mixture demonstrated the highest toxicity to the exposed model organisms. Nevertheless, when studying the toxicity of synthetic wastewater, no increase in toxicity was seen for wastewater spiked with sulfamethoxazole and diclofenac. Consequently, complex interactions likely occur between matrix components in environmental samples, thus affecting the wastewater toxicities that are observed. The concentrations in the eight Scanian WWTPs varied from a few to 281 ng/L, with average concentrations of 134 ng/L for sulfamethoxazole and 52 ng/L for trimethoprim. Both compounds were detected at Henriksdal at 60 ng/L (RSD = 67%) and 35 ng/L (RSD = 45%), and at Bromma with concentrations of 52 ng/L (RSD = 54%) and 186 ng/L (RSD = 29%), respectively, by Stockholm Vatten [24]. National Screening [34] quantified sulfamethoxazole between 30 and 290 ng/L and trimethoprim between 60 and 510 ng/L. Pharmaceutical emissions from Scanian WWTPs are similar to Stockholm Vatten [24] but lower than National Screening [34]. The outlet concentration of sulfamethoxazole and trimethoprim in five Swedish WWTPs ranged between <LOQ and 304 ng/L and between 66 and 1340 ng/L, respectively [49]. In the EU Monitoring, two different results were reported for sulfamethoxazole [39]: (I) detection frequency of 83% with maximum and average concentrations of 1691 and 280 ng/L, respectively; and (II) detection frequency of 81% with maximum and average concentrations of 1147 and 142 ng/L, respectively. For trimethoprim the detection frequency was 93%, with a maximum concentration of 800 ng/L and an average concentration of 229 ng/L. In general, these concentrations were higher than those reported in Scania.

Antibiotics and size of WWTP: The analyzed antibiotics did not show any obvious relation between concentrations and size of WWTP. However, a concrete difference was observed between the two WWTPs in Gärds Köpinge and Sankt Olof (the smallest WWTPs) and the six other WWTPs (Figure 6).

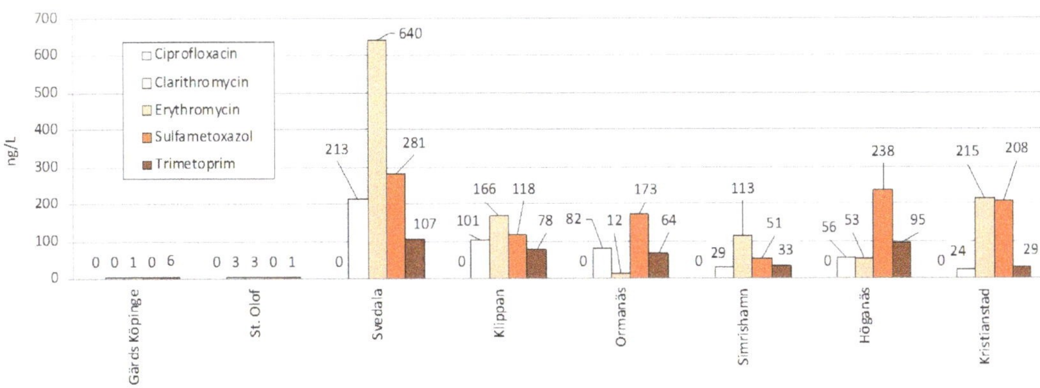

Figure 6. Measured outlet water concentrations (ng/L) of the five antibiotics, namely ciprofloxacin, clarithromycin, erythromycin, sulfamethoxazole and trimethoprim, in the 8 Scanian WWTPs.

These two WWTPs are a factor 5–30 smaller compared to the three in order (Figure 2b), and the number of PE is only 425 in Gärds Köpinge and 600 in Sankt Olof. This can possibly be explained by antibiotics not being consumed as regularly as, for example, heart medication. Instead, these antibiotics are taken as specific courses of treatment for a limited time. Small WWTPs seldom have enough parallel antibiotic treatments to allow detection.

Tramadol, Citalopram, Fluconazole and Sertraline

The four pharmaceuticals, i.e., tramadol, citalopram, fluconazole and sertraline, had average concentrations in the range of 24–147 ng/L for the eight Scanian WWTPs (Figure 3 and Supplementary Materials, Table S2).

Tramadol is an opioid and classed as a narcotic in Sweden. Tramadol has previously been investigated in 33 Swedish WWTPs' influent wastewater, together with 23 other illicit drugs [52]. Tramadol was detected in 100% of the samples, showing prevalent usage. In Scanian WWTPs, the concentrations were between 81 and 208 ng/L, with an average concentration of 147 ng/L. This differs from Henriksdal and Bromma, with concentrations of 571 ng/L (RSD = 49%) and 474 ng/L (RSD = 50%), respectively, Stockholm Vatten [24]. Even higher concentrations were observed in National Screening between 730 and 3000 ng/L [34]. It is noteworthy that tramadol showed one of the largest measurement uncertainties among five laboratories in a recent inter-calibration exercise that was headed by us on assignment of the Swedish Agency for Marine and Water Management [53]. This might explain part of this large variation. The EU Monitoring [39] reported a detection frequency of 100% with maximum and average concentrations of 1166 and 256 ng/L, respectively. Finally, tramadol was present in lake Mälaren up to 18 ng/L, showing occurrences also in Swedish aquatic ecosystems [35].

Citalopram is an antidepressant of the SSRI type. Swedish researchers have shown partially inhibited feeding [54] and long-lasting behavioral effects after exposure during development of three-spine stickleback (*Gasterosteus aculeatus*) to relevant environmental concentrations [55]. The Scanian WWTPs had concentrations between 80 and 217 ng/L, with an average concentration of 128 ng/L. This was close to data from Stockholm Vatten for Henriksdal and Bromma WWTPs, with concentrations of 196 ng/L (RSD = 44%) and 140 ng/L (RSD = 56%), respectively [24]. Slightly higher concentrations were found in National Screening [34] ranging between 170 and 480 ng/L. In the EU Monitoring, the detection frequency was 83%, maximum reported concentration 189 ng/L and average concentration 34 ng/L [39]. Overall, the Scanian concentrations were on par with previously reported concentrations. The highest concentration quantified in Swedish lake Mälaren was 4.1 ng/L [35].

Fluconazole is an antifungal medication, and the concentrations varied between 3 and 105 ng/L, with an average concentration of 48 ng/L. An analysis of fluconazole was not performed by Stockholm Vatten [24]. National Screening, however, reported much higher concentrations ranging from 72 to 1100 ng/L [34]. What might cause this large difference is not known. The effluents from 11 Swedish WWTPs showed a detection frequency of 97% with maximum and average concentrations of 170 and 60 ng/L, respectively [47] which is very close to our findings. Previous Swedish studies of five WWTPs had similar concentrations ranging between <LOQ and 140 ng/L [56]. The EU Monitoring [39] had a detection frequency of 98% and maximum and average concentrations of 598 and 108 ng/L, respectively, which is somewhere between the Scanian values and the National Screening values.

Sertraline is also an SSRI-type antidepressant, as citalopram above. Swedish researchers have studied the behavior of freshwater snails (*Radix balthica*) exposed to the pharmaceutical, observing a general lack of effects on the snail's activity [57]. In contrast, the same researchers found that the fish species Eurasian perch (*Perca fluviatilis*) presented an exposure-dependent decrease in feeding habits with increasing sertraline concentrations at a μg/L level [58]. The concentrations in the Scanian WWTPs varied in the interval 4–47 ng/L. The average concentration of 24 ng/L was similar to Henriksdal and Bromma, with concentrations of 26 ng/L (RSD = 59%) and 21 ng/L (RSD = 72%), respectively, Stockholm Vatten [24]. In National Screening [34], concentrations were between 0 and 32 ng/L, while in the EU Monitoring [39] the reported maximum and average concentrations were 38 and 2 ng/L, respectively, but with a detection frequency of only 12%. The sertraline concentrations in Scania are close to those observed in other studies. Sertraline could not be identified in lake Mälaren, Sweden [35].

Hormones

Two hormones were analyzed, namely estrone (a natural hormone) and levonorgestrel (a synthetic hormone), as shown in Figure 3 and Supplementary Materials, Table S2.

Estrone and other estrogenic substances in WWTP effluents have since long been known to induce estrogenic effects in fish [59]. In Scania, estrone occurred in low concentrations in outlet wastewater between 1 and 18 ng/L, except for Ormanäs WWTP, at 63 ng/L. The average concentration, with and without the Ormanäs value, was 13 and 6 ng/L, respectively. This is similar to Stockholm Vatten [24], where estrone analyses were performed as discrete "special analyses" and found concentrations of 4.2 ng/L (RSD = 89%) and 0.5 ng/L (RSD = 46%) at Henriksdal and Bromma, respectively. Estrone was not included in the National Screening [34], while in the EU Monitoring [39] it was analyzed by a multi-compound hormone screening method. However, no hormones were detected in any of the samples above the LOQ of 10 ng/L.

Levonorgestrel has been investigated by Swedish researchers and shown to disrupts the seasonal breeding cycle in male three-spine stickleback (*Gasterosteus aculeatus*) at ng/L levels [60]. Levonorgestrel could not be detected in any Scanian wastewaters, while Stockholm Vatten [24] did not analyze it. No levonorgestrel could be quantified in the water samples in National Screening [34], while it was not included in the EU Monitoring [39].

Ketoconazole, Zolpidem and Methotrexate

These pharmaceuticals are all present in low concentrations (Figure 3 and Supplementary Materials, Table S2).

Ketoconazole is an antifungal compound not quantified in any of the Scanian WWTPs. Stockholm Vatten [24] found 8 ng/L (RSD = 77%) at Henriksdal and 9 ng/L (RSD = 81%) at Bromma WWTP. In National Screening [34], it was only be found in one of 12 samples, but with the remarkably high concentration of 120 ng/L. In the 11 Swedish WWTPs investigated, very high levels of ketoconazole were found in sludge, with an average concentration of 2835 ng/g dry weight and a detection frequency of 85%. Consequently, it binds to sludge [47]. In the same study, the effluent sample had maximum and average concentrations of 41 and 36 ng/L, respectively, which is slightly higher than previous Swedish studies. Ketoconazole was also recently quantified by us in sludge, showing concentrations as high as 3009 ng/kg [48]. Ketoconazole was not analyzed in the EU Monitoring [39].

Zolpidem is administered for difficulties with sleeping. The average concentration was 3 ng/L, ranging from 1 to 4 ng/L. The concentrations at Henriksdal and Bromma were 5.1 ng/L (RSD = 46%) and 4.8 ng/L (RSD = 55%), respectively, according to Stockholm Vatten [24]. In National Screening [34], zolpidem concentrations varied between 3 and 41 ng/L. In the EU Monitoring [39], it had a detection frequency of 58%, with maximum and average concentrations of 43 and 1.5 ng/L, respectively, which is relatively close to the Swedish studies.

Methotrexate is used against rheumatic and inflammatory diseases, and in cancer treatment. It was not detected in any of the Scanian WWTP outlet water samples. No analysis of methotrexate was performed in the two previous Swedish studies by Stockholm Vatten [24] and National Screening [34]. It was not analyzed in the EU Monitoring, either [39].

2.2.2. Chemical Emissions of Pharmaceuticals from Eight WWTPs in Kilograms

The basic parameters of the WWTPs comprised the annual volume of treated wastewater (Supplementary Materials, Table S1a). By multiplying this volume with the outlet concentrations of each pharmaceutical at each WWTP (Supplementary Materials, Table S2), an estimate of the annually emitted amounts of pharmaceuticals in grams could be calculated (Supplementary Materials, Table S4). Thereafter the pharmaceutical emissions from each WWTP in kilograms could be estimated. The results are seen in Figure 7a.

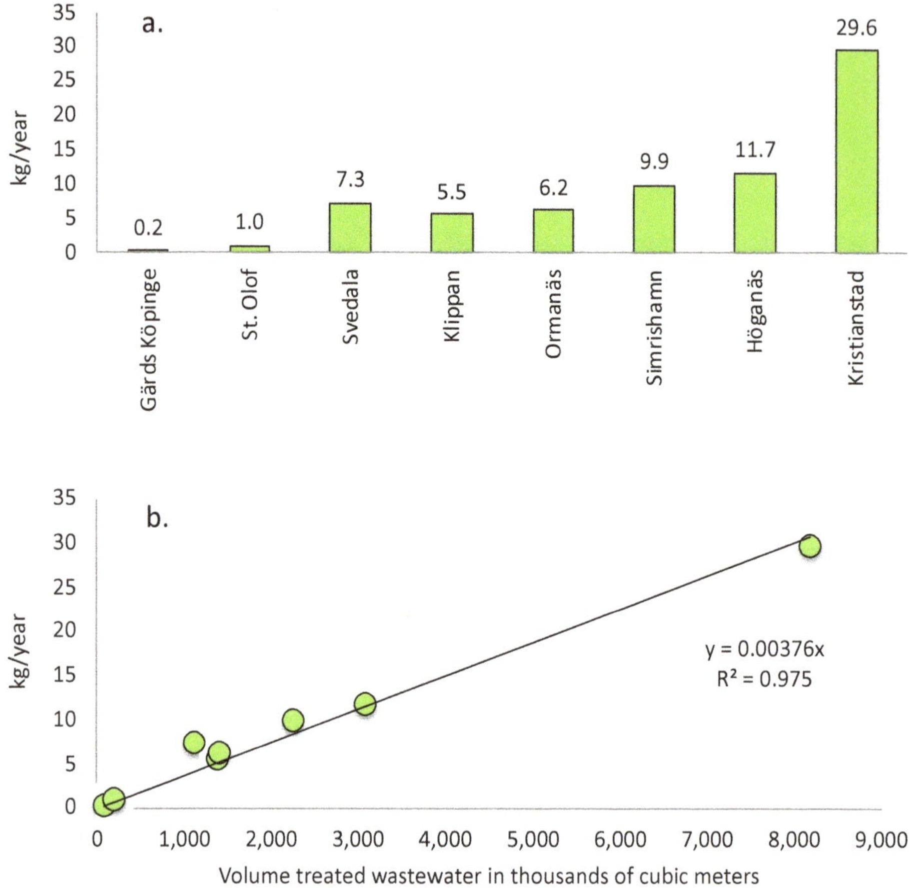

Figure 7. (**a**) Total emission in kg/year of the 21 pharmaceuticals in the 8 Scanian WWTPs. (**b**) Relationship between the emitted amount of pharmaceuticals as the sum of 21 pharmaceuticals in kg/year (y-axis) and the treated wastewater volume in thousands of cubic meters per year. In both panels, the order of the WWTPs is based on size according to Figure 2b.

The emissions from the eight WWTPs added up to a total of 71,409 which equals ca. 71 kg. These 71 kg only includes those 21 pharmaceuticals analysed. The smallest WWTP in Gärds Köpinge released only 0.2 kg while the largest WWTP in Kristianstads released as much as 30 kg. By relating the released amount of drugs per year (kg/year) to the treated volume of water in thousands of cubic meters (m^3) the relationship between the volume wastewater and the emitted amount of pharmaceuticals could be estimated (Figure 7b). From this, it could be seen that, when a WWTP in Scania releases 1 million cubic meter (1,000,000 m^3) of treated wastewater, roughly 3.76 kg $\approx$ 4 kg of the 21 pharmaceuticals passes through into the recipient simultaneously.

2.3. Estimate of Total Pharmaceutical Emissions in All of Scania Based on Open Data of Amount Treated Wastewater from Scanian WWTP Operators

Scania has 33 municipalities and a population of 1,322,193 persons (as of 2016) [61]. The WWTP operators most often release data on the total volume of treated wastewater. As part of this study, a large effort was put forth to identify as many of these data as possible to estimate the total amount of treated wastewater in Scania. An overview of the 92 identified WWTPs and their yearly volume of treated wastewater is shown in Figure 8, where the great diversity in size of WWTPs becomes clear. The figure also shows that the eight WWTPs in this study (marked in green) covers this diversity. Not all data could be identified for all 92 WWTPs, and, in such cases, the volume of wastewater was estimated (as discussed below). These WWTPs were, however, few (marked in gray in Figure 8). In some cases, several WWTPs were reported as a sum of treated wastewater (indicated to the right of Figure 8 in light red). Summing all reported volumes gave a total volume of 152,887 thousand m^3 of wastewater. This can be compared to the total treated volume of wastewater in Sweden, which, in 2016, was 1,078,652 thousand m^3. Consequently, Scania has ca. 14.2% of the wastewater in Sweden. This is realistic, considering that the population of Scania was 1,322,193 persons in 2016, compared to Sweden's total population of 9,995,153 the same year [61], corresponding to 13.2% of the total Swedish population. The population of Scania is not evenly distributed, and especially the western side of Scania has several larger cities. In Southern and Eastern Scania, the size of the population is slightly lower with a few small- and medium-sized cities. The identified treated wastewater volumes were grouped into six different geographical regions: Northwest, Southwest, South, Southeast, Northeast and Central Scania.

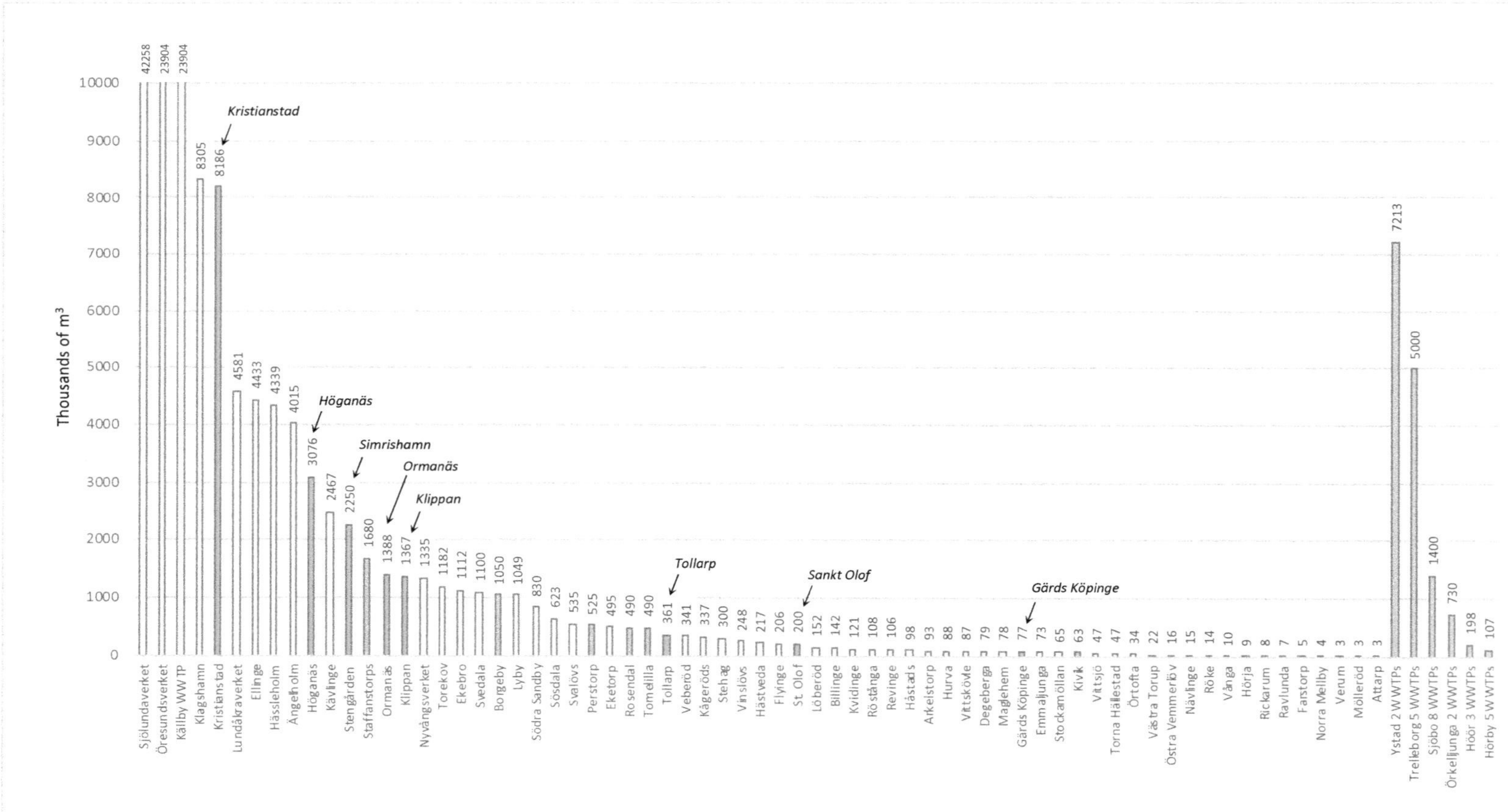

Figure 8. Yearly volume treated wastewater in thousands of m^3 in County Scania. For explanations, see text.

2.3.1. Northwestern Scania

Nordvästra Skånes Vatten och Avlopp AB (NSVA) is the organization that treats a majority of the wastewater in this area in six municipalities: Bjuv, Båstad, Helsingborg, Landskrona, Svalöv and Åstorp. In total, 10 WWTPs are operated with differing size, according to the available environmental reports (Supplementary Materials, Table S5). The total volume is 33,710,670 m^3, resulting in 126.8 kg. Additionally, Höganäs Municipality and its WWTP releases 11.7 kg according to our study, while the Klippan Municipality WWTP discharges 5.6 kg. The WWTP in Ängelholm Municipality treats around 11,000 m^3 of wastewater/day, according to their information, corresponding to 4,015,000 m^3 wastewater/year resulting in 15.1 kg. The total emission is estimated to be as follows (Equation (1)):

$$126.8 + 11.7 + 5.6 + 15.1 = 159.2 \text{ kg} \approx 160 \text{ kg/year} \quad (1)$$

2.3.2. Southwestern Scania

The largest organization in this part is VA SYD AB, which, all together, treats wastewater from more than 500,000 people. One of Sweden's largest WWTPs is Sjölundaverket, which treats wastewater from the City of Malmö and Burlöv, Lomma, Staffanstorp and Svedala Municipalities. Klagshamn WWTP receives wastewater from some areas of Malmö City and Vellinge Municipality. Additionally, VA SYD operates several other WWTPs as seen in Supplementary Materials Table S5, with a total volume of 68,694,617 m^3 resulting in 258.3 kg. For Borgeby WWTP in Lomma Municipality the exact volume is not known, but the WWTP is dimensioned for 15,000 persons. Based on similar-sized WWTPs this would give roughly 1,050,000 m^3 of wastewater and 3.9 kg. Likewise, Staffanstorp Municipality treats wastewater from 14,000 persons, corresponding to 980,000 m^3 wastewater and 3.7 kg. For Kävlinge Municipality the stated volume treated wastewater was 6758 m^3/day, corresponding to 2,466,670 m^3/year and 9.3 kg. In addition, Svedala Municipality discharges 7.3 kg, according to our own study. The total emission is estimated to as follows (Equation (2)):

$$258.3 + 3.9 + 3.7 + 9.3 + 7.3 = 282.5 \text{ kg} \approx 283 \text{ kg/year} \quad (2)$$

2.3.3. Southern Scania

Here Trelleborg Municipality operates five WWTPs in Trelleborg, Smygehamn, Västra Alstad, Sjörup and Grönalund, which annually treat at total of 5,000,000 m^3 wastewater, corresponding to 18.8 kg. Ystad WWTP treats wastewater from Ystad and Skurup Municipality, in total 7,212,600 m^3 of wastewater/year, resulting in 27.1 kg. The total emission is estimated to be as follows (Equation (3)):

$$18.8 + 27.1 = 45.9 \text{ kg} \approx 46 \text{ kg/year} \quad (3)$$

2.3.4. Southeastern Scania

Simrishamn Municipality has >19,000 residents; Stengården WWTP is the municipality's largest WWTP and was included in this study, with a discharge of 9.9 kg. Simrishamn Municipality operates a few smaller WWTPs: Kivik, Sankt Olof, Östra Vemmerlöv, and Ravlunda. The treated volume wastewater was not stated, but their dimensions were 3000, 1000, 250 and 140 PE. Sankt Olof treatment plant was measured in this study to discharge 1.0 kg. The additional three WWTPs would give 3400 PE and a volume of 238,000 m^3 wastewater and 0.9 kg pharmaceuticals. Tomelilla Municipality has Rosendal WWTP which treats water from 7000 persons, corresponding to 490,000 m^3 and 1.8 kg. Sjöbo Municipality has ca. 19,000 residents, and the treatment of wastewater is performed at Sjöbo WWTP and in seven smaller WWTPs. In total, 1,400,000 m^3 of wastewater is treated yearly, resulting in 5.3 kg. The total emission is estimated to be as follows (Equation (4)):

$$9.9 + 1.0 + 0.9 + 1.8 + 5.3 = 18.9 \text{ kg} \approx 19 \text{ kg/year} \quad (4)$$

2.3.5. Northeastern Scania

Kristianstad Municipality has the largest WWTP, which was also included in this study, discharging 29.6 kg. Kristianstad Municipality operates 11 other WWTPs, where Gärds Köpinge was also part of this study, with 0.2 kg. The additional seven WWTPs are Arkelstorp, 92,501 m^3; Vittskövle, 86,989 m^3; Degeberga, 79,000 m^3; Maglehem, 78,283 m^3; Vånga, 9790 m^3; and Rickarum, 8264 m^3, with a total volume of 431,365 m^3 and 1.6 kg pharmaceuticals. For the two smallest WWTPs, no volume could be found. The municipalities of Bromölla, Östra Göinge and Osby are located north of Kristianstad, and wastewater treatment is performed by Skåne Blekinge Vattentjänst AB (SBVT). The volume of treated water in Bromölla Municipality is 1,270,500 m^3/year, corresponding to 4.8 kg. Östra Göinge Municipality has a number of WWTPs in Knislinge, Broby, Sibbhult, Immeln, Östanå, Boalt and Kräbbleboda, which treats 1,600,000 m^3 wastewater per year, corresponding to 6.0 kg. Five WWTPs are operated in Osby Municipality: Osby town, Lönsboda, Killeberg, Hökön and Visseltofta. These correspond to 1,600,000 m^3/year and 6.0 kg of pharmaceuticals. The total emission is estimated to be as follows (Equation (5)):

$$29.6 + 0.2 + 1.6 + 4.8 + 6.0 + 6.0 = 48.2 \text{ kg} \approx 48 \text{ kg/year} \tag{5}$$

2.3.6. Central Scania

Örkelljunga Municipality has around 10,000 residents and two WWTPs in Skånes Fagerhult and Örkelljunga. The municipality stated that, every day, between 2000 and 5000 m^3 of wastewater is treated, which gives a minimum of 730,000 m^3 of wastewater treated yearly, corresponding to 2.7 kg of pharmaceuticals. Perstorp Municipality has 7000 residents, but no information about WWTPs could be found, so they were excluded from calculations. Hässleholm Municipality has 52,000 residents, and the wastewater is treated by Hässleholms Vatten AB. The treatment plants are listed in Supplementary Materials, Table S5. Hässleholm Vatten treats a total volume of 5,623,878 m^3/year, corresponding to 21.1 kg. Höör and Hörby Municipalities have a total of 32,000 residents. Wastewater treatment is performed by the organization Mittskåne Vatten, which operates 10 WWTPs. Hörby Municipality has six WWTPs, and Höör Municipality consequently has four. In our study, Ormanäs WWTP was included, discharging 6.2 kg. Apart from this, 1,048,604 m^3 wastewater is treated in Lyby WWTP, releasing 3.9 kg. The additional eight WWTPs together generate 304,997 m^3, corresponding to 1.1 kg. The total emission is estimated to be as follows (Equation (6)):

$$2.7 + 21.1 + 6.2 + 3.9 + 1.1 = 35.0 \text{ kg} \approx 35 \text{ kg/year} \tag{6}$$

2.3.7. Total Estimated Emission of Pharmaceuticals in Scania

The calculated chemical burdens above are estimates that build on information available from the municipalities. Additionally, some WWTPs are not part of the calculations, as information was lacking. Furthermore, in some cases, the WWTPs must brim their wastewater; thereby, it is released untreated directly to the recipient, thus increasing the burden. Furthermore, the study does not take into account private sewage systems, so-called on-site sewage facilities (OSSFs), which are known to be less efficient and release an unknown number of pharmaceuticals [62]. Around 10% of the Swedish population (1 million persons) is connected to such OSSFs. One should also consider that only a very limited number of pharmaceuticals, out of several hundred active medications on the market, were included in this study. Nonetheless, this study adds up to the following (Equation (7)):

$$159.2 + 282.5 + 45.9 + 18.9 + 48.2 + 35.0 = 589.7 \text{ kg} \approx 590 \text{ kg/year} \tag{7}$$

This total load in Scania and the distribution between the various areas are illustrated in Supplementary Materials, Figure S4. Still, it is not unlikely that the total pharmaceutical load to Scanian waters is many times larger, meaning one or several tons yearly.

2.4. Occurrence of Pharmaceuticals in Scanian Streams and Lakes Downstream WWTPs

A critical topic is whether the concentrations of drugs in the recipient are increasing downstream the WWTPs. For study this, water sampling was performed in the recipient both upstream and downstream of the WWTPs. The sampling of these environmental waters depended on the geographical placement of the various WWTPs. The results of the analyses are seen in Supplementary Materials, Table S6 and visualized for the four rivers in Figure 9. Based on this information, the pharmaceutical occurrence could be compared between different types of recipients.

2.4.1. Gärds Köpinge WWTP and Vramsån River

Gärds Köpinge WWTP discharges ca. 8.75 m^3/h or 0.0024 m^3/s of wastewater into the Vramsån River which has an annual flow of approximately 4 m^3/s [63]. The sampling sites in Vramsån river are seen in Supplementary Materials, Figure S5. The results show that this WWTP has a limited impact on the concentrations of pharmaceuticals in the river (Figure 9), since they basically are at the same low level before and after the WWTP, and always less than 6 ng/L. This can be explained by the tiny contribution from the WWTP to the total river flow. The ratio between the average annual flow (m^3/s) of the river and the daily flow from the WWTP (m^3/s) gives a value of 1667 (4/0.0024), revealing extensive dilution. However, it is noteworthy that the concentrations upstream Gärds Köpinge WWTP for some pharmaceuticals (diclofenac, carbamazepine, losartan, metoprolol and oxazepam) also occur at a few ng/L (Figure 9). The reason for this might be that the larger WWTP in Tollarp is located only a few kilometers upstream. Studies are underway to determine the contribution from Tollarp WWTP to the chemical load of the Vramsån River.

2.4.2. Klippan WWTP and Bäljane Å River

Klippan WWTP releases 156 m^3/h, or 0.043 m^3/s, wastewater into the Bäljane Å river, which has an average annual flow of 2.4 m^3/s [64]. The average dilution factor in the river has been reported to be 40-fold; however, in extremely low flow periods, it may drop to 2-fold, meaning that, at times, the river can comprise one-third of the treated wastewater [64]. The sampling points are shown in Supplementary Materials, Figure S5. From Figure 9, it can be seen that the concentrations downstream Klippan WWTP are higher than upstream: diclofenac, 41 vs. 2.3 ng/L; carbamazepine, 27 vs. 6.7 ng/L; losartan, 12 vs. 1.9 ng/L; metoprolol, 52 vs. 7.7 ng/L; naproxen, 13 ng/L vs. <LOQ; oxazepam, 23 vs. 5.6 ng/L; sulfamethoxazole, 11 vs. 4.0 ng/L; and tramadol, 11 ng/L vs. <LOQ. The Bäljane Å river runs into the Rönne Å river, which thereafter feeds into the sea at Skälderviken Bay on the western coast of Scania. In previous studies, seawater was sampled from this coast in a Baltic Sea Expedition [41]. Of the 92 pharmaceuticals analyzed, carbamazepine was found in all samples at concentrations between 1.2 and 2.4 ng/L, fluconazole at 1.0 ng/L (1 sample), ketoconazole at 5.5 ng/L (one sample) and metoprolol at 1.0 and 2.3 ng/L (two samples). Accordingly, there is some agreement between those pharmaceuticals found at the highest concentration in the Bäljane Å River and those quantified in the western coast of the Baltic sea. By dividing the concentrations in the river downstream the WWTP with concentrations upstream the WWTP it was shown that the downstream concentrations are roughly six times higher at Klippan WWTP for diclofenac, erythromycin, fluconazole, carbamazepine, losartan, metoprolol, oxazepam and sulfamethoxazole. Likewise, by dividing the outlet concentrations in the WWTP with the downstream river concentrations, it is shown that the outlet concentrations are 22 times higher on average. This is in relatively good agreement with the average dilution factor of 40 [64]. The ratio between the average annual flow of the river and the daily flow of wastewater of the WWTP yields a value of 56 (2.4/0.043), which is also fairly close to 40. Overall, this indicates that dilution plays a major role in the reduction of pharmaceuticals in the river, though not as large as in the Vramsån river above.

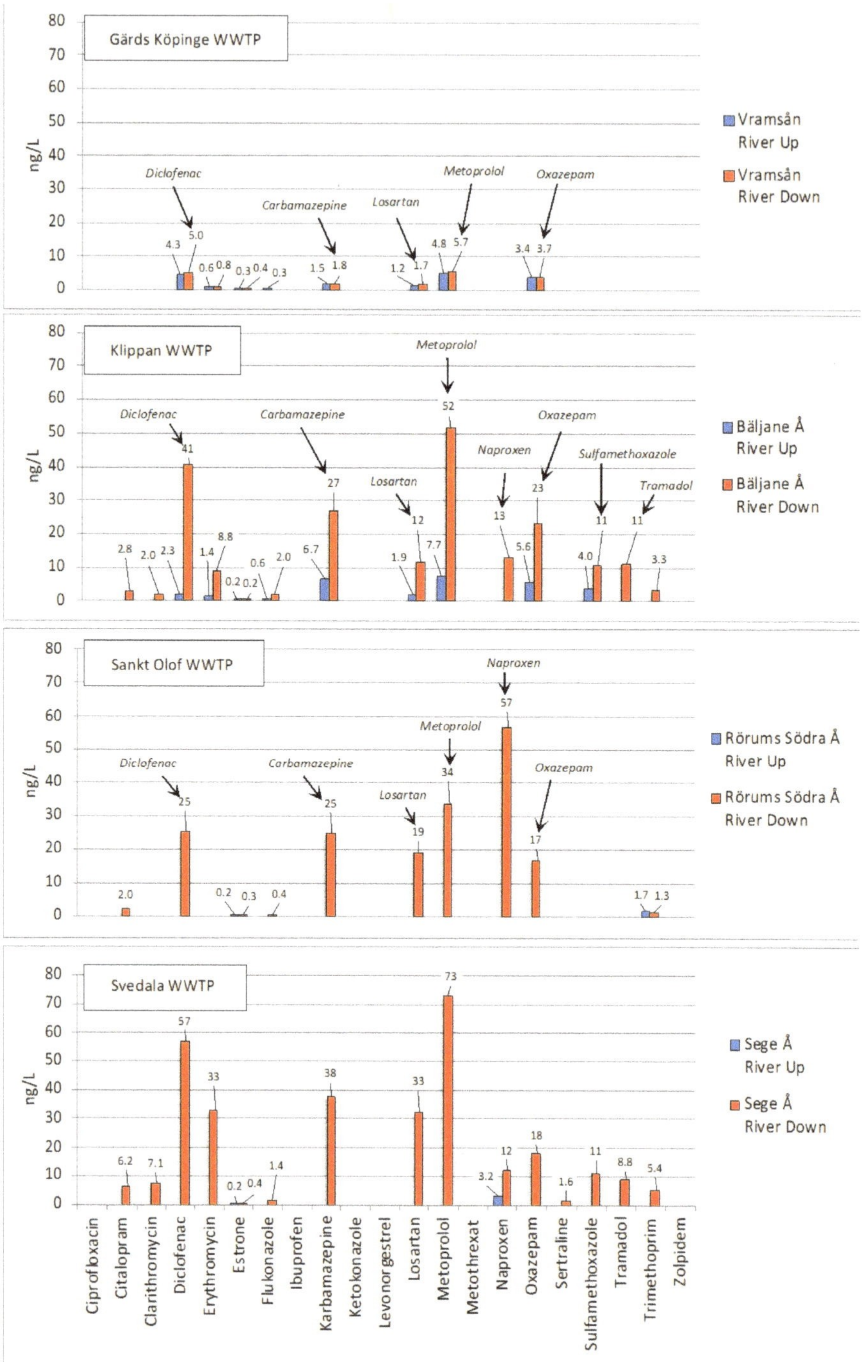

Figure 9. Measured concentrations of pharmaceuticals in the 4 investigated rivers, upstream (Up) and downstream (Down) of the WWTPs.

2.4.3. Sankt Olof WWTP and Rörums Södra Å River

Sankt Olof WWTP releases its water into Rörums Södra Å river. Flow data were not available; however, a realistic estimate would be 20 m^3/h or 0.0056 m^3/s. The Rörums Södra Å river has a reported flow of 0.4 m^3/s [65]. The sampling sites are shown in Supplementary Materials Figure S5, while the concentrations of pharmaceuticals are shown in Figure 9, revealing elevated concentrations downstream Sankt Olof WWTP. Six pharmaceuticals exceed 10 ng/L: diclofenac, 25 ng/L; carbamazepine, 25 ng/L; losartan, 19 ng/L; metoprolol, 34 ng/L; naproxen, 57 ng/L; and oxazepam, 17 ng/L. The river runs eastwards and feeds directly into Hanöbukten bay of the Baltic Sea. In the Baltic Sea Expedition [41], no samples were taken directly in the Hanöbukten Bay, but two samples were taken in the Southern Baltic proper, where carbamazepine was quantified in two samples at 2.1 and 3.3 ng/L, and fluconazole in one sample at 1.9 ng/L. Looking deeper into the river data, a few things can be noted in concerning the observations made for Gärds Köpinge WWTP in the Vramsån river and Klippan WWTP in the Bäljane Å river (above). The size of Gärds Köpinge WWTP is about the same as Sankt Olof WWTP, even so higher concentrations of pharmaceuticals was not observed in the Vramsån river as dilution was higher. Calculating the relationship between the annual average river flows and the daily wastewater flows at the WWTPs in Gärds Köpinge and Sankt Olof generated values of 1667 (above) and 71 (0.4/0.0056), respectively. Consequently, the dilution of wastewater in the Vramsån river is 23 times larger than in the Rörums Södra Å river. This results in considerably lower concentrations in the Vramsån river. Turning to Klippan WWTP, it can be seen that the relation between the flow of wastewater per day and the Bäljane Å river annual average flow is 56 (above). This relation is similar to that observed in the Rörums Södra Å river, with a corresponding value of 71. The similarities in concentrations in the two rivers is therefore not startling (Figure 9). By looking at upstream concentration data in the Rörums Södra Å river, we see that Sankt Olof WWTP seems to be the main pharmaceuticals source in this river.

2.4.4. Svedala WWTP and Sege Å River

Svedala WWTP releases 125 m^3/h or 0.035 m^3/s wastewater into the Sege Å River. The average river flow has been reported to 2.7 m^3/s [66], and the sampling sites in are shown in Supplementary Materials, Figure S5. The concentrations of pharmaceuticals downstream Svedala WWTP are shown in Figure 9 which reveals an increased occurrence. In total, eight pharmaceuticals had concentrations exceeding 10 ng/L: diclofenac, 57 ng/L; erythromycin, 33 ng/L; carbamazepine, 38 ng/L; losartan, 33 ng/L; metoprolol, 73 ng/L; naproxen, 12 ng/L; oxazepam, 18 ng/L; and sulfamethoxazole, 11 ng/L. The river feeds into the Southern Lommabukten bay of the western coast of Scania (just as the Bäljane Å river above), and correspondingly there is some correlation between the compounds released into the Sege Å river and those identified on the western coast, i.e., carbamazepine, fluconazole, ketoconazole and metoprolol [41]. The ratio between the average annual river flow and the daily wastewater for the WWTP results in a value of 77 (2.7/0.035). This value is resembling both the value 56 for Bäljane Å river and the value 71 for Rörums Södra Å river at. It is therefore reasonable that the concentrations downstream Svedala WWTP in the Sege Å river are similar to those reported in the Bäljane Å river, as well as the Rörums Södra Å river above.

2.4.5. Kristianstad WWTP and Hammarsjön Lake

Kristianstad WWTP releases 958 m^3/h or 0.27 m^3/s wastewater into the Hammarsjön lake with a volume of 782,000 m^3. The upstream sampling point is situated in the Helge Å River (Supplementary Materials, Figure S6), and has an approximate flow of 39 m^3/s [63]. The Helge Å river enters the Hammarsjön lake in its northwestern corner. The location of the sampling point downstream Kristianstad WWTP is ca. 2 km from the WWTP at a place named "Ekenabben" situated in the northeastern corner of the Hammarsjön lake. From Figure 10a it is clear that the occurrence of pharmaceuticals downstream at "Ekenabben" are higher than those found at the upstream sampling point in the river. On average the lake concentrations are 13 times higher than the river concentrations upstream the WWTP. Similarly, Kristianstad WWTP outlet concentrations are 11 times higher (on average) than concentrations downstream the WWTP in the Hammarsjön lake. Consequently, the pharmaceutical concentrations are diluted 11 times in the Hammarsjön lake before reaching "Ekenabben" inlet. Simultaneously, a minor calculation reveals that the treated wastewater contributes quite substantially to the total lake volume considering that the volume of the lake is ca. 782,000 m^3 and Kristianstad WWTP releases 958 m^3 water per hour. Through the following basic calculation, 782,000 $m^3/958\ m^3/h = 816\ h = 34$ days, it can be seen that, in only 1 month, the released volume of treated wastewater equals the entire volume of the Hammarsjön lake. Despite this, the lake concentrations are still not extremely high, and this probably can be explained by the fact that the Helge Å river flows through the lake at a rate of ca. 39 $m^3/s = 140,400\ m^3/h$. It can therefore be estimated that the Helge Å river renews the lake water in ca. 6 h. No identification of the flow profile of the lake water has been found, but even though the flow of water through the lake is high, rather high concentrations are still observed at the inlet "Ekenabben". This is probably because this is a more stagnant part of the lake.

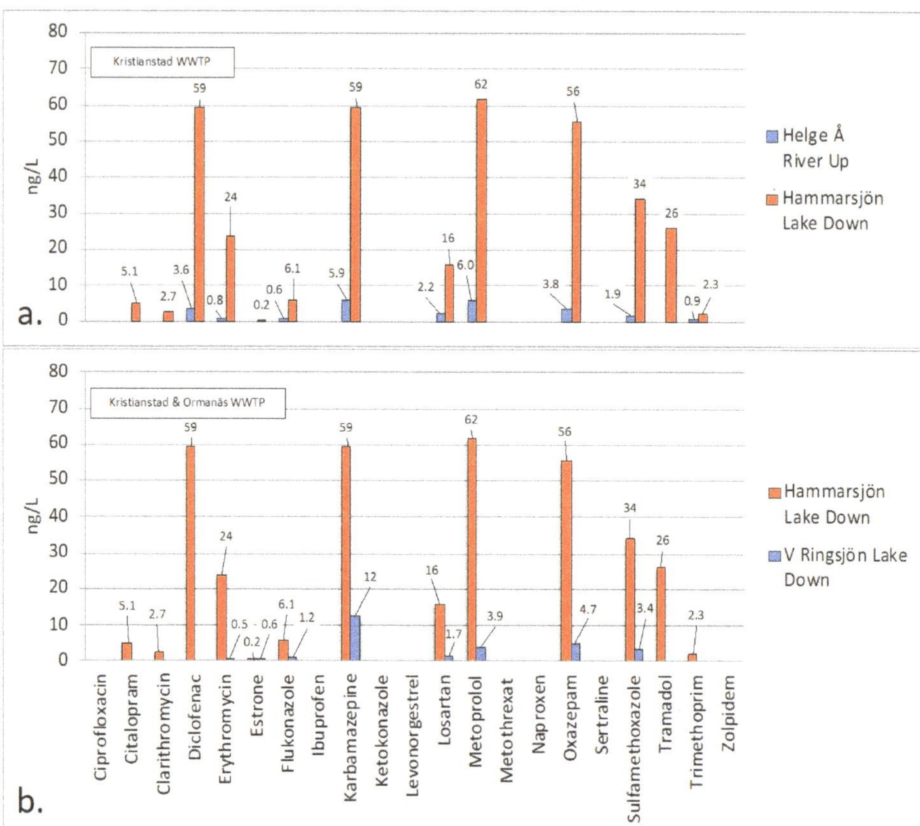

Figure 10. (**a**) Measured concentrations of pharmaceuticals in the Helge Å river upstream and the Hammarsjön lake downstream of Kristianstad WWTP. (**b**) Comparison of measured concentrations of pharmaceuticals in the Västra Ringsjön lake downstream Ormanäs WWTP and the Hammarsjön lake downstream Kristianstad WWTP.

2.4.6. Ormanäs WWTP and Västra Ringsjön Lake

Ormanäs WWTP releases 184 m^3/h, or 0.051 m^3/s, wastewater into the Västra Ringsjön lake. The downstream sampling point is shown in Supplementary Materials Figure S6, while no upstream sampling point could be located. The exact position of the outlet of the wastewater in the lake was not known, but most likely some distance from the shore. Even so, lake water was sampled 2 m from the shore in a southwesterly direction from the WWTP. The volume of the Västra Ringsjön lake is ca. 39,110,000 m^3 and a comparison of the concentrations of pharmaceuticals in the Hammarsjön lake and the Västra Ringsjön lake is shown in Figure 10b. The Hammarsjön lake had concentrations that were 13 times higher than the Västra Ringsjön lake. Even though the point of release of wastewater in the Västra Ringsjön lake is not exactly known and the lake currents were not identified, we can still estimate that the Västra Ringsjön lake volume is around 50 times greater than the Hammarsjön lake volume. Consequently, the dilution is thereby much bigger. Additionally, Ormanäs WWTP only makes a minor contribution of water compared to the total volume of the lake. Through the following simple calculation, 39,110,000 (m^3)/184 (m^3/h) = 212,554 h = 8856 days, it takes 24 years until the discharged volume treated wastewater equals the water volume of the Västra Ringsjön lake. It should be noted though, that the Västra Ringsjön lake lacks the flow of a river as large as the Helge Å river and therefore the turnover of water in the Västra Ringsjön lake is therefore most likely not as high.

2.4.7. Höganäs WWTP and Öresund and Simrishamn WWTP and the Baltic Sea

Höganäs WWTP and Simrishamn WWTP discharge their wastewater into a sea environment and their location are shown in Supplementary Materials Figure S7. For both WWTPs the volume of the recipient is unknown and no downstream samples were taken. The recipient Öresund (western coast) is likely not as sensitive as the Baltic Sea (eastern coast), since the latter is a closed brackish water sea. At the same time, the released amount of pharmaceuticals (kg) into Öresund is much larger than into the Baltic Sea (Supplementary Materials Figure S4). Based on the fact that fresh water is becoming a limited resource during the summer period it might be worthwhile to consider to stop releasing the wastewater into the salty water of the sea, since the costs of turning it back into fresh water is very costly. Additionally, there is a shortage of fresh water in the summer season in many parts of Sweden, and therefore increased reuse and circulation of fresh water may be a way to improve water deficiency in these areas.

3. Materials and Methods

3.1. *Questionnaire, Wastewater Treatment Plants (WWTPs) and Recipients*

Eight WWTPs in Scania (the most southern county in Sweden) were included in the study, covering several different scenarios. This constituted a suitable basis for model studies of pharmaceutical emissions from the roughly 90 WWTPs in Scania. All participating WWTPs were given a questionnaire and asked to provide basic information about the WWTPs, such as size and volumes. The answers to these questions are shown in Supplementary Materials Table S1. The 8 WWTPs chosen are spread out geographically all over Scania and discharge their wastewater in several differing recipients, as seen in Figure 2a.

In Northwest Höganäs, WWTP discharges directly into the salty water of the Öresund coastal sea area, Kattegatt. In inland Northwest Scania, Klippan WWTP has a point of discharge in the Bäljane Å river that continues to Rönne Å river and thereafter reaches Ängelholm and the Skälderviken bay of the western coast of Scania. In Southwest Scania, Svedala WWTP has a point of discharge in the Sege Å river. This river ends in the Southern Lommabukten bay on the western coast. In Central Scania, Ormanäs WWTP discharges wastewater in the northern part of the Västra Ringsjön lake. In Northeastern Scania, Kristianstad WWTP and Gärds Köpinge WWTP were both included. Both are situated in the UNESCO Biosphere Reserve Area Kristianstads Vattenrike [67] and connected to the basin of the Helge Å river, as described in more detail in our previous paper [68]. Kristianstad WWTP discharges wastewater into a 1500 m long canal, that ends in the Hammarsjön lake. The Helge Å river is the largest river in Scania, and both the inlet and outlet of the river are located in the Hammarsjön lake. The point of discharge of Gärds Köpinge WWTP is in the Vramsån river, which runs into the Helge Å river. The Helge Å river thereafter continues on eastwards to Hanöbukten bay (Baltic Sea with brackish water), with its largest exit in Gropahålet (close to the village Yngsjö) and a minor exit in Åhus harbor. Sampling was also performed at Sankt Olof WWTP and Simrishamn WWTP, both located in the southeast corner of Scania. Sankt Olof WWTP releases its wastewater into the Rörums Södra Å river, which continues east and ends in the Hanöbukten Bay north of Vik. Simrishamn WWTP discharges its wastewater directly into the Hanöbukten bay.

3.2. *Sampling and Analysis*

In spring 2017, contact was taken with the WWTP organizations before sampling. The actual sampling campaign was performed through visits at the 8 WWTPs on the 4/5 April 2017. All samples underwent chemical analysis at Kristianstad University during April 2017. All sampling was conducted in collaboration with the WWTP staff. All samples at the WWTP were grab samples taken in 100 mL HDPE bottles. Surface-water samples from rivers and lakes were all collected by the authors as grab samples, using 500 mL HDPE bottles. Samples were taken 0.2 m below surface for all rivers and lake samples. All sample types were kept frozen at -18 °C until analysis. For pharmaceutical determination, 50 and 500 mL of the collected sample volume were extracted for wastewaters and surface waters, respectively. For extraction, HLB cartridges (hydrophilic–lipophilic balance 200 mg sorbent, 6 mL cartridge) were purchased from Waters Oasis (Milford, MA, USA). In this project, a robust and flexible method developed at Kristianstad University, Sweden, was used [30,31]. The method has been validated according to a method established by the United States Environmental Protection Agency (US EPA) in 2007 for analysis of pharmaceuticals and personal hygiene products in water, soil, sediment and biomaterial, using HPLC/MS/MS [69]. The method has also been validated in a recent intercalibration exercise funded by the Swedish Agency for Marine and Water Management, including 5 well-established laboratories (4 Swedish and 1 Danish), demonstrating good overall performance [53]. For analysis, a Waters Acquity UPLC H-Class with the following components was used: a Quaternary Solvent Manager (QSM), a Sample Manager with Flow-Through Needle (SM-FTN) and a Column Manager (CM), enabling fast column switching (Waters, Milford, MA, USA). The chromatographic column was a Waters Acquity UPLC BEH C18 column (2.1 mmID × 50 mm, 1.8 µm). For final detection, a Xevo TQ-S™ triple quadrupole mass spectrometer (Waters Micromass, Manchester, UK) equipped with a Z-spray electro-spray interface was used.

4. Conclusions

The concentrations of pharmaceuticals released by the eight WWTPs in County Scania are, in general, on par with those reported in previous Swedish monitoring studies. There is also some agreement with previous European screening programs. Of the 21 analyzed substances, metoprolol and diclofenac have the highest average outlet concentrations of 946 and 680 ng/L, respectively. This is followed by carbamazepine, losartan, naproxen and oxazepam, all of which are above 250 ng/L.

A separate investigation on seasonal variation in outlet concentration from the largest WWTP in Kristianstad showed a peak in January and a dip in July, but still with a consistent flow of pharmaceuticals to the recipient all year around.

Despite the fact that the concentrations are low (ng/L to µg/L levels), the total chemical burden from these eight WWTPs is substantial on a yearly basis and adds up to 71 kg, as a consequence of the large total volumes of treated wastewater. The total treated wastewater volume for the >90 WWTPs in Scania was estimated to be 152,887,000 m^3, which is ca. 14% of the total Swedish wastewater volume. Based on a correlation between the volume of treated wastewater and the total amount released of the 21 pharmaceuticals, an estimate of the total chemical burden to the Scanian aquatic systems could be performed. This showed a total release of 590 kg/year.

An analysis of river samples upstream and downstream the WWTPs showed that the dilution factor of the WWTP wastewater flow in the receiving river flow is central for the concentrations found downstream of WWTPs. In short, small WWTPs in large rivers causes less concern. Still, the need for implementation of advanced treatment should be judged on a case-by-case basis, as there might be specific ecological values that need protection also from minor discharges. For example, the Vramsån river has fine specimens of the very rare freshwater pearl mussel (*Margaritifera margaritifera*) [70] which can live for more than 250 years [71]. Being exposed to even minute amounts of contaminants for centuries may cause unknown effects, especially considering that these mussels are sensitive to environmental pollution [72].

In summary, regional monitoring studies will aid in prioritizing where the need for advanced treatment is highest, as requested by the Swedish EPA [28]. Even a few measurements can make a large difference in dialogue with politicians and stakeholders. Considering the large financial undertakings associated with construction of modern treatment facilities that should last for decades, the time and resources spent on a sampling campaign is quite insignificant. It will provide essential information in taking knowledge-based decisions that will save time and resources and, at the same time, make sure that the investments made will maximize the protection of our aquatic ecosystems.

Supplementary Materials: The following supporting information can be downloaded. Figure S1: Outlet concentrations of diclofenac and metoprolol. Figure S2: Outlet concentrations of oxazepam. Figure S3: Seasonal concentrations of 4 pharmaceuticals at Kristianstad WWTP. Figure S4: Estimated release of pharmaceuticals in kg. Figure S5: River sampling points. Figure S6: Lake sampling points. Figure S7: WWTPs by the coast. Table S1.a.: Information about the 8 WWTPs. Table S1.b. Complementary information about the eight WWTPs. Table S2: WWTP outlet concentrations. Table S3: Seasonal concentrations at Kristianstad WWTP. Table S4: Outlet masses from WWTPs. Table S5: Outlet wastewater volumes. Table S6: Upstream and downstream concentrations.

Author Contributions: Conceptualization, E.B. and O.S.; methodology, E.B. and O.S.; formal analysis, E.B. and O.S.; investigation, E.B. and O.S.; resources, E.B. and O.S.; data curation, E.B. and O.S.; writing—original draft preparation, E.B.; writing—review and editing, O.S.; visualization, E.B.; project administration, O.S.; funding acquisition, E.B. All authors have read and agreed to the published version of the manuscript.

Funding: This study was funded by Region Skåne, Sweden, through MILJÖVÅRDSFONDEN 2015—Insatsområde friskt och livskraftigt hav och vatten i Skåne; Projekt M219 (THE ENVIRONMENTAL CONSERVATION FUND 2015—Application area of healthy and vital seas and waters in Scania, Project 2019).

Institutional Review Board Statement: Not applicable.

Informed Consent Statement: Not applicable.

Data Availability Statement: The chromatograms of the analyses are stored on a server at Kristianstad University, Sweden.

Acknowledgments: The staff at six different wastewater treatment organizations in County Scania (Region Skåne), namely Höganäs Municipality, Klippan Municipality, Kristianstad Municipality, Mittskåne Vatten (Höör and Hörby Municipalities), Simrishamn Municipality and Svedala Municipality, are gratefully acknowledged for aiding in providing information about the investigated WWTPs and assisting in sampling at the WWTPs and in the recipients. Peter Askman at Region Skåne is gratefully acknowledged for constructive discussions on the occurrence and strategies for reduction of pharmaceuticals in the aquatic environment of Scania.

Conflicts of Interest: The authors declare no conflict of interest. The funders had no role in the design of the study; in the collection, analyses or interpretation of data; in the writing of the manuscript; or in the decision to publish the results. The data and part of the discussion were previously published as a Swedish report to Region Skåne called "*LUSKA—Läkemedelsutsläpp från Skånska Avloppsreningsverk 2017*" by Ola Svahn and Erland Björklund. The report was later translated into English in a report with the title "*Pharmaceutical emissions from Scanian wastewater treatment plants in 2017*". The cost of translation was funded by Interreg South Baltic as part of the project European project "*Model Areas for Removal of Pharmaceutical Substances in the South Baltic—MORPHEUS*", Project No. STHB.02.02.00-SE-0038/16".

Sample Availability: The physical wastewater samples and surface water sample are not available, as they were used during the sample preparation.

References

1. Halling-Sørensen, B.; Nors Nielsen, S.; Lanzky, F.; Ingerslev, F.; Holten Lützhøft, S.; Jørgensen, S. Occurrence, fate and effects of pharmaceutical substances in the environment—A review. *Chemosphere* **1998**, *36*, 357–393. [CrossRef]
2. Cleuvers, M. Mixture toxicity of the anti-inflammatory drugs diclofenac, ibuprofen, naproxen, and acetylsalicylic acid. *Ecotoxicol. Environ. Saf.* **2004**, *59*, 309–315. [CrossRef]
3. Zhang, Y.; Geißen, S.-U.; Gal, C. Carbamazepine and diclofenac: Removal in wastewater treatment plants and occurrence in water bodies. *Chemosphere* **2008**, *73*, 1151–1161. [CrossRef] [PubMed]
4. Runnqvist, H.; Bak, S.; Hansen, M.; Styrishave, B.; Halling-Sørensen, B.; Björklund, E. Determination of pharmaceuticals in environmental and biological matrices using pressurised liquid extraction—Are we developing sound extraction methods? *J. Chromatogr. A* **2010**, *1217*, 2447–2470. [CrossRef]
5. Santos, L.; Araújo, A.; Fachini, A.; Pena, A.; Delerue-Matos, C.; Montenegro, M. Ecotoxicological aspects related to the presence of pharmaceuticals in the aquatic environment. *J. Hazard. Mater.* **2010**, *175*, 45–95. [CrossRef] [PubMed]
6. Verlicchi, P.; Galletti, A.; Petrovic, M.; Barceló, D. Hospital effluents as a source of emerging pollutants: An overview of micropollutants and sustainable treatment options. *J. Hydrol.* **2010**, *389*, 416–428. [CrossRef]
7. Vulliet, E.; Cren-Olivé, C. Screening of pharmaceuticals and hormones at the regional scale, in surface and groundwaters intended to human consumption. *Environ. Pollut.* **2011**, *159*, 2929–2934. [CrossRef]
8. Martín, J.; Camacho-Munoz, D.; Santos, J.; Aparicio, I.; Alonso, E. Occurrence of pharmaceutical compounds in wastewater and sludge from wastewater treatment plants: Removal and ecotoxicological impact of wastewater discharges and sludge disposal. *J. Hazard. Mater.* **2012**, *239–240*, 40–47. [CrossRef]
9. Arnold, K.; Boxall, A.; Ross Brown, A.; Cuthbert, R.; Gaw, S.; Hutchinson, T.; Jobling, S.; Madden, J.; Metcalfe, C.; Naidoo, V.; et al. Assessing the exposure risk and impacts of pharmaceuticals in the environment on individuals and ecosystems. *Biol. Lett.* **2013**, *9*, 20130492. [CrossRef]
10. Li, W. Occurrence, sources, and fate of pharmaceuticals in aquatic environment and soil. *Environ. Pollut.* **2014**, *187*, 193–201. [CrossRef]
11. Beek, T.; Weber, F.-A.; Bergmann, A.; Hickmann, S.; Ebert, I.; Hein, A.; Küster, A. Pharmaceuticals in the environment—Global occurrence and perspectives. *Environ. Toxicol. Chem.* **2016**, *35*, 823–835. [CrossRef]
12. Daughton, C. Pharmaceuticals in the environment (PiE): Evolution and impact of the published literature revealed by bibliometric analysis. *Sci. Total Environ.* **2016**, *562*, 391–426. [CrossRef] [PubMed]
13. Tlili, I.; Caria, G.; Ouddane, B.; Ghorbel-Abid, I.; Ternane, R.; Trabelsi-Ayadi, M.; Net, S. Simultaneous detection of antibiotics and other drug residues in the dissolved and particulate phases of water by an off-line SPE combined with on-line SPE-LC-MS/MS: Method development and application. *Sci. Total Environ.* **2016**, *563–564*, 424–433. [CrossRef]
14. Wang, J.; Wang, S. Removal of pharmaceuticals and personal care products (PPCPs) from wastewater: A review. *J. Environ. Manag.* **2016**, *182*, 620–640. [CrossRef] [PubMed]
15. Xie, Z.; Lu, G.; Li, S.; Nie, Y.; Ma, B.; Liu, J. Behavioral and biochemical responses in freshwater fish Carassius auratus exposed to sertraline. *Chemosphere* **2018**, *135*, 146–155. [CrossRef]
16. Brodin, T.; Fick, J.; Jonsson, M.; Klaminder, J. Dilute Concentrations of a Psychiatric Drug Alter Behavior of Fish from Natural Populations. *Science* **2013**, *339*, 814–815. [CrossRef]
17. Whitlock, S.; Glória Pereira, M.; Shore, R.; Lane, J.; Arnold, K. Environmentally relevant exposure to an antidepressant alters courtship behaviours in a songbird. *Chemosphere* **2018**, *211*, 17–24. [CrossRef]
18. Saaristo, M.; Brodin, T.; Balshine, S.; Bertram, M.; Brooks, B.; Ehlman, S.; McCallum, E.; Sih, A.; Sundin, J.; Wong, B.; et al. Direct and indirect effects of chemical contaminants on the behaviour, ecology and evolution of wildlife. *Proc. R. Soc. B* **2018**, *285*, 20181297. [CrossRef]
19. Escher, B.; Stapleton, H.; Schymanski, E. Tracking complex mixtures of chemicals in our changing environment. *Science* **2020**, *367*, 388–392. [CrossRef]
20. Pedrero, F.; Kalavrouziotis, I.; Alarcóna, J.; Koukoulakis, P.; Asanoc, T. Use of treated municipal wastewater in irrigated agriculture—Review of some practices in Spain and Greece. *Agric. Water Manag.* **2010**, *97*, 1233–1241. [CrossRef]

21. Christou, A.; Agüera, A.; Maria Bayona, J.; Cytryn, E.; Fotopoulos, V.; Lambropoulou, D.; Manaia, C.; Michael, C.; Revitt, M.; Schröder, P.; et al. The potential implications of reclaimed wastewater reuse for irrigation on the agricultural environment: The knowns and unknowns of the fate of antibiotics and antibiotic resistant bacteria and resistance genes—A review. *Water Res.* **2017**, *123*, 448–467. [CrossRef]
22. Moghaddam, V.; Changani, F.; Mohammadi, A.; Hadei, M.; Ashabi, R.; Ebrahimi Majd, L.; Hossein Mahvi, A. Sustainable development of water resources based on wastewater reuse and upgrading of treatment plants: A review in the Middle East. *Desalination Water Treat.* **2017**, *65*, 463–473. Available online: https://www.deswater.com/DWT_abstracts/vol_65/65_2017_463.pdf (accessed on 21 December 2021).
23. SMHI—Sveriges Meteorologiska och Hydrologiska Institut. Available online: https://www.smhi.se/klimat/klimatet-da-och-nu/arets-vader/sommaren-2018-extremt-varm-och-solig-1.138134 (accessed on 31 January 2021).
24. Wahlberg, C.; Björlenius, B.; Paxéus, N. Läkemedelsrester I Stockholms vattenmiljö—Förekomst, förebyggande åtgärder och rening av avloppsvatten. *Stockholm Vatten* **2010**, *2010*, 140. Available online: https://www.stockholmvattenochavfall.se/globalassets/pdf1/rapporter/avlopp/avloppsrening/lakemedelsrapport_slutrapport.pdf (accessed on 21 December 2021).
25. MistraPharma—Identification and Reduction of Environmental Risks Caused by Human Pharmaceuticals, MistraPharma Research 2008–2015. *Final. Rep.* **2016**, *2016*, 86.
26. Ågerstrand, M.; Berg, C.; Björlenius, B.; Breitholtz, M.; Brunström, B.; Fick, J.; Gunnarsson, L.; Larsson, J.; Sumpter, J.; Tysklind, M.; et al. Improving Environmental Risk Assessment of Human Pharmaceuticals. *Environ. Sci. Technol.* **2015**, *49*, 5336–5345. [CrossRef]
27. Cimbritz, M.; Mattsson, A. Reningstekniker för läkemedel och mikroföroreningar i avloppsvatten. *Havs-Och Vattenmyndighetens Rapp.* **2018**, *2018*, 60.
28. Sundin, A.-M.; Linderholm, L.; Hedlund, B.; Bly Joyce, K.; Klingspor, K. Avancerad rening av avloppsvatten för avskiljning av läkemedelsrester och andra oönskade ämnen-Behov, teknik och konsekvenser. *Nat. Rep.* **2017**, *2017*, 88.
29. Mattson, B.; Andersson, A.; Ovesjö, M.-L. Miljöindikatorer inom ramen för nationella läkemedelsstrategin (NLS). In *Rapport Från CBL-Kansliet, Läkemedelsverket*; 2015; Volume 7. Available online: https://docplayer.se/16152094-Miljoindikatorer-inom-ramen-for-nationella-lakemedelsstrategin-nls.html (accessed on 21 December 2021).
30. Svahn, O. Applied Environmental Analytical Chemistry for Monitoring and Measures against Antibiotics and Drug Residues in Vattenriket. Ph.D. Thesis, Centre for Environmental and Climate Change, Lund University, Lund, Sweden, 2016.
31. Svahn, O.; Björklund, E. Increased electrospray ionization intensities and expanded chromatographic possibilities for emerging contaminants using mobile phases of different pH. *J. Chromatogr. B* **2016**, *1033*, 1–10. [CrossRef]
32. Läkemedelsrester i Avloppsvatten. Länsstyrelsen Skåne. *TVL-Info* **2014**, *6*. Available online: https://www.lansstyrelsen.se/download/18.4df86bcd164893b7cd9361a4/1535032536259/TVL-info%202014:12%20L%C3%A4kemedelsrester%20i%20avloppsvatten.pdf (accessed on 21 December 2021).
33. Wick, A.; Fink, G.; Joss, A.; Siegrist, H.; Ternes, T. Fate of beta blockers and psycho-active drugs in conventional wastewater treatment. *Water Res.* **2009**, *123*, 448–467. [CrossRef]
34. Fick, J.; Lindberg, R.; Kaj, L.; Brorström-Lundén, E. Results from the Swedish National Screening Programme 2010. *Pharmaceuticals* **2011**, *2011*, 56.
35. Golovko, O.; Rehrl, A.-L.; Köhler, S.; Ahrens, L. Organic micropollutants in water and sediment from Lake Mälaren, Sweden. *Chemosphere* **2020**, *258*, 127293. [CrossRef]
36. Zou, H.; Radke, M.; Kierkegaard, A.; McLachlan, M. Temporal Variation of Chemical Persistence in a Swedish Lake Assessed by Benchmarking. *Environ. Sci. Technol.* **2015**, *4*, 9881–9888. [CrossRef]
37. Näslund, J.; Fick, J.; Asker, N.; Ekman, E.; Larsson, J.; Norrgren, L. Diclofenac affects kidney histology in the three-spined stickleback (*Gasterosteus aculeatus*) at low µg/L concentrations. *Aquat. Toxicol.* **2017**, *189*, 87–96. [CrossRef]
38. Ringbom, T.; Salin, K.; Scholz, B.; Hillver, S.-E.; Ljung, R. Tonvis med diklofenak i våra vatten–regeländring behövs. *Läkartidningen* **2017**, *47*. Available online: https://lakartidningen.se/opinion/debatt/2017/11/sverige-slapper-ut-flera-ton-diklofenak-i-miljon/ (accessed on 31 January 2021).
39. Loos, R.; Carvalho, R.; António, D.; Comero, S.; Locoro, G.; Tavazzi, S.; Paracchini, B.; Ghiani, M.; Lettieri, T.; Blaha, L.; et al. EU-wide monitoring survey on emerging polar organic contaminants in wastewater treatment plant effluents. *Water Res.* **2013**, *47*, 6475–6487. [CrossRef]
40. Durán-Alvarez, J.; Prado, B.; González, D.; Sánchez, Y.; Jiménez-Cisneros, B. Environmental fate of naproxen, carbamazepine and triclosan in wastewater, surface water and wastewater irrigated soil—Results of laboratory scale experiments. *Sci. Total Environ.* **2015**, *538*, 350–362. [CrossRef]
41. Björlenius, B.; Ripszámb, M.; Haglund, P.; Lindberg, R.; Tysklind, M.; Fick, J. Pharmaceutical residues are widespread in Baltic Sea coastal and offshore waters–Screening for pharmaceuticals and modelling of environmental concentrations of carbamazepine. *Sci. Total Environ.* **2018**, *633*, 496–1509. [CrossRef]
42. Daneshvar, A.; Aboulfadl, K.; Viglino, L.; Broséus, R.; Sauvé, S.; Madoux-Humery, A.-S.; Weyhenmeyer, G.; Prévost, M. Evaluating pharmaceuticals and caffeine as indicators of fecal contamination in drinking water sources of the Greater Montreal region. *Chemosphere* **2012**, *88*, 131–139. [CrossRef]
43. Sanzi Cortez, F.; da Silva Souza, L.; Lopes Guimarãe, L.; Emanoel Almeida, J.; Hermes Pusceddu, F.; Alves Maranho, L.; Gonçalves Mota, L.; Rodrigues Nobre, C.; Barbosa Moreno, B.; Moledo de Souza Abessa, D.; et al. Ecotoxicological effects of losartan on the

brown mussel *Perna perna* and its occurrence in seawater from Santos Bay (Brazil). *Sci. Total Environ.* **2018**, *637–638*, 1363–1371. [CrossRef]

44. Näslund, J.; Asker, N.; Fick, J.; Larsson, J.; Norrgren, L. Naproxen affects multiple organs in fish but is still an environmentally better alternative to diclofenac. *Aquat. Toxicol.* **2020**, *227*, 105583. [CrossRef]
45. Brodin, T.; Nordling, J.; Lagesson, A.; Klaminder, J.; Hellström, G.; Christensen, B.; Fick, J. Environmental relevant levels of a benzodiazepine (oxazepam) alters important behavioral traits in a common planktivorous fish, (*Rutilus rutilus*). *J. Toxicol. Environ. Health Part A* **2017**, *80*, 16–18. [CrossRef]
46. Fick, J.; Brodin, T.; Heyne, M.; Klaminder, J.; Jonsson, M.; Grabicova, K.; Randa, T.; Grabic, R.; Kodes, V.; Slobodnik, J.; et al. Screening of benzodiazepines in thirty European rivers. *Chemosphere* **2017**, *176*, 324–332. [CrossRef]
47. Östman, M.; Lindberg, R.; Fick, J.; Björn, E.; Tysklind, M. Screening of biocides, metals and antibiotics in Swedish sewage sludge and wastewater. *Water Res.* **2017**, *115*, 318–328. [CrossRef]
48. Svahn, O.; Björklund, E. Extraction Efficiency of a Commercial Espresso Machine Compared to a Stainless-Steel Column Pressurized Hot Water Extraction (PHWE) System for the Determination of 23 Pharmaceuticals, Antibiotics and Hormones in Sewage Sludge. *Appl. Sci.* **2019**, *9*, 1509. [CrossRef]
49. Lindberg, R.; Wennberg, P.; Johansson, M.; Tysklind, M.; Andersson, B. Screening of Human Antibiotic Substances and Determination of Weekly Mass Flows in Five Sewage Treatment Plants in Sweden. *Environ. Sci. Technol.* **2005**, *39*, 3421–3429. [CrossRef]
50. Thiebault, T. Sulfamethoxazole/Trimethoprim ratio as a new marker in raw wastewaters: A critical review. *Sci. Total Environ.* **2020**, *715*, 136916. [CrossRef]
51. Drzymała, J.; Kalka, J. Ecotoxic interactions between pharmaceuticals in mixtures: Diclofenac and sulfamethoxazole. *Chemosphere* **2020**, *259*, 127407. [CrossRef]
52. Östman, M.; Fick, J.; Näsström, E.; Lindberg, H. A snapshot of illicit drug use in Sweden acquired through sewage water analysis. *Sci. Total Environ.* **2014**, *715*, 136916. [CrossRef]
53. Svahn, O.; Björklund, E. *Interkalibrerad Läkemedelsanalys 2017—Ett Samarbetsprojekt för Ökad Analyskvalité*; Kristianstad University: Kristianstad, Sweden, 2017.
54. Kellner, M.; Porseryd, T.; Porsch-Hällström, I.; Hansen, S.; Olsén, K. Environmentally relevant concentrations of citalopram partially inhibit feeding in the three-spine stickleback (*Gasterosteus aculeatus*). *Aquat. Toxicol.* **2015**, *158*, 165–170. [CrossRef]
55. Kellner, M.; Porseryd, T.; Porsch-Hällström, I.; Borg, B.; Hansen, S.; Roufidou, C.; Olsén, K. Developmental exposure to the SSRI citalopram causes long-lasting behavioural effects in the three-spined stickleback (*Gasterosteus aculeatus*). *Ecotoxicology* **2018**, *27*, 12–22. [CrossRef]
56. Lindberg, R.; Fick, J.; Tysklind, M. Screening of antimycotics in Swedish sewage treatment plants—Waters and sludge. *Water Res.* **2010**, *44*, 649–657. [CrossRef]
57. Hedgespeth, M.; Karasek, T.; Ahlgren, J.; Berglund, O.; Brönmark, C. Behaviour of freshwater snails (*Radix balthica*) exposed to the pharmaceutical sertraline under simulated predation risk. *Ecotoxicology* **2018**, *27*, 144–153. [CrossRef]
58. Hedgespeth, M.; Nilsson, A.; Berglund, O. Ecological implications of altered fish foraging after exposure to an antidepressant pharmaceutical. *Aquat. Toxicol.* **2014**, *151*, 84–87. [CrossRef] [PubMed]
59. Parkkonen, J.; Larsson, J.; Adolfsson-Erici, M.; Pettersson, M.; Berg, A.; Olsson, P.; Förlin, L. Contraceptive pill residues in sewage effluent are estrogenic to fish. *Mar. Environ. Res.* **2000**, *50*, 191–199. [CrossRef]
60. Svensson, J.; Fick, J.; Brandt, I.; Brunström, B. Environmental concentrations of an androgenic progestin disrupts the seasonal breeding cycle in male three-spined stickleback (*Gasterosteus aculeatus*). *Aquat. Toxicol.* **2014**, *147*, 84–91. [CrossRef] [PubMed]
61. Statistics Sweden. Available online: https://www.scb.se/hitta-statistik/statistik-efter-amne/befolkning/befolkningens-sammansattning/befolkningsstatistik/pong/tabell-och-diagram/helarsstatistik--forsamling-landskap-och-stad/folkmangd-i-landskapen-den-31-december-2016/ (accessed on 2 February 2021).
62. Olshammar, M.; Ek, M.; Rosenquist, L.; Ejhed, H.; Sidvall, A.; Svanström, S. *Uppdatering Av Kunskapsläget Och Statistik för Små Avloppsanläggningar*; Report 166; Swedish Meteorological and Hydrological Institute: Norrköping, Sweden, 2015; p. 34.
63. Kommittén för Samordnad Kontroll av Helgeån. *Med Långtidsdiagram 1973–2011*; Rapport 2012-05-25; Kommittén för Samordnad Kontroll av Helgeån: Kristianstad, Sweden, 2012; 224p.
64. Klippans Läderfabrik. Kompletterande undersökningar av Bäljane Å 2005. In *Rapport Klippans Kommun*; 2005; Volume 60. Available online: https://www.klippan.se/download/18.126f9670127a4c8435b800017151/1358344559597/Unders%C3%B6kningar%20av%20B%C3%A4ljane%20%C3%A5,%20ALcontrol%202005.pdf (accessed on 21 December 2021).
65. Miljöövervakning Länsstyrelsen i Kristianstads Län. Klammersbäck, Mölleån, Rörums norra å, Rörums södra å, Kvarnbybäcken. In *Österlen-Åar–Underl.*; 1995; Volume 111. Available online: http://www.osterlensvattenrad.se/wp-content/uploads/2012/08/Osterlenaar_rapport_1995-Klammersback-Mollean-Rorums_Norra_a-Rorums_Sodra_a-Kvarnbybacken.pdf (accessed on 21 December 2021).
66. Available online: http://www.segea.se/Om-Segea.html (accessed on 31 October 2018).
67. Available online: https://vattenriket.kristianstad.se/other-languages/english/ (accessed on 3 February 2021).
68. Björklund, E.; Svahn, O.; Bak, S.; Oppong Bekoe, S.; Hansen, M. Pharmaceutical Residues Affecting the UNESCO Biosphere Reserve Kristianstads Vattenrike Wetlands: Sources and Sinks. *Arch. Environ. Contam. Toxicol.* **2016**, *71*, 423–436. [CrossRef]

69. U.S. Environmental Protection Agency; Office of Water; Office of Science and Technology Engineering and Analysis Division. *US EPA Method 1694: Pharmaceuticals and Personal Care Products in Water, Soil, Sediment, and Biosolids by HPLC/MS/MS*; EPA: Washington, DC, USA, 2007; Volume 2007, p. 72.
70. Available online: https://vattenriket.kristianstad.se/vramsan-vattendrag/ (accessed on 3 February 2021).
71. Dunca, E.; Söderberg, H.; Norrgrann, O. Shell growth and age determination in the freshwater pearl mussel Margaritifera margaritifera in Sweden: Natural versus limed streams. *Ferrantia* **2011**, *64*, 48–58.
72. Hartmut, F.; Gerstmann, S. Declining Populations of Freshwater Pearl Mussels (Margaritifera margaritifera) Are Burdened with Heavy Metals and DDT/DDE. *AMBIO: A J. Hum. Environ.* **2007**, *36*, 571–574. [CrossRef]

molecules MDPI

Article

Occurrence of Disinfection By-Products in Swimming Pools in the Area of Thessaloniki, Northern Greece. Assessment of Multi-Pathway Exposure and Risk

Akrivi Sdougkou, Kyriaki Kapsalaki, Argyri Kozari, Ioanna Pantelaki and Dimitra Voutsa *

Environmental Pollution Control Laboratory, School of Chemistry, Aristotle University, 541 24 Thessaloniki, Greece; akrivisd@chem.auth.gr (A.S.); kapsalaki@chem.auth.gr (K.K.); akozaria@chem.auth.gr (A.K.); ipantelaki@chem.auth.gr (I.P.)

* Correspondence: dvoutsa@chem.auth.gr; Tel.: +30-23-1099-7858

Abstract: This study investigated the occurrence of disinfection by-products (DBPs) (trihalomethanes (THMs), haloacetic acids (HAAs), halonitriles (HANs), halonitromethane (TCNM) and haloketones (HKs)) in different type of swimming pools in the area of Thessaloniki, northern Greece by employing the EPA methods 551.1 and 552.3. Moreover, general water quality parameters (pH, residual chlorine, dissolved organic carbon, UV_{254} absorption, total nitrogen, alkalinity and conductivity) were also measured. The concentrations of DBPs showed great variability among swimming pools as well as within the same pool between sampling campaigns. HAAs exhibited the highest concentrations followed by THMs, HANs, TCNM and HKs. Exposure doses for four age groups (3–<6 y, 6–<11 y, 11–<16 y and adults) were calculated. Route-specific exposures varied among DBPs groups. Inhalation was the dominant exposure route to THMs and TCNM (up to 92–95%). Ingestion and dermal absorption were the main exposure routes to HAAs (40–82% and 18–59%, respectively), depending on the age of swimmers. HANs contributed up to 75% to the calculated cytotoxicity of pool water. Hazard indices for different exposure routes were <1, suggesting non-carcinogenic risk. Inhalation posed the higher carcinogenic risk for THMs, whereas risk via oral and dermal routes was low. Ingestion and dermal contact posed the higher risk for HAAs. Risk management strategies that minimise DBPs exposure without compromising disinfection efficiency in swimming pools are necessary.

Keywords: DBPs; haloacetic acids; halonitriles; trihalomethanes; swimming pool; carcinogenic risk; cytotoxicity; dermal; inhalation; ingestion

Citation: Sdougkou, A.; Kapsalaki, K.; Kozari, A.; Pantelaki, I.; Voutsa, D. Occurrence of Disinfection By-Products in Swimming Pools in the Area of Thessaloniki, Northern Greece. Assessment of Multi-Pathway Exposure and Risk. *Molecules* **2021**, *26*, 7639. https://doi.org/10.3390/molecules26247639

Academic Editor: Esteban Alonso

Received: 9 November 2021
Accepted: 13 December 2021
Published: 16 December 2021

1. Introduction

Disinfection is a necessary water treatment process used in swimming pools to inactivate pathogens and prevent outbreaks of infectious diseases. Chlorine is the most common disinfectant for this purpose. However, this practice also results in the formation of undesirable disinfection by-products (DBPs) from the reaction of chlorine with the organic matter present in filling waters (natural organic matter) and released from swimmers (human body fluids, sweat, sebum, skin particles, personal care products etc). Differences in operation conditions and disinfection methods affect the levels and speciation of the DBPs in swimming pools [1–3].

A large number of DBPs are considered as cytotoxic, neurotoxic, mutagenic, genotoxic, carcinogenic and teratogenic [4–6]. Concerns have been raised regarding potential negative effects on human health from water disinfectants used in swimming pools. The exposure to DBPs through different intake routes (inhalation, dermal absorption, water ingestion) may pose health risk for swimmers and pool staff. Links between exposure to DBPs and several health issues have been investigated [7–9]. Overall, the available knowledge suggests that the health benefits of swimming outweigh the potential health risks and, positive effects of swimming should be enhanced by minimizing potential risks. Although many studies reported the occurrence of commonly investigated DBPs (mainly trihalomethanes and haloacetic acids) in swimming pools, there are only a few studies that investigated emerging DBP classes, such as nitrogenous DBPs (halonitromethanes, haloacetonitriles etc.).

The aim of this study was to investigate for the first time the presence of various priority and emerging DBP groups (trihalomethanes, haloacetic acids, halonitriles, halonitromethanes and haloketones) under swimming pool operation conditions as practiced in the area of Thessaloniki, Northern Greece. DBPs as well as other water quality parameters (pH, residual chlorine, dissolved organic carbon, UV absorption at 254 nm, alkalinity, conductivity, total nitrogen) were determined during the period July 2019–February 2020. Specifically, the main objectives of this article are to:

(a) Examine the occurrence of different DBPs groups in swimming pools;
(b) Find out possible correlations among DBPs and physicochemical water parameters;
(c) Estimate the contribution of different exposure routes to DBPs and;
(d) Present a multi-pathway risk assessment for four age-groups of swimmers.

2. Experimental

2.1. Analytical Standards and Reagents

Target DBPs classes and individual species are shown in Table 1. Analytical standards for the studied DBPs were used for the purpose of this study. The analytical standard for THMs EPA 501/601 trihalomenthanes calibration mix certified reference material, 100 μg/mL each component in methanol, was obtained from Sigma-Aldrich and is a mixture of four compounds (TCM, DCBM, CDBM and TBM). Analytical standard EPA 551B Halogenated Volatiles Mix certified reference material (2000 μg/mL each compnentin methyl tert-butyl ether) purchased from Sigma–Aldrich contains four haloacetonitriles (DCAN, TCAN, BCAN, DBAN), one halonitromethane (TCNM) and two haloketones (DCP, TCP). The analytical standard EPA 552.2 Haloacetic Acids Mix certified reference material 2000 μg/mL each component in methyl tert-butyl ether was obtained from Supelco and is a mixture of nine haloacetic acids (BCAA, BDCAA, CDBAA, DBAA, DCAA, MBAA, MCAA, TBAA, TCAA). The internal standard 1,2-dibromopropane was obtained from Chem. Service (West Chester, U.S.A). The surrogate standard 2,3 dibromoproprionic acid was provided from Supelco. Methyl tert-butyl ester (MtBE), Na_2HPO_4, KH_2PO_4, Na_2SO_4, NH_4Cl and methanol were provided from Sigma-Aldrich. Milli-Q water provided by the Simplicity UV Ultrapure Water System (Millipore, Molsheim, France).

Table 1. DBPs groups determined in this study.

DBPs Groups	DBPs Species	Abbreviation	Chemical Formula
THMs Trihalomethanes	trichloromethane	TCM	$CHCl_3$
	dichlorobromomethane	DCBM	$CHBrCl_2$
	chlorodibromo	CDBM	$CHBr_2Cl$
	methanetribromomethane	TBM	$CHBr_3$
HANs Haloacetonitriles	dichloroacetonitrile	DCAN	$CHCl_2CN$
	trichloroacetonitile	TCAN	CCl_3CN
	bromochloroacetonitrile	BCAN	$CHBrClCN$
	dibromoacetontrile	DBAN	$CHBr_2CN$
HNMs Halonitromethanes	trichloronitromethane	TCNM	CCl_3NO_2
HKs Haloketones	1,1-dichloropropanone	DCP	$C_3H_4Cl_2O$
	1,1,1-trichloropropanone	TCP	$C_3H_3Cl_3O$

Table 1. *Cont.*

DBPs Groups	DBPs Species	Abbreviation	Chemical Formula
HAAs Haloacetic acids	monochloroacetic acid	MCAA	$CH_2ClCOOH$
	monobromoacetic acid	MBAA	$CH_2BrCOOH$
	dichloroacetic acid	DCAA	$CHCl_2COOH$
	trichloroacetic acid	TCAA	CCl_3COOH
	bromochloroacetic acid	BCAA	$CHBrClCOOH$
	dibromoacetic acid	DBAA	$CHBr_2COOH$
	dibromochloroacetic acid	DBCAA	$CBr_2ClCOOH$
	chlorodibromo acetic acid	CDBAA	$CBr_2ClCOOH$
	tribromoacetic acid	TBAA	CBr_3COOH

2.2. Water Samples

A range of swimming pool types (indoor, outdoor, only for children and for children/adults) located in the area of Thessaloniki, Northern Greece were included in this study. The type of pools and the disinfectant used are described in Table 2. In order to ensure confidentiality, samples have been identified by applying codes (SP1 to SP14).

Table 2. Description of swimming pools.

Code	Indoor/Outdoor	Children/Adults	Disinfection
SP-1	Indoor	Children [1]	NaOCl
SP-2	Indoor	Children/Adults [2]	NaOCl
SP-3	Outdoor	Children/Adults	NaOCl
SP-4	Outdoor	Children/Adults	NaOCl
SP-5	Indoor	Children	NaOCl
SP-6	Indoor	Children/Adults	NaOCl
SP-7	Outdoor	Children/Adults	NaOCl
SP-8	Outdoor	Children/Adults	NaOCl
SP-9	Outdoor	Children/Adults	NaOCl
SP-10	Indoor	Children	Electrolysis NaCl
SP-11	Indoor	Children/Adults	Electrolysis NaCl
SP-12	Indoor	Children/Adults	Electrolysis NaCl
SP-13	Outdoor	Children	NaOCl
SP-14	Outdoor	Children/Adults	NaOCl

[1] swimming pools only for children ≤7 years old, [2] Swimming pools for children >7 and adults.

Water samples were collected from 14 swimming pools, consisting of 7 outdoor and 7 indoor pools, 4 for children (≤7 years) and the rest for adults/children >7 year during the period July 2019 to February 2020 (5 sampling were conducted). Unfortunately, it was not possible to have access to detailed information (e.g., number of swimmers, average daily attendance rate, frequency of pool water replacement, regularity of filter backwashing and occurrence of shock chlorination procedures).

Sampling was conducted in evening, between 18.00 and 20.00 p.m. Equal volumes of water samples collected at the four corners of pool, from approximately 50 cm from the side walls and 30 cm below the water's surface, were combined in order to obtain a composite sample. Prior to sampling different reagents were added to the sample vials in order to quench any chlorine residual or standardise pH values. At each pool, three bottles were filled: (1) one 60 mL headspace free amber glass bottle for determination of THMs-HANs-HNM-HKs. This vial contained NH_4Cl as dechlorination agent and $KH_2PO_4/NaHPO_4$ as buffer to lower pH 4.8–5.5 in order to inhibit base catalysed degradation of HANs and standardise the pH in all samples, (2) one 60 mL headspace free amber glass bottle contained NH_4Cl as dechlorination agent for determination of HAAs and, (3) one 500 mL glass bottle for determination of water quality parameters. Water samples were immediately placed in a cooler, transported to the laboratory within 2 h and then stored at 4 °C until analysis (within 1–4 days). Tap water from the main drinking water supply system of the city was used as filling water for swimming pools. Tap water samples were also collected at selected facilities to assess their impact on the formation of DBPs in the pools.

All glassware used for sampling and analysis was meticulously cleaned with detergent and tap water, rinsed with tap water followed by Milli-Q water and solvents.

2.3. Analytical Methods

DBPs were recovered from water samples by employing various extraction methods. THMs, HANs, TCNM and HKs were extracted with MtBE according to the EPA 551.1 method [10]. HAAs were analysed according to the EPA 552.3 [11] that includes liquid–liquid extraction with MtBE followed by derivatisation (methylation) via acidic methanol. GC/ECD (Trace GC Ultra, Thermo Scientific) with an analytical column AT-5 (30 m, id 0.25 mm, 0.25 μm, Grace) was used for determination of DBPs. THM, HANs, TCNM, HKs were analysed in the same run and the GC temperature program was: 35 °C for 9 min, 1 °C/min to 40 °C (3 min), 6 °C/min to 120 °C (5 min) and then 10 °C/min to 220 (5 min). For HAAs, the GC temperature program was: 37 °C for 21min, 5 °C/min to 136 °C (3 min), and then 20 °C/min to 240 (5 min). The injector and detector temperatures were set at 210 °C and 280 °C, respectively.

Procedural standard calibration curves were used to quantify DBPs. Internal standards as per individual method requirements were employed and quantification for each analyte was based on relative response ratios. The recovery of DBPs ranged from 80–112%, precision was better than 11% and detection limits ranged from 0.14–0.50 μg/L for THMs, 0.15–0.34 μg/L for NDBPs and HKs and 0.04–0.15 μg/L for HAAs (Table S1).

Free chlorine was measured on site upon the addition of N,N-diethylp-phenylenediamine (DPD) reagent using a portable Pocket Colorimeter (HANNA instruments (HI 96, 710 and free chlorine reagent Hi37101-01). pH and conductivity were also measured on site using a portable multi-meter (Dr Lange, ECM). DOC (as nonpurgeable organic carbon) was measured and filtrated through 0.45-μm samples, using a TOC-Vcsh analyser (Shimadzu). UV absorbance at 254 nm was also measured in filtrated samples by a spectrophotometer (Hitachi U-2001). SUVA was calculated as the ratio of UV_{254} to DOC. Total nitrogen (TN) was measured after digestion of unfiltered sample by persulfate method and alkalinity by titration with sulfate acid according to standard methods [12].

2.4. Risk Assessment

2.4.1. Assessment of Cytotoxicity

The chronic cytotoxicity of pool waters due to the presence of the studied DBPs was evaluated based on effective concentrations of EC_{50} values (a measure of the minimum concentration of a particular compound that induces a 50% reduction in density of Chinese hamster ovary cells after 72 h) [13]. The molar concentration of each DBP was divided by its EC_{50} value for those DBP species that were available, resulting in a dimensionless cytotoxicity value. The sum of these values represent the DBP-derived cytotoxic nature of pool water.

2.4.2. Exposure Doses

Swimmers are exposed to DBPs through various routes such as incidental oral exposure, dermal exposure, inhalation buccal, aural and nasal orbital exposures according to Swimmer Exposure Assessment Model (SWIMODEL) [14].

The average daily dose (ADD, mg/kg-event) though ingestion (ADD_{Ing}), dermal absorption(ADD_{Abs}) and inhalation (ADD_{Inh}) was estimated according to the following equations [15]:

Average Daily Dose through ingestion (ADD_{ing}, mg/kg-day):

$$ADD_{Ing} = (Cw \times IngR \times EF \times ED)/(BW \times AT)$$

Average Daily Dose absorbed through dermal (ADD_{Abs}, mg/kg-day):

$$ADD_{Abs} = (Cw \times Kp \times t \times SA \times EF \times ED)/(BW \times AT)$$

Average Daily Intake through inhalation (ADD_{Inh}, mg/kg-day):

$$ADD_{Inh} = (Ca \times InhR \times EF \times ED)/(BW \times AT)$$

where: Cw: concentration of DBPs in water (mg/L), Ca: Concentration of DBPs in air (mg/m^3), IingR: Ingestion rate (L/day), EF: Exposure frequency (days/year), ED: Exposure duration (years), BW: Body weight (kg), AT: Averaging time (days), Kp: Permeability coefficient (cm/h), t: Time of contact (h/event, days), SA: Skin surface area available for contact (cm^2), InhR: Inhalation rate (m^3/h).

When both water and air concentrations of DBPs are measured in swimming pool facilities the actual exposure doses are obtained. However, in case that air concentrations are not determined, they can be calculated from the measured water concentrations through different approaches i.e., using Henry's Law or Raoult's Law as in SWIMODEL or through fugacity model [14,16,17]. Dyke et al. [16] who compared measured air concentrations of THMs with those calculated by different methods reported that the Henry's Law approach appeared to overestimate two-three orders of magnitude the experimental data. Lourencetti et al. [17] also found that the actual THMs concentrations measured in air were significantly lower than those expected from the Henry's Law and the ratios of measured to calculated concentrations of THMs ranged from 0.6 to 5.6% for individual species (0.9–1.4% for chloroform). In our study the Henry's Law based on equilibrium with water was used to calculate air concentrations. This approach overestimates the actual concentrations occurred in a natatorium where ventilation is continuously driving the system away from its equilibrium [16,17]. In addition, in order to estimate exposure through inhalation and possible risks, we made the assumption that actual THMs air concentrations are 2% of the calculated values, based on the findings of Lourencetti et al. [17]. The efficiency of the absorption through exposure routes was assumed to be 100%. The exposure factors employed in this study and the properties of the studied DBPs are shown in Tables 3 and 4.

Table 3. Exposure factors for non-competitive swimmers [18].

Abr	Exposure Factors	Age Groups of Non-Competitive Swimmers			
		Children (3–<6 y)	**Children (6–<11 y)**	**Children (11–<16 y)**	**Adults (>18 y)**
BW	Body weight (kg)	19	32	57	80
SA	Surface area (m^2)	0.76	1.08	1.59	1.94
InhR	Inhalation rate (m^3/h)	0.66	0.66	0.78	0.74
IngR	Ingestion rate (L/h)	0.049	0.049	0.049	0.025
EF	Exposure frequency (min/month)	137	151	139	181
ED	Exposure duration (years)	4	5	5	30
AT	Average Time	78	78	78	78

Table 4. Properties and risk factors used for assessing exposure and health risk.

DBPs	H (atm m^3/mol) [19,20]	Kp (cm/h) [19,20]	EC_{50} (M) [13]	Rfd (mg/kg-day) [21]	IARC [22]	SF (mg/kg-day) [21,23]		
						Oral	Dermal *	Inhalation *
THMs								
TCM	3.67×10^{-3}	6.83×10^{-3}	9.62×10^{-3}	1×10^{-2}	2B	1×10^{-2}	1×10^{-2}	1.9×10^{-2}
BDCM	2.12×10^{-3}	4.02×10^{-3}	1.15×10^{-2}	2×10^{-2}	2B	6.2×10^{-2}	6.2×10^{-2}	1.3×10^{-1}
CDBM	7.83×10^{-4}	2.89×10^{-3}	5.36×10^{-3}	2×10^{-2}	3	8.4×10^{-2}	8.4×10^{-2}	8.4×10^{-2}
TBM	5.35×10^{-4}	2.35×10^{-3}	3.96×10^{-3}	2×10^{-2}	3	7.9×10^{-2}	7.9×10^{-2}	3.9×10^{-3} **
HAAS								
DCAA	8.38×10^{-9}	1.21×10^{-3}	7.3×10^{-3}	4×10^{-3}	2B	5×10^{-2}	5×10^{-2}	5×10^{-2}
TCAA	1.35×10^{-8}	1.45×10^{-3}	2.4×10^{-3}	2×10^{-3}	2B	7×10^{-2}	5×10^{-2}	5×10^{-2}
NDBPs ***								
TCAN	1.34×10^{-6}	7.6×10^{-3}	1.601×0^{-4}		3			
DCAN	3.79×10^{-6}	6.5×10^{-4}	5.73×10^{-5}		3			
BCAN	1.24×10^{-6}	4.1×10^{-4}	8.46×10^{-6}		3			
DBAN	4.06×10^{-7}	2.5×10^{-4}	2.85×10^{-6}		2B			
TCNM	2.05×10^{-3}	5.8×10^{-3}	5.36×10^{-4}					
HKs ***								
DCA	6.15×10^{-6}	4.4×10^{-4}						
TCA	2.17×10^{-6}	1.2×10^{-3}						

* Oral SFs were used for dermal exposure and inhalation when data were not available. ** SF was derived from inhalation unit risk. *** Kp was calculated according to the equation $\log Kp = -2.72 + 0.71 \log K_{ow} - 0.006161MW$.

2.4.3. Non-Carcinogenic Risk

The non-carcinogenic risk for an individual i DBP specie was assessed as hazard quotient (HQ) according to the equation $HQ_i = \frac{ADDi}{RfDi}$ where ADD_i (mg/kg-day) is average daily intake considering AT = ED and RfD_i is the reference daily dose for the i DBP specie. RfDs values for oral exposure are shown in Table 4. The same values were also used for assessing hazard indices through dermal and inhalation routes. Hazard Index (HI) represents the sum of hazard quotients for all DBPs.

2.4.4. Carcinogenic Risk

Lifetime cancer risk (CR_i) for individual i DBP specie for a specific exposure route was calculated according to the equation: $CR_i = LADD_i \times SF_i$, where $LADD_i$: is the Lifetime Average daily dose (mg/kg-day) calculated from ADD, considering AT as lifetime (in days), and SF_i is the cancer slope factor for individual i DBP specie. SF_i values are shown in Table 4. Cancer slope factors through oral exposure were used also for the other routes that SFs are not available.

2.5. Statistical Evaluation

Statistical analysis was performed using SPSS Statistics version 24 software (IBM, Armonk, NY, USA). Non detectable concentrations were treated as half of LOD. Correlation among DBPs and water quality parameters was examined through non-parametric Spearman's correlation analysis, since data are not follow normal distribution.

3. Results and Discussion

3.1. Water Quality Parameters

The general water quality parameters are shown in Figure 1. The pH values of waters ranged from 6.4–8.4. The majority of samples (95%) meet the local operational guideline for a range of 7.2 to 8.2 for waters treated with chlorine. Free chlorine concentrations were found to vary greatly among the investigated waters (0.3–5 mg/L). According to national legislation, free residual chlorine range is between 0.4 and 0.7 mg/L [24]. Recently, for precautionary reasons, a concentration of 1.5 mg/L is suggested for protection of public health to SARS-CoV-2 [25]. WHO suggests that a range of residual chlorine from 1 to 3 mg/L [26]. In our study most samples were within the acceptable limits proposed by WHO and only 15% of samples exceed the upper WHO guideline. The alkalinity in pool water ranged from 35–335 mg/L, and in general meet the local guideline for concentrations above 50 mg/L (only 5% were below this limit). Many studies worldwide reported deviations from national guidelines either for pH values or for free chlorine values [3].

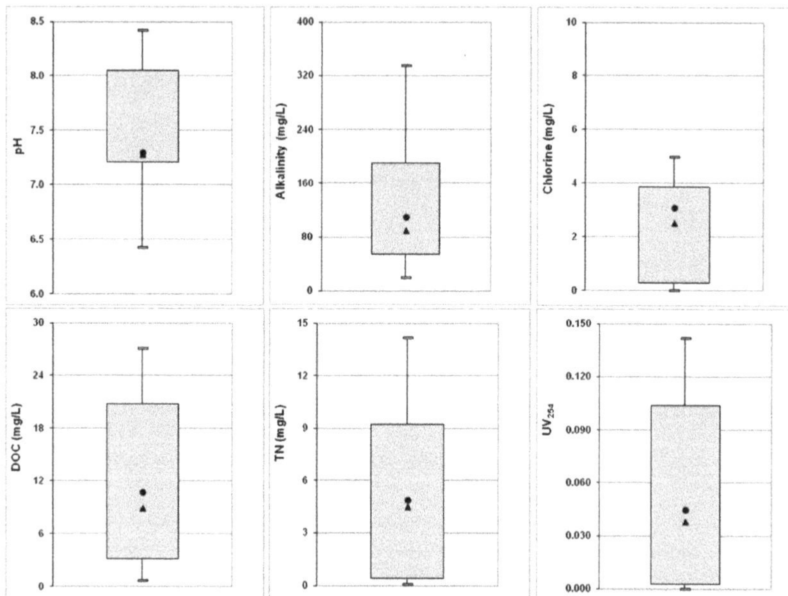

Figure 1. Water quality parameters in swimming pools (boxplots represent 10th and 90th percentiles, min and max values; circle and triangle represent mean and median values, respectively).

DOC varied among pools and ranged from 0.7 to 27 mg/L. These values are within the range of values reported in other studies [1,3,27]. Total nitrogen (TN) ranged from <0.1 to 14 mg/L. Yeh et al. [28] reported an increasing accumulation of TN in swimming pool. The organic and nitrogen loads are highly variable and depend on their presence in filling water but mainly from the anthropogenic releases. Human body fluids, sweat, sebum, skin particles and hair, and personal care products contribute to both carbon and nitrogen load in pools [28–31].

3.2. Occurrence of Disinfection By-Products

The range of various DBPs groups determined in pool water as well as of the individual species in each group is shown in Figure 2.

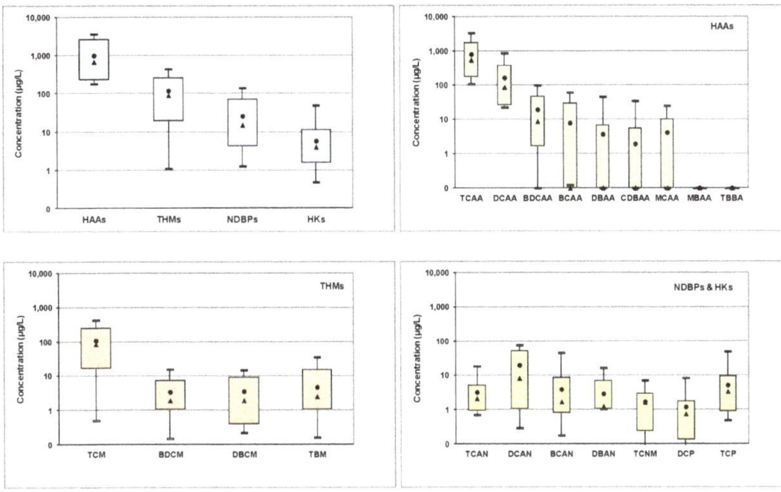

Figure 2. Water concentrations of DBPs in swimming pools (boxplots represent 10th and 90th percentiles, min and max values; circle and triangle represent mean and median values, respectively).

HAAs exhibited the highest concentrations followed by THMs, HANs, TCNM and HKs. HAAs as sum of nine compounds ranged from 178–3640 μg/L (median 680 μg/L) (Figure 2). The dominant compounds were TCAA and DCAA with relative contributions 79 ± 15% and 18 ± 14%, respectively. BCAA, DBAA, CDBAA were determined in few samples. Yeh et al. [28] reported that due to high abundance of HAAs (especially DCAA and TCAA) these compounds could be used as indicator chemicals to define guideline values for monitoring pool water quality. The concentrations of HAAs are below ECHA proposed limits for swimming pools that ranged from 800 to 8000 μg/L for individual HAAs, and particularly 1500 μg/L for DCAA and 8000 μg/L for TCAA [32].

THMs, as sum of four compounds, was the second most abundant group with concentrations ranging from 1–410 μg/L (median 89 μg/L), with TCM being the dominant compound (84 ± 22%). Some studies compared concentrations of DBPs with drinking water limits 100 μg/L for THMs [33]. National limits have been proposed in various countries for THMs, i.e., 20 μg/L in Germany, 50 in Netherlands and Hungary [34]. ECHA proposed a limit of 50 μg/L for THMs, as chloroform equivalent, in swimming pools [32].

HANs as sum of four acetonitriles ranged from 0.9–130 μg/L (median 15 μg/L). DCAN (median 8.1 μg/L) and TCAN (2.1 μg/L) were the dominant compounds with relative contribution 51 + 32% and 21 + 17%, respectively, followed by BCAN, DBAN. The majority of samples exhibited concentrations below the ECHA limits for swimming pools (20 μg/L for DCAN, 70 μg/L for DBAN and 20 μg/L BCAN) [32]. TCNM ranged from <0.2–7 μg/L (median 1.6 μg/L). The haloacetone 1,1,1-TCP was the dominant propanone with concentrations 0.5–48 μg/L (median 3.4 μg/L).

The concentrations of DBPs varied among swimming pools (Figure S1) as well as within the same pool between sampling campaigns. This variation could be attributed to different conditions occurred in each swimming pool i.e., regarding low or heavy load of swimmers, replacement of pool water, filter backwashing or shock chlorination. Several outliers were found to originate mainly from two swimming pools that also exhibited high DOC values, low residual chlorine concentrations and employed electrolysis of sodium chloride as disinfection process. They also showed relatively elevated concentrations of brominated DBP species (1–12% for Br-HAAs, 13–40% for Br-THMs and 3–30% for Br-HANs). Organic load is a significant precursor of DBPs. Moreover, it has been reported that electrolysis of salt solution could result in higher formation of DBPs as well as brominated analogues due to impurities in salt [1].

Table 5 summarises the concentrations of DBPs in swimming pools worldwide. DBPs varied both quantitatively and in terms of speciation since their occurrence in pool water depends on operational and environmental conditions (pH, temperature, concentration and origin of organic carbon, concentration of chlorine, management conditions etc.,). The concentrations of DBPs in our study are within the reported range.

Table 5. Concentrations of DBPs (μg/L) in swimming pool water worldwide.

Country	HAAs	THMs	NDBPs	HKs	References
Australia	366–5126 230–2400 (DCAA) 110–2600 (TCAA)	65–84 (TCM)	4.9–8.9 (DCAN) nd–2.3 (TCNM)		[28]
Canada	155–2224				[35]
Canada		21–132 6.7–125 (TCM)	3.4–78.6 (HANs) 4.5 (TCNM)	0.3–7.3 (TCP)	[36]
China	1.2–1889	25.7 ± 33.1	12.3 ± 15.5		[37]
France		80 70 (TCM)	75 (DCAN) nd–4.5 (TCNM)	72 (TCP)	[38]
Greece	7.7–653.7	8.1–57.4	0.8–20.6 (HANs)	nd–15.3	[39]
USA	70–3980 50–2040 (DCAA) 20–2970 (TCAA)				[40]
USA		26–213 25–207 (TCM)	4–47 (DCAN)		[41]
Singapore	45–828 (DCAA) 114–1020 (TCAA)	32–170 30–167 (TCM)			[42]

nd: non detected.

Tap water from drinking water distribution system is used to fill and regularly top-up the pools at each facility. DBPs in tap water was therefore investigated. The concentrations of DBPs were low, THMs ranged from 0.4 to 17 μg/L, NDBPs <1–5 μg/L and HAAs were not found at detectable concentrations whereas DOC ranged from 0.1–1.2 mg/L. The majority of pool waters (90%) contained higher concentration of DBPs and DOC compared to filling waters.

Correlation coefficients among the investigated DBP classes and other general water quality parameters are shown in Table S1. Significant correlations were observed among individual species in each DBP group (HAAs-DCAA-TCAA-BCAA-DBCAA-CBAA and THMs-TCM-TBM), between studied DBPs classes (THMs-HANs, HAAs-TCNM). UV was the water quality parameter that showed significant correlation with all DBPs species. Various studies reported correlation between various DBPs species and organic content or free chlorine [35,43]. However, other studies did not report significant correlations [3,40]. The origin of dissolved organic carbon in swimming pool may affect the levels and speciation of DBPs. It has been reported [27] that human body fluids exhibited higher formation potentials of HAAs than THMs, whereas the opposite was observed for the natural organic matter that produce more THMs.

3.3. Exposure Routes

Ingestion, dermal contact and inhalation are the main exposure pathways to DBPs. The exposure to DBPs depends on the physical activity of swimmers and level of their effort, average time of swimming, body surface area, inhalation rate and rate of inadvertent ingestion of pool water [15,44]. The relative contribution of exposure routes to DBPs for different age-groups is shown in Figure 3.

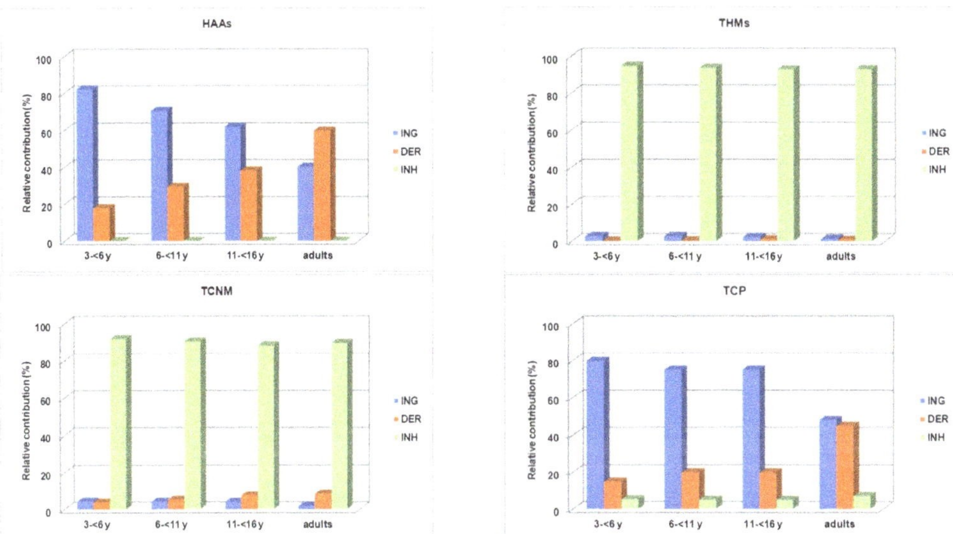

Figure 3. Relative contribution of exposure routes to DBPs (ING: ingestion, DER: dermal and INH: inhalation).

Route specific exposures varied among DBPs groups. Inhalation was the dominant exposure route for THMs (93–95%) and TCNM (88–92%). These compounds are highly volatile and occur in the air of natatorium. The presence of DBPs in air depends on their volatility, water concentrations, temperature of water, height above surface of swimming pool, water turbulence, humidity and air ventilation rates [16,34,45]. Inhalation was also found to be the main route of exposure to 1,1,1-TCP for children (75–80%), whereas adults exposed almost equally through ingestion and dermal absorption. Ingestion and dermal absorption were the main exposure routes for HAAs, 40–82% and 18–59%, respectively. Their relative contributing is age dependant; ingestion was the dominant exposure pathway for children, with a decreasing trend toward adults where dermal absorption became the main exposure pathway. This is because the inadvertent water intake varies with the age of swimmers, their skill and experience and type of activity.

3.4. Assessment of Possible Risks

The presence of DBPs in water and air is of major human health concern because a number of DBPs species are cytotoxic, others are carcinogenic, mutagenic or have reproductive and developmental effects [4].

3.4.1. Cytotoxicity of Pool Water

The assessment of cytotoxicity of pool water was based on the measured DBP concentrations in water and their effective concentration (EC_{50}) values [13]. The relative contribution of specific DBPs groups to the total calculated cytotoxicity of pool waters is shown in Figure 4.

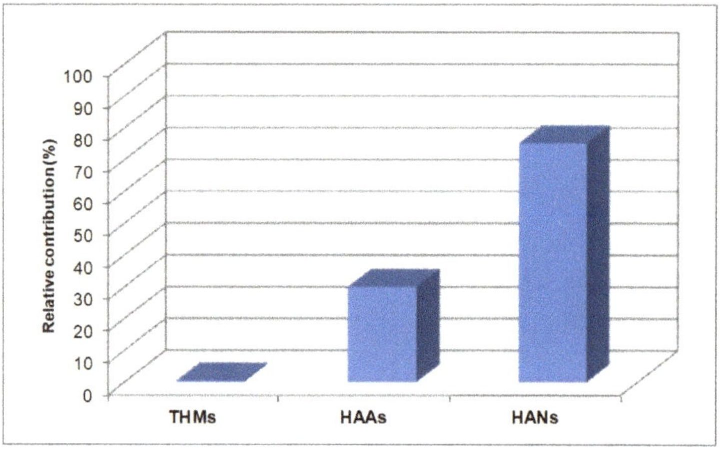

Figure 4. Relative contribution of DBPs groups to calculated overall cytotoxicity of water in swimming pools.

THMs and HAAs contributed only to 0.5% and 24.5% of the total calculated cytotoxicity, respectively, although were the predominant DBP classes. NDBPs contributed significantly (up to 75%) to the calculated cytotoxicity, although determined at lower concentrations, Carter et al. [3] also reported low contribution of HAAs and THMs to the overall cytotoxicity of water in swimming pools whereas other DBP species such haloacetaldeydes and nitrogenous species (haloacetonitriles and haloacetamides) were the major forcing agents of toxicity. Yeh et al. [28] who actually measured cytotoxicity of pool water reported that HAAs, although the dominant DBPs, explained less than 4% of the observed cytotoxicity.

Therefore, the occurrence of THMs and HAAs commonly measured in swimming pools cannot interpret the cytotoxicity of pool water and other DBPs species, at relatively lower concentrations, significantly contribute to this risk.

3.4.2. Non-Carcinogenic and Carcinogenic Risk

Non-carcinogenic and carcinogenic risk from the exposure to THMs and HAAs was calculated since these DBPs were the most prominent compounds and their toxicity data are available (Table 4).

Hazard indices for different exposure routes are illustrated in Figure 5. HI values for both THMs and HAAs from ingestion, dermal absorption and inhalation ranged from 7×10^{-8} to 3×10^{-1}, well below the acceptable maximum value of 1, suggesting that non-carcinogenic risk was not of apparent concern.

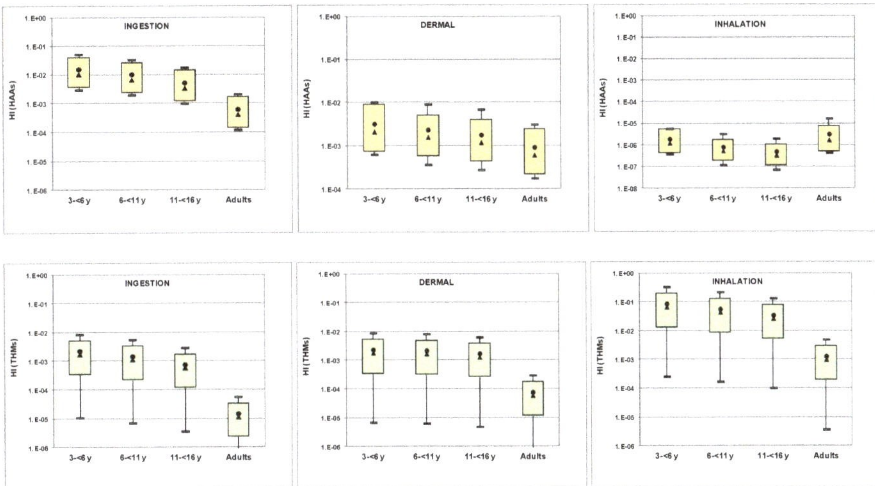

Figure 5. Hazard indices through different exposure routes for THMs and HAAs (boxplots represent 10th and 90th percentiles, min and max values; circle and triangle represent mean and median values, respectively).

Carcinogenic risks due to exposure to THMs and HAAs through ingestion, dermal contact and inhalation are shown in Figure 6. Inhalation posed the higher risk for THMs (4×10^{-9}–4×10^{-6}), with values occasionally exceeding the negligible risk limit of 10^{-6} in some facilities, whereas risk via oral and dermal routes was low. Ingestion and dermal contact posed the higher risk (3×10^{-8}–3×10^{-6}) for HAAs. Other studies also reported high risk posed by THMs through inhalation, often exceeding the limit of 10^{-6} [34,39,46,47].

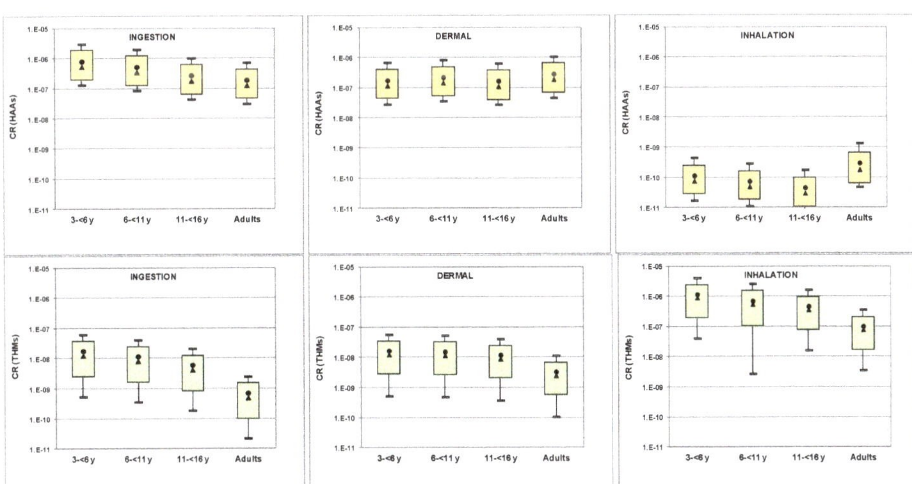

Figure 6. Carcinogenic risk through different exposure routes) for THMs and HAAs (boxplots represent 10th and 90th percentiles, min and max values; circle and triangle represent mean and median values, respectively).

There is a lot of discussion regarding calculated risks and the realistic DBPs-related risks in swimming pools. Risks could be significantly under- or over- estimated due to a number of uncertainties and assumptions that may affect the outcome. These uncertainties arise: (a) from the absence of RfDs and SFs values for each exposure route. In these cases, oral values commonly used to assess the risk via dermal exposure or inhalation. However, this extrapolation might introduce a level of uncertainty due to differences on route-specific rate and magnitude of absorption, (b) air concentrations in the natatorium calculated through different approaches, in case that are not actually measured, differ by several orders of magnitude and introduce a relevant uncertainty and, (c) other DBPs also occurred in swimming pools that may significantly contribute to realistic DBPs-related risks [34,47]. All these assumptions and uncertainties significantly affect the final risk evaluation. This urges for a common, approved methodology for risk assessment in swimming pools, for monitoring requirements of DBPs in water and indoor air quality in swimming pools facilities and for regulatory values for DBPs.

4. Conclusions

This study investigated the occurrence of various disinfection by-products, DBPs (THMs, HAAs, HANs, TCNM and HKs) in different types of swimming pools in the area of Thessaloniki during the period July 2019–February 2020. Moreover, water quality parameters (pH, residual chlorine, dissolved organic carbon, UV_{254} absorption, total nitrogen, alkalinity and conductivity) were measured.

The concentrations of DBPs varied among swimming pools as well as within the same pool between sampling campaigns. HAAs exhibited the highest concentrations in pool water followed by THMs, HANs, TCNM and HKs. The dominant species were TCAA, DCAA for haloacetic acids, TCM for trihalomethanes, DCAN, BCAN, DBAN and TCAN for haloacetoniriles. The majority of pool waters contained higher concentrations of DBPs and dissolved organic carbon and total nitrogen compared to filling waters. Exposure doses by ingestion, dermal absorption and inhalation for four age groups (3–<6 y, 6–<11 y, 11–<16 y and adults) were calculated. Inhalation was the dominant exposure route for THMs (93–95%) and TCNM (88–92%). Ingestion and dermal absorption were the main exposure routes for HAAs, 40–82% and 18–59%, respectively. Their relative contribution was age dependant; ingestion was the dominant exposure pathway for children, with a decreasing trend towards adults where dermal absorption became the main exposure pathway.

HANs, although present at lower concentrations, significantly contributed, up to 75%, to calculated cytotoxicity of pool waters. The estimation of DBPs-related health risk was based on the measured water concentrations and the respective calculated air concentrations. Hazard indices for different exposure routes were very low suggesting non-carcinogenic risk. Inhalation posed the higher carcinogenic risk for THMs, with values occasionally exceeding the negligible risk limit of 10^{-6}, whereas risk via oral and dermal routes was low. Ingestion and dermal contact posed the higher risk for HAAs although at lower levels. Due to uncertainties and assumptions in the risk assessment process, further studies are needed to comprehensively evaluate the extent and the acceptability of risks to DBPs in swimming pools. Risk management strategies that minimise the exposure to DBPs without compromising disinfection efficiency as well as the development of health-based guidelines are necessary.

Supplementary Materials: The following are available online. Figure S1: Concentrations of DBPs (mean ± sd) in swimming pools. Table S1: Spearman's correlation coefficients (values in bold were significant at 0.01 level, values in italics were significant at 0.05 level).

Author Contributions: A.S.: data curation; investigation; formal analysis; methodology; validation; writing—original draft preparation. K.K.: data curation; investigation; formal analysis; methodology; validation; writing—original draft preparation. A.K.: formal analysis; methodology; validation. I.P., formal analysis; methodology; validation. D.V.: conceptualisation, supervision methodology, validation, writing—review and editing, project administration, funding acquisition. All authors have read and agreed to the published version of the manuscript.

Funding: This research received no external funding.

Conflicts of Interest: The authors declare no conflict of interest.

References

1. Carter, R.A.A.; Joll, C.A. Occurrence and formation of disinfection by-products in the swimming pool environment: A critical review. *J. Environ. Sci.* **2017**, *58*, 19–50. [CrossRef] [PubMed]
2. Lyas, H.; Masih, I.; van der Hoek, J.V. Disinfection Methods for Swimming Pool Water: Byproduct Formation and Control. *Water* **2018**, *10*, 797. [CrossRef]
3. Carter, R.A.A.; Allard, S.; Croué, J.P.; Joll, C.A. Occurrence of disinfection by-products in swimming pools and the estimated resulting cytotoxicity. *Sci. Total Environ.* **2019**, *664*, 851–864. [CrossRef] [PubMed]
4. Richardson, S.D.; Plewa, M.J.; Wagner, E.D.; Schoeny, R.; DeMarine, D.M. Occurrence, genotoxicity, and carcinogenicity of regulated and emerging disinfection by-products in drinking water: A review and roadmap for research. *Mutat. Res.* **2007**, *636*, 178–242. [CrossRef]
5. Richardon, S.D.; Postigo, C. Discovery of New Emerging DBPs by High-Resolution Mass Spectrometry. *Compreh. Anal. Chem.* **2016**, *71*, 335–355. [CrossRef]
6. Stalter, D.; O'Malley, E.; von Gunten, U.; Escher, B.J. Fingerprinting the reactive toxicity pathways of 50 drinking water disinfection by-products. *Water Res.* **2016**, *91*, 19–30. [CrossRef] [PubMed]
7. Villanueva, C.M.; Cordier, S.; Font-Ribera, L.; Salas, L.A.; Levallois, P. Overview of Disinfection By-products and Associated Health Effects. *Curr. Environ. Health. Rpt.* **2015**, *2*, 107–115. [CrossRef]
8. Font-Ribera, L.; Marco, E.; Grimalte, J.O.; Pastor, S.; Marcos, R.; Abramsson-Zetterberg, L.; Pedersen, M.; Grummt, T.; Junek, R.; Barreiro, E.; et al. Exposure to disinfection by-products in swimming pools and biomarkers of genotoxicity and respiratory damage–The PISCINA2 Study. *Environ. Int.* **2019**, *131*, 104998. [CrossRef] [PubMed]
9. Couto, M.; Bernard, A.; Delgado, L.; Drobnic, F.; Kurowski, M.; Moreira, A.; Rodrigues-Alves, R.; Rukhadze, M.; Seys, S.; Wiszniewska, M.; et al. Health effects of exposure to chlorination by-products in swimming pools. *Allergy* **2021**, *76*, 3257–3275. [CrossRef]
10. U.S. EPA. *Method 551.1 Determination of Chlorination Disinfection Byproducts, Chlorinated Solvents, and Halogenated Pesticides/Herbicides In Drinking Water By Liquid-Liquid Extraction And Gas Chromatography with Electron-Capture Detection*; Revision 1.0.; U.S. EPA: Cincinnati, OH, USA, 1995. Available online: https://www.epa.gov/sites/default/files/2015-06/documents/epa-551.1.pdf (accessed on 6 September 2021).
11. U.S. EPA. *Method 552.3 Determination of Haloacetic Acids and Dalapon In Drinking Water by Liquid-Liquid Microextraction, Derivatization and Gas Chromatography with Electron Capture Detection*; Revision 1.0.; U.S. EPA: Cincinnati, OH, USA, 2003. Available online: https://nepis.epa.gov/Exe/ZyPDF.cgi/901V0400.PDF?Dockey=901V0400.PDF (accessed on 6 September 2021).
12. APHA. *Standard Methods for the Examination of Water and Waste Water*, 22nd ed.; AWWA, WEF: Washington, DC, USA, 2012.
13. Wagner, E.D.; Plewa, M.J. CHO cell cytotoxicity and genotoxicity analyses of disinfectionby-products: An updated review. *J. Environ. Sci.* **2017**, *58*, 64–76. [CrossRef]
14. U.S. EPA. *Swimmer Exposure Assessment Model (SWIMODEL)*; Version 3.0; U.S. EPA: Cincinnati, OH, USA, 2003.
15. U.S. EPA. Exposure Assessment Tools by Routes. Available online: https://www.epa.gov/expobox/exposure-assessment-tools-routes (accessed on 6 September 2021).
16. Dyck, R.; Sadiq, R.; Rodriguez, M.J.; Simard, S.; Tardif, R. Trihalomethane exposures in indoor swimming pools: A level III fugacity model. *Water Res.* **2011**, *45*, 5084–5098. [CrossRef]
17. Lourencetti, C.; Grimalt, J.O.; Marco, E.; Fernandez, P.; Font-Ribera, L.; Villanueva, C.M.; Kogevinas, M. Trihalomethanes in chlorine and bromine disinfected swimming pools: Air-water distributions and human exposure. *Environ. Int.* **2012**, *45*, 59–67. [CrossRef]
18. U.S. EPA. *Exposure Factors Handbook*; EPA/600/R-09/052F; U.S. EPA: Cincinnati, OH, USA, 2011.
19. R.A.I.S. Risk Assessment Information System (RAIS) Online Database. Available online: https://rais.ornl.gov/cgi-bin/tools/TOX_search?select=chemspef (accessed on 6 September 2021).
20. PubChem. National Center for Biotechnology Information. U.S. National Library of Medicine, USA. Available online: https://pubchem.ncbi.nlm.nih.gov/compounds (accessed on 6 September 2021).
21. U.S. EPA. The USEPA Integrated Risk Information System (IRIS) Online Database. Washington, DC, USA. 2014. Available online: http://www.epa.gov/iris/subst/index.html (accessed on 6 September 2021).
22. I.A.R.C. International agency for Research of Cancer. Agents Classified by the IARC Monographs, Volumes 1–128. Available online: https://monographs.iarc.fr/list-of-classifications (accessed on 6 September 2021).
23. O.E.H.H.A. California Office of Environmental Health Hazard Assessment. Available online: https://oehha.ca.gov/chemicals/chloroform (accessed on 6 September 2021).
24. *Greek National Legislation Γ1/443/1973 On Swimming Pools Following Instructions for Their Construction and Operation*; Greek National Legislation: Athens, Greece, 1973.
25. *Greek National Legislation F.E.K. 21.5.2020, (Δ1(δ)/ΓΠοικ 32179/2020) Regarding Measures for Protection of Public Health In The Context Of Avoiding Spread Of SARS-Cov-2 After Commissioning Of Swimming Pools*; Greek National Legislation: Athens, Greece, 2020.
26. WHO. *Guidelines for Safe Recreational Water Environments*; Volume 2: Swimming Pools and Similar Environments; WHO: Geneva, Switzerland, 2006.
27. Kanan, A.; Karanfil, T. Formation of Disinfection By-Products in Indoor Swimming Pool Water: The Contribution from Filling Water Natural Organic Matter and Swimmer Body Fluids. *Water Res.* **2011**, *45*, 926–932. [CrossRef]

28. Yeh, R.Y.; Farre, M.J.; Stalter, D.; Tang, J.Y.; Molendijk, J.; Escher, B.I. Bioanalytical and chemical evaluation of disinfection by-products in swimming pool water. *Water Res.* **2014**, *59*, 172–184. [CrossRef]
29. Keuten, M.G.A.; Peters, M.C.F.M.; Daanen, H.A.M.; de Kreuk, M.K.; Rietveld, L.C.; van Dijk, J.C. Quantification of continual anthropogenic pollutants released in swimming pools. *Water Res.* **2014**, *53*, 259–270. [CrossRef] [PubMed]
30. Tsamba, L.; Cimetière, N.; Wolbert, D.; Correc, O.; Le Cloirec, P. Body fluid analog chlorination: Application to the determination of disinfection byproduct formation kinetics in swimming pool water. *J. Environ. Sci.* **2020**, *87*, 112–122. [CrossRef]
31. Wang, J.; Gong, T.; Xian, J. Formation of haloacetic acids from different organic precursors in swimming pool water during chlorination. *Chemosphere* **2020**, *247*, 125793. [CrossRef] [PubMed]
32. European Chemical s Agency. *Guidance on The BPR: Vol V Disinfection By-Products. ECHA-17-G-01-EN.*; European Chemical s Agency: Telakkakatu, Finland, 2017.
33. European Parliament. *EU Directive 2020/2184 Of the European Parliament and of The Council of 16 December 2020 On the Quality Of Water Intended For Human Consumption. OJ. L 435/1, 23.12.2020*; European Parliament: Strasbourg, France, 2020.
34. Pándics, T.; Hofer, A.; Dura, G.; Vargha, M.; Szigeti, T.; Tóth, E. Health risk of swimming pool disinfection by-products: A regulatory perspective. *J. Water Health* **2018**, *16*, 947–957. [CrossRef] [PubMed]
35. Simard, S.; Tardif, R.; Rodriguez, M.J. Variability of chlorination by-product occurrence in water of indoor and outdoor swimming pools. *Water Res.* **2013**, *47*, 763–1772. [CrossRef]
36. Tardif, R.; Catto, C.; Haddad, S.; Simard, S.; Rodriguez, M. Assessment of air and water contamination by disinfection by-products at 41 indoor swimming pools. *Environ. Res.* **2016**, *148*, 411–420. [CrossRef]
37. Peng, F.; Peng, J.; Li, H.; Li, Y.; Wang, B.; Yang, Z. Health risks and predictive modeling of disinfection byproducts in swimming pools. *Environ. Int.* **2020**, *139*, 105726. [CrossRef]
38. Manasfi, T.; De Méo, M.; Coulomb, B.; Di Giorgio, C.; Boudenne, J.L. Identification of disinfection by-products in freshwater and seawater swimming pools and evaluation of genotoxicity. *Environ. Int.* **2016**, *88*, 94–102. [CrossRef] [PubMed]
39. Kargaki, S.; Iakovides, M.; Stephanou, E. Study of the occurrence and multi-pathway health risk assessment of regulated and unregulated disinfection by-products in drinking and swimming pool waters of Mediterranean cities. *Sci. Total Environ.* **2020**, *739*, 1398890. [CrossRef] [PubMed]
40. Wang, X.; Garcia, L.; Zhang, H.; Yang, H.; Yuefeng, X. Haloacetic acids in swimming pool and spa water in the United States and China. *Front Environ. Sci. Eng.* **2014**, *8*, 820–824. [CrossRef]
41. Kanan, A.; Selbes, M.; Karanfil, T. Occurrence and formation of disinfection by-products in indoor US swimming pools. *ACS Symp. Ser.* **2015**, *1190*, 405–430. [CrossRef]
42. Yang, L.; Schmalz, C.; Zhou, J.; Zwiener, C.; Chang, V.W.; Ge, L.; Pun, M. An insight of disinfection by-product (DBP) formation by alternative disinfectants for swimming pool disinfection under tropical conditions. *Water Res.* **2016**, *101*, 535–546. [CrossRef]
43. Zhang, X.; Yang, H.; Wang, X.; Zhao, Y.; Wang, X.; Xie, Y. Concentration levels of disinfection by-products in 14 swimming pools of China. *Environ. Sci. Eng.* **2015**, *9*, 995–1003. [CrossRef]
44. Font-Ribera, L.; Kogevinas, M.; Schmalz, C.; Zwiener, C.; Marco, E.; Grimaltf, J.O.; Liu, J.; Zhang, X.; Mitch, W.; Critelli, R.; et al. Environmental and personal determinants of the uptake of disinfection by-products during swimming. *Environ. Res.* **2016**, *149*, 206–215. [CrossRef]
45. Nitter, T.B.; Svendsen, K.H. Modelling the concentration of chloroform in the air of a Norwegian swimming pool facility—A repeated measures study. *Sci. Total Environ.* **2019**, *664*, 1039–1044. [CrossRef] [PubMed]
46. Chowdhury, S. Predicting human exposure and risk from chlorinated indoor swimming pool: A case study. *Environ. Monit. Assess.* **2015**, *187*, 502. [CrossRef]
47. Gouveia, P.; Felgueiras, F.; Mourão, Z.; Fernandes, E.D.O.; Moreira, A.; Gabriel, M.F. Predicting health risk from exposure to trihalomehtanes in an Olympic-size indoor swimming pool among elite swimmers and coaches. *J. Toxicol. Environ. Health A* **2019**, *82*, 577–590. [CrossRef] [PubMed]

molecules

MDPI

Article

Multicriteria Decision Analysis and Grouping of Analytical Procedures for Phthalates Determination in Disposable Baby Diapers

Magdalena Fabjanowicz [1], **Justyna Płotka-Wasylka** [2,*] and **Marek Tobiszewski** [3]

[1] Department of Analytical Chemistry, Chemical Faculty, Gdańsk University of Technology (GUT), 11/12 G. Narutowicza St., 80-233 Gdańsk, Poland; magfabja@student.pg.edu.pl
[2] Department of Analytical Chemistry, Chemical Faculty and BioTechMed Center, Gdańsk University of Technology (GUT), 11/12 G. Narutowicza St., 80-233 Gdańsk, Poland
[3] Department of Analytical Chemistry, Chemical Faculty and EcoTech Center, Gdańsk University of Technology (GUT), 11/12 G. Narutowicza St., 80-233 Gdańsk, Poland; martobis@pg.edu.pl
* Correspondence: juswasyl@pg.edu.pl

Abstract: This study presents the application of one of the tools from the multicriteria decision analysis set (MCDA), the Technique for Order Preference by Similarity to Ideal Solution (TOPSIS). Selected green analytical chemistry metrics were used to rank analytical procedures for the phthalate determination in disposable baby diapers. Nine analytical procedures were assessed in order to find one that has the lowest environmental impact and the best analytical figures of merit. Nine different criteria, where weighting was based on the experts' evaluation, were used in the procedures' assessment. With the use of TOPSIS, an easy and straightforward technique, selection of the most appropriate procedure was made.

Keywords: disposable baby diapers; phthalate; analytical procedures; TOPSIS; MCDA; green analytical chemistry; environmental impact

Citation: Fabjanowicz, M.; Płotka-Wasylka, J.; Tobiszewski, M. Multicriteria Decision Analysis and Grouping of Analytical Procedures for Phthalates Determination in Disposable Baby Diapers. *Molecules* **2021**, *26*, 7009. https://doi.org/10.3390/molecules26227009

Academic Editors: Victoria Samanidou, Eleni Deliyanni and Dimitra Voutsa

Received: 7 October 2021
Accepted: 16 November 2021
Published: 19 November 2021

1. Introduction

Phthalates are a group of compounds derived from the esters of 1,2-dibenzene dicarboxylic acid. Due to their properties, they are widely applied in the industry as plasticizers (addition of phthalates increases the flexibility and workability of high molecular weight polymers), heat-transfer fluids and carriers (thanks to their low melting point and high boiling point) [1].

The most commonly used phthalates are di(2-ethylhexyl) phthalate (DEHP) and the primary metabolite mono-(2-ethylhexyl) phthalate (MEHP). However, di(n-butyl) phthalate (DBP), diethyl phthalate (DEP), butyl-benzyl phthalate (BBP) and di(n-octyl) phthalate (DnOP) are also abundant [2]. Phthalates are present in ink, paint, adhesives, vinyl flooring, food packaging, furniture, toys, cosmetics and pharmaceuticals, as well as in sanitary products [1,3].

There is a growing concern regarding the adverse effect of phthalates on human health. There are several studies presenting the link between human exposure and certain disease occurrences, e.g., cancer, abnormal development and thyroid function, fertility disturbances, type 2 diabetes and cardiovascular disorders. Moreover, several phthalate metabolites can increase the risk of obesity [3].

When it comes to exposure to phthalates, newborns and toddlers are one of the most vulnerable groups of people due to the fact that up to the age of four, their skin is at risk of prolonged exposure to the compounds present in disposable diapers [4]. Nowadays, disposable baby diapers are filled with superabsorbent material, increasing their softness and overall comfort of use. However, because of that, they can also be a source of volatile organic compounds (VOCs) and endocrine-disrupting chemicals, such as phthalates [5]. It is very important to control and monitor the safety of disposable baby diapers in order to reduce the adverse effect that studied compounds can have on the well-being of diaper-wearing children [5].

To the best of the authors' knowledge, only a few studies dealing with phthalate determination in disposable baby diapers have been published within the last 10 years, as can be seen in Table 1. However, there has been a significant increase in the number of publications on this topic in the last two years [6–8], which indicates the importance of this topic and increased interest in it.

Table 1. Dataset characteristics.

Criteria	Analytical Procedures								
	1 [9]	2 [10]	3 [11]	4 [6]	5 [7]	6 [12]	7 [5]	8 [13]	9 [8]
Analytical methodology abbreviation	DART-MS/MS	MSPE/GC-FID	GC-MS	GC-MS	GC-MS	PLE/GC-MS	LC-MS/MS	GC-MS	GC-MS
LOQ [mg/L]	0.001485	0.00066	0.1	0.7689	0.33	0.02376	0.0001	0.002	0.001188
LOD [mg/L]	0.00045	0.0002	NI	0.233	0.1	0.0072	NI	NI	0.00036
Reagent (list of all reagents used in the procedure)	direct analysis	6 mL HCl	~10 mL DCM	83 mL hexane, 47 mL acetone	0.5 mL Milli-Q water. 10 mL hexane. 3 mL MeOH	100 mL ethyl acetate. 20 mL methanol	6 mL methanol	5 mL hexane	102.2 mL DCM, 154.8 mL hexane
Solvent hazards	0	0	598	6878.4	861.1	1044	94.2	407	18,712.28
Reagent price (euro)	-	0.18	0.22	3.61	0.37	5.56	0.15	0.14	8.78
Amount of sample (g)	0.9985	0.1	0.05	0.5	0.3	5.25	0.435	0.3	100
Number of other analytes	21	4	7	15	8	54	4	24	57
Time of analysis (min)	few sec.	48	NI about the extraction time + 37.5	86	138.5	31	36	118.66	992
Sample throughput	few sec.	3	1	2	1	2	2	1	1
Procedure steps	2	3	3	3	2	3	2	3	4
Energy consumption (kWh)	3.2	1.6	2.55	1.77	4.03	1.77	3.2	2.62	2.6

Gas chromatography coupled to mass spectrometry (GC-MS) is the most widely applied technique for phthalate determination in disposable baby diapers and other sanitary products of similar matrices. GC-MS was applied in seven out of nine research articles gathered in Table 2. Due to sufficient volatility of phthalates, there is no derivatization needed. However, due to the complicity of the matrix, there is a need for extraction prior to the chromatographic analysis. According to the published scientific reports, solvent extraction is preferred.

Table 2. Weights applied for TOPSIS according to the scenario.

	Scenario
LOQ	0.01
Solvent hazards	0.28
Reagent price	0.05
Amount of sample	0.01
Number of other analytes	0.2
Time of analysis	0.15
Sample throughput	0.15
Procedure steps	0.05
Energy consumption	0.1

Although there is an increasing number of research articles depicting phthalate determination in different sanitary products, there is a lack of critical comparisons of developed methodologies in terms of both the parameters of analytical figures of merits as well as their green character. It is obvious that modern analytical procedures should be designed keeping in mind both the requirements of green analytical chemistry and the necessity of sustainable development. Multicriteria decision analysis (MCDA) can be applied in the assessment of the environmental impact of different approaches to phthalates determination. In MCDA, a set of tools such as the Preference Ranking Organization METHOD for Enrichment of Evaluations (PROMETHEE) [14] or Technique for Order of Preference by Similarity to Ideal Solution (TOPSIS) [15] are used to rank procedures.

This work aims to fill the gap of knowledge in the field of simultaneous assessment of validation parameters and environmental impact of available procedures for phthalate determination in baby sanitary products. Moreover, the applicability of TOPSIS as the appropriate method selection is described. This case study was aimed at finding a green analytical procedure for phthalate determination in diapers and related product samples. The present study stands as a starting point for future research concerning the safety of newborns and toddlers exposed to the chemicals released from disposable baby diapers. We believe that this work will benefit those who research phthalates in sanitary products and will enable the development of methods with minimal influence on the environment.

2. Result and Discussion

2.1. Clustering of Variables

Cluster analysis facilitates finding possible correlations between variables and data dimensionality reduction. According to the results (Figure 1), three clusters can be identified (the cut-off was set to 60% of the distance, to a maximum distance ratio from 10 criteria).

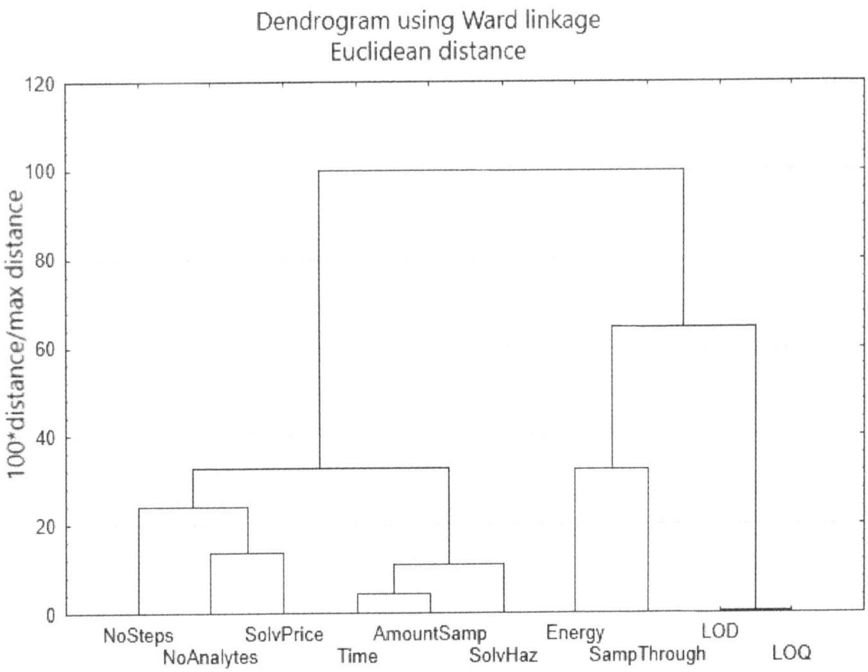

Figure 1. Results of cluster analysis grouping of variables.

The obtained results from the clustering of variables are as follows:

i. There are several correlations between selected variables observed: LOD and LOQ can be further treated as a single variable. The relation between LOD and LOQ is the following: LOQ = 3 × LOD, which was calculated either by the authors of the investigated procedures or by the authors of this study. Therefore, the resulting determination coefficient $R^2 = 1$.
ii. Time and amount of sample are correlated, which indicates that more time is needed for the preparation and analysis of a larger sample. Correlation is affected mostly by Procedure 9, which is an outlier, described in more detail later;
iii. Solvent hazard variability is correlated with two former variables. It indicates that a large number of solvents are needed to extract analytes from large samples;
iv. The price of solvents is correlated with the number of analytes, which is an indication that multianalyte procedures require larger inputs in terms of solvents;
v. The number of analytical steps is correlated with two former variables. Thus, it can be concluded that more analytical steps are needed to deal with multianalyte situations. On the other hand, a higher number of steps requires higher solvent contribution;
vi. Both groups (time, amount of sample and solvent hazards, as well as solvent price, number of analytes and number of steps) are loosely correlated;
vii. The last cluster consists of energy and sample throughput. Sample throughput measures the number of analyses that can be performed within one hour, which is related to the energy consumption. The higher the sample throughput, the less energy is needed for a single analysis.

2.2. Clustering of Analytical Procedures

According to the cluster analysis (Figure 2), the clustering of objects shows the following patterns:

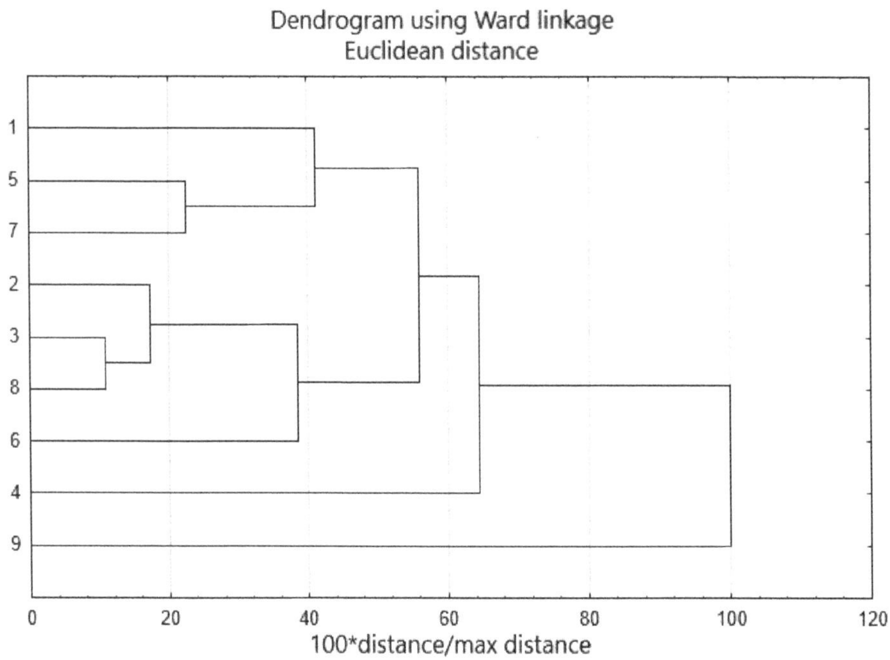

Figure 2. Cluster analysis results of analytical procedures grouping.

i. Procedures 3 and 8 are more loosely related to 2;
ii. Procedures 5, 7 and 1 are also somehow related;

iii. Procedure 9 is an outlier.

Going in-depth, the first cluster, which gathers procedures 3, 8 and 2, clusters methodologies based on the same analytical technique; procedures 3 and 8 use gas chromatography coupled to mass spectrometry (GC-MS), while procedure 2 uses gas chromatography with flame ionization detector (GC-FID). However, all of them are characterized by a similar amount of solvent used, the same number of procedures steps and similar energy consumption.

The second cluster grouped procedures 5 (based on the liquid chromatography coupled to tandem mass spectrometry (LC-MS/MS), 7 (based on GC-MS) and 1 (DART-MS/MS). All of them are characterized by the same number of procedure steps and similar energy consumption.

Procedure 9 is an outlier since the amount of solvent used during the analysis was ten times higher than in the rest of the analytical procedures described before. This procedure significantly differs from the others in terms of solvent hazards and time of analysis (six times longer).

2.3. Ranking

The application of TOPSIS requires setting the ranking of the weights of all the variables (gathered in Table 1). The weights were applied based on the variable's importance in the context of green chemistry and analytical figures of merit. Thus, LOQ is given a very low weight as all analytical procedures have satisfactory LOQ. The number of samples is of little importance as diapers are not very expensive and easy to obtain. Reagent price is given relatively low weight as it is within a satisfactory range. The number of steps is given the same weight of 0.05 due to little variability. The highest weight is given to solvent hazards as it is the most influential factor of the analytical procedure's greenness assessment. The number of analytes and two time-related criteria are given high weight values, as all criteria describe the amount of analytical information obtained in a reasonable unit of time. Energy consumption is given a moderate weight as it should not be minimized but is not critical in terms of optimization of analytical methodologies.

The ideal methodology is characterized by the best parameters that could be achieved [16]. After the application of TOPSIS ranking (Table 3), it can be seen that procedure 1, DART-MS/MS, is the closest to the ideal analytical procedures. Direct analysis in real-time does not require the sample preparation step and there is no solvent usage, which is a significant advantage, resulting in the highest place in the ranking. Second, achieving a 0.677 score, is a procedure based on the liquid chromatography coupled to tandem mass spectrometry (procedure 6). The least ideal solution and the last place in the ranking is dedicated to the outlier—procedure 9. This is not a surprising result since the use of this procedure takes six times longer than the others. Furthermore, it obtained the highest solvent hazard score.

Table 3. Ranking results of analytical procedures according to a given scenario.

Scenario	
Procedure Number	**Similarity to an Ideal Solution**
1	0.799
6	0.677
8	0.638
2	0.616
7	0.612
3	0.609
5	0.599
4	0.521
9	0.268

3. Materials and Methods

3.1. Dataset Creation

The dataset used for TOPSIS analysis was collected from the research papers published in the Web of Science, Scopus, Science Direct and Nature databases. Specific keywords like phthalate, disposable baby diapers, and newborns were applied in order to find publications of interest. Due to the matrix similarity, sanitary pads, wipes and toilet tissue paper were also taken under consideration.

3.2. Cluster Analysis

Cluster Analysis (CA) belongs to a set of multivariate statistical tools that allow for grouping objects and variables according to their similarity. Briefly, the unsupervised algorithm finds an internal pattern in the set of data with no previous considerations on the dataset. The similarity (or dissimilarity) of objects or variables is calculated on the basis of correlation or distance. The grouping of both variables and objects is performed with the Euclidean distance and cluster formation method after Ward. After standardization of the initial dataset, CA calculations are performed with Statistica12 software. More details about the algorithm may be found elsewhere [17].

3.3. TOPSIS

One of the frequently applied MCDA algorithms is TOPSIS. This is an expert system that is applied for decision making and was developed in 1981 by Hwang and Yoon [18]. The analysis leads to the final ranking of all alternatives; as a result, the selection of the best option among all available is easily possible. The alternative that has the shortest distance to the positive ideal solution and the farthest distance to the negative ideal solution at the same time is the winner.

The assessment procedure for MCDA can be performed in a few simple steps. In the beginning, the main aim of the analysis should be clearly stated. This case study is aimed at finding a green analytical procedure for phthalate determination in diapers and related product samples. The dataset prepared for application in MCDA methodology consists of criteria and alternatives, which should be clearly defined. Generally, criteria are parameters that describe the set of alternatives and make the evaluation easy and systematic. Alternatives are available options that allow reaching the goal stated in the aim of the MCDA study. All the alternatives must be described by criteria expressed with numerical values or at least they must be easily transformable into numbers. In the next step of MCDA assessment, the assigning of proper weight values to each criterion is done. This means that there is a possibility to differentiate the importance of criteria.

The last step is the application of the TOPSIS algorithm. Its mechanism can be described in several steps, as described in other papers [19]. All TOPSIS calculations were performed in the Excel program (Microsoft 2016).

3.4. The Dataset

Based on the literature reports available in the Web of Science and Scopus databases, nine analytical methodologies were found and applied to a given research problem. All of them were described by certain variables, which were used to set the criteria for clustering analysis in order to find parameters that are the most significant in choosing the best methodology. In this context, the term "the best" corresponds to the methodology, which is described by the most relevant analytical figures of merit as well as meets the requirements of green analytical chemistry. Thus, ten criteria were established in order to perform the first grouping:

- number of steps necessary to perform the analysis;
- number of analytes determined in the single run;
- the solvent prize, taken from the website for the Polish market, with prices recalculated based on the price for 1 L;

- the time needed to perform the analysis (extraction time + time of chromatographic analysis);
- amount of sample used for the analysis;
- solvent hazards, which was calculated according to [20];
- energy consumption during a single run;
- sample throughput calculated as 60 min divided by the time of chromatographic analysis;
- limit of detection (LOD) and limit of quantification (LOQ), calculated as 3.3 × LOD.

Such a selection of variables comprehensively covers many aspects of the procedure assessment. They deal with the most basic metrological parameters but also refer to many green analytical chemistry aspects (solvents applied, energy consumed) and general performance (number of analytes, analysis time, number of procedural steps, analytical throughput) of the procedure.

All the data gathered for the MCDA and CA analyses are presented in Table 3.

4. Conclusions

In this study, TOPSIS was applied for the selection of the most relevant analytical procedure for the phthalate determination in disposable baby diapers. According to data analysis and procedures rank, the procedure that gained the best result was the direct analysis in real-time (DART) coupled to mass spectrometry (MS) technique. The lack of sample preparation is a crucial advantage of the given technique. Apart from the low impact on the environment, good metrological parameters were observed. Multicriteria decision analysis is a useful approach for the selection of procedures meeting the green analytical chemistry requirements.

Author Contributions: Conceptualization, J.P.-W. and M.T.; bibliographic research, M.F.; application of methodology, M.T.; writing—original draft, M.F.; writing—review & editing, M.T. and J.P.-W.; supervision, J.P.-W. All authors have read and agreed to the published version of the manuscript.

Funding: This work was supported by the National Science Center granted by the decision number DEC-2020/37/B/ST4/02886 project number UMO-2020/37/B/ST4/02886.

Institutional Review Board Statement: Not applicable.

Informed Consent Statement: Not applicable.

Data Availability Statement: Data is contained within this article.

Acknowledgments: The authors would like to thank Kaja Kalinowska for her assistance and support.

Conflicts of Interest: The authors declare that the research was conducted in the absence of any commercial or financial relationships that could be construed as a potential conflict of interest.

References

1. Peijnenburg, W.J.G.M. Phthalates. In *Encyclopedia of Ecology*; Elsevier: Amsterdam, The Netherlands, 2008.
2. Kim, H.S.; Lee, B.M. Endocrine Disrupting Chemicals and Human Cancer. In *Encyclopedia of Environmental Health*; Elsevier: Amsterdam, The Netherlands, 2011.
3. Muscogiuri, G.; Colao, A. Phthalates: New Cardiovascular Health Disruptors? *Arch. Toxicol.* **2017**, *91*, 1513–1517. [CrossRef] [PubMed]
4. WHO Summary of Principles for Evaluating Health Risks in Children Associated with Exposure to Chemicals. WHO Library Cataloguing-in-Publication Data. 2011. Available online: https://www.who.int/ceh/health_risk_children.pdf (accessed on 15 September 2021).
5. Park, C.J.; Barakat, R.; Ulanov, A.; Li, Z.; Lin, P.-C.; Chiu, K.; Zhou, S.; Perez, P.; Lee, J.; Flaws, J.; et al. Sanitary Pads and Diapers Contain Higher Phthalate Contents than Those in Common Commercial Plastic Products. *Reprod. Toxicol.* **2019**, *84*, 114–121. [CrossRef] [PubMed]
6. Tang, Z.; Chai, M.; Cheng, J.; Wang, Y.; Huang, Q. Occurrence and Distribution of Phthalates in Sanitary Napkins from Six Countries: Implications for Women's Health. *Environ. Sci. Technol.* **2019**, *53*, 13919–13928. [CrossRef] [PubMed]
7. Gao, C.-J.; Wang, F.; Shen, H.-M.; Kannan, K.; Guo, Y. Feminine Hygiene Products—A Neglected Source of Phthalate Exposure in Women. *Environ. Sci. Technol.* **2020**, *54*, 930–937. [CrossRef] [PubMed]

8. Adjei, J.K.; Essumang, D.K.; Twumasi, E.; Nyame, E.; Muah, I. Levels and Risk Assessment of Residual Phthalates, Polycyclic Aromatic Hydrocarbons and Semi-Volatile Chlorinated Organic Compounds in Toilet Tissue Papers. *Toxicol. Rep.* **2019**, *6*, 1263–1272. [CrossRef] [PubMed]
9. Kuki, Á.; Zelei, G.; Nagy, L.; Nagy, T.; Zsuga, M.; Kéki, S. Rapid Mapping of Various Chemicals in Personal Care and Healthcare Products by Direct Analysis in Real Time Mass Spectrometry. *Talanta* **2019**, *192*, 241–247. [CrossRef] [PubMed]
10. Razavi, N.; Es'haghi, Z. Employ of Magnetic Polyaniline Coated Chitosan Nanocomposite for Extraction and Determination of Phthalate Esters in Diapers and Wipes Using Gas Chromatography. *Microchem. J.* **2018**, *142*, 359–366. [CrossRef]
11. Ishii, S.; Katagiri, R.; Minobe, Y.; Kuribara, I.; Wada, T.; Wada, M.; Imai, S. Investigation of the Amount of Transdermal Exposure of Newborn Babies to Phthalates in Paper Diapers and Certification of the Safety of Paper Diapers. *Regul. Toxicol. Pharmacol.* **2015**, *73*, 85–92. [CrossRef] [PubMed]
12. Celeiro, M.; Lamas, J.P.; Garcia-Jares, C.; Llompart, M. Pressurized Liquid Extraction-Gas Chromatography-Mass Spectrometry Analysis of Fragrance Allergens, Musks, Phthalates and Preservatives in Baby Wipes. *J. Chromatogr. A* **2015**, *1384*, 9–21. [CrossRef] [PubMed]
13. Gao, C.-J.; Kannan, K. Phthalates, Bisphenols, Parabens, and Triclocarban in Feminine Hygiene Products from the United States and Their Implications for Human Exposure. *Environ. Int.* **2020**, *136*, 105465. [CrossRef] [PubMed]
14. Tobiszewski, M.; Orłowski, A. Multicriteria Decision Analysis in Ranking of Analytical Procedures for Aldrin Determination in Water. *J. Chromatogr. A* **2015**, *1387*, 116–122. [CrossRef] [PubMed]
15. Al-Hazmi, H.; Namieśnik, J.; Tobiszewski, M. Application of TOPSIS for Selection and Assessment of Analytical Procedures for Ibuprofen Determination in Wastewater. *Curr. Anal. Chem.* **2016**, *12*, 261–267. [CrossRef]
16. Pavić, Z.; Novoselac, V. Clustering and Outlier Detection by the EM Algorithm Based on the Restriction Principle View Project Notes on TOPSIS Method. *Int. J. Eng. Sci.* **2013**, *1*, 5–12.
17. Kettenring, J.R. The Practice of Cluster Analysis. *J. Classif.* **2006**, *23*, 3–30. [CrossRef]
18. Behzadian, M.; Khanmohammadi Otaghsara, S.; Yazdani, M.; Ignatius, J. A State-of the-Art Survey of TOPSIS Applications. *Expert Syst. Appl.* **2012**, *39*, 13051–13069. [CrossRef]
19. Kwok, P.K.; Lau, H.Y.K. Hotel Selection Using a Modified TOPSIS-Based Decision Support Algorithm. *Decis. Support Syst.* **2019**, *120*, 95–105. [CrossRef]
20. Tobiszewski, M.; Namieśnik, J. Scoring of Solvents Used in Analytical Laboratories by Their Toxicological and Exposure Hazards. *Ecotoxicol. Environ. Saf.* **2015**, *120*, 169–173. [CrossRef] [PubMed]

molecules MDPI

Article

Hexosomal Dispersion: A Nano-Based Approach to Boost the Antifungal Potential of Citrus Essential Oils against Plant Fungal Pathogens

Mohamed S. Sedeek [1,*], Abdulaziz M. Al-Mahallawi [2,3], Rania A. A. Hussien [4], Ahmed M. Abdelhaleem Ali [5], Ibrahim A. Naguib [6] and Mai K. Mansour [7]

1 Pharmacognosy Department, Faculty of Pharmacy, Cairo University, Kasr-El-Aini Street, Cairo 11562, Egypt
2 Department of Pharmaceutics and Industrial Pharmacy, Faculty of Pharmacy, Cairo University, Cairo 11562, Egypt; abdulaziz.mohsen@pharma.cu.edu.eg
3 School of Life and Medical Sciences, University of Hertfordshire Hosted by Global Academic Foundation, New Administrative Capital, Cairo 11562, Egypt
4 Fungicide, Bactericide and Nematicide Department, Central Agricultural Pesticides Lab (CAPL), Agriculture Research Center (ARC), Giza 11835, Egypt; raniahussien187@gmail.com
5 Department of Pharmaceutics and Industrial Pharmacy, College of Pharmacy, Taif University, P.O. Box 11099, Taif 21944, Saudi Arabia; a.mali@tu.edu.sa
6 Department of Pharmaceutical Chemistry, College of Pharmacy, Taif University, P.O. Box 11099, Taif 21944, Saudi Arabia; i.abdelaal@tu.edu.sa
7 Department of Medicinal Plants and Natural Products, Egyptian Drug Authority, Giza 11553, Egypt; mai.khaled@pharma.cu.edu.eg or mkm107@live.com
* Correspondence: drmohamedsedeek@live.com or mohamed.sedeek@pharma.cu.edu.eg

Citation: Sedeek, M.S.; Al-Mahallawi, A.M.; Hussien, R.A.A.; Ali, A.M.A.; Naguib, I.A.; Mansour, M.K. Hexosomal Dispersion: A Nano-Based Approach to Boost the Antifungal Potential of Citrus Essential Oils against Plant Fungal Pathogens. *Molecules* **2021**, *26*, 6284. https://doi.org/10.3390/molecules26206284

Academic Editors: Victoria Samanidou, Eleni Deliyanni and Dimitra Voutsa

Received: 29 September 2021
Accepted: 14 October 2021
Published: 17 October 2021

Abstract: The demand for natural fungicides to replace synthetic ones has surged since toxic residues persist in soils, causing environmental contamination and posing a serious threat to worldwide public health. In the context of crop protection and enhancing the efficiency and safety of fungicides, nanotechnology is an eco-friendly strategy in managing fungal pathogens. In the present study, essential oils were isolated from the peels of four citrus fruits (*Citrus lemon*, *Citrus aurantifolia*, *Citrus maxima*, and *Citrus sinensis*) and were investigated using gas chromatography-mass spectrometric analysis. Monoterpene hydrocarbon was the most predominant group and limonene was the most abundant in the four oils. The antifungal potential of the oils was investigated, and the most active oil (*Citrus lemon*) was loaded into hexosomal dispersion, and its antifungal potential was retested against the same fungi. The structurally unique nano-based formulation showed great potency for fungal control. To the best of our knowledge, it is the first time the oil of *Citrus lemon* in nano-hexosomes has been formulated and its fungicidal activity examined. The data collected suggest that citrus essential oils (CEOs), especially when nano-formulated, could be successfully used in integrated fungus management programs.

Keywords: fungicides; citrus; essential oil; nanotechnology; hexosomes; environmental contamination

1. Introduction

Fungal plant pathogens play a crucial role in plant production [1]. These pathogens can significantly reduce the productivity and quality of field crops and this is becoming a more pressing concern for human health and the global economy and costs billions of dollars annually [1,2].

Agricultural crops are exposed to more than 10,000 pathogenic fungi that are known to cause important plant diseases [3], resulting in a significant loss in agricultural crops and reduction of world food production. Farmers generally rely on the use of synthetic fungicides to control plant diseases caused by pathogenic fungi; however, misuse of these fungicides can cause serious health and environmental problems [3].

One of the most important fruit crops in the world is citrus [4]. Citrus is grown in over 100 nations throughout the world, primarily in tropical and subtropical regions. As a byproduct of citrus fruit processing, a vast quantity of residual peels is generated, which add no value to the product even though they are discarded or dumped, presenting an environmental problem [5]. Citrus peels have received much attention for their potential use as value-added products because they contain numerous biologically active compounds, including natural antioxidants and essential oils [6,7].

Citrus essential oils (CEOs) have been utilized for therapeutic and health purposes in numerous cultures since ancient times for antibacterial, antiviral, antifungal, anticarcinogenic, antimutagenic, anti-inflammatory, and antioxidant properties [8]. The numerous health benefits linked to the use of CEO have been well-documented. In addition to medicinal and health applications, CEO is increasingly being used in the food sector, food packaging, and agriculture. Synthetic chemical compounds that are more hazardous or to which pests, bacteria, or fungi have proven resistance are being replaced with essential oils [9]. Essential oils have been shown to be more effective than chemical preservatives in reducing pathogen growth and delaying food spoiling in several studies. *Citrus trifoliata* L. essential oil's insecticidal and fungicidal activities against *Spodoptera littoralis*, *Fusarium oxysporum*, and *Fusarium solani* are examples of citrus essential oils' potential activity [10,11]. Furthermore, they are free of the negative health hazards connected with synthetic pesticides and fungicides.

Despite their promising features, essential oil-based insecticides have significant limitations linked to their chemical nature (e.g., volatility, poor water solubility, and environmental degradation) [12], which can limit their application. Because of the small size of the particles, encapsulating essential oils inside nanoparticles could alleviate these issues by further enhancing efficacy [11,13]. Essential oils as well as the other botanical products have become a prevalent option for searching new fungicidal agents to be incorporated into the fungus management programs. Nanoencapsulation of fungicidal essential oils would optimize the fungus control system through the effective protection of active ingredients, reducing the need for high doses, the toxicity, and offering protection against environmental degradation and loss. Nanoencapsulation offers significant potential for increasing agricultural productivity while reducing environmental and human health impacts [14,15].

Herein, the fungicidal activity of the essential oil obtained from the peels of *Citrus lemon*, *Citrus aurantifolia*, *Citrus maxima*, and *Citrus sinensis* were assessed against seven pathogenic fungi that have been reported as pathogenic to humans too [16–19], namely *Rhizoctonia solani*, *Sclerotium rolfsii*, *Fusarium solani*, *Fusarium oxysporum*, *Fusarium semtectium*, *Botrytis cinerea*, and *Alternaria alternata*. In the presented study, a nano-based valorization approach is reported to establish a novel hexosomal system containing the most active essential oil, allowing for an enhancement of the efficiency of natural fungicides, reducing costs and increasing competitiveness.

2. Results and Discussion

2.1. Chemical Composition of Different Citrus Peels' Essential Oils

The results obtained by GC-MS analysis of *C. lemon*, *C. aurantifolia*, *C. maxima*, and *C. sinensis* peel essential oils are presented in Table 1. Twenty-four compounds were identified, constituting 97.66%, 94.93%, 99.99%, and 99.93% of the citrus peels essential oils under investigation, respectively. Analysis of C. maxima and *C. sinensis* oil revealed that most of the compounds were monoterpenes, where limonene *represented* the major one, with the percentages (97.51% and 96.71%) in agreement with previously published data [11,20]. *C. lemon* and *C. aurantifolia* essential oils' composition showed that the percentage of monoterpenes is 64.13% and 80.34%, respectively. The oxygenated hydrocarbon composition is higher in *C. lemon* and *C. aurantifolia*, with percentages of 33.53% and 14.59%, respectively. High *concentrations* of oxygenated hydrocarbons in *C. lemon* and *C. aurantifolia* could be attributed to the strong antifungal activities of these oils [21]. They could synergistically increase the effect of limonene and other monoterpene hydrocarbons [22].

Table 1. Essential oil composition of *C. lemon*, *C. aurantifolia*, *C. maxima*, and *C. sinensis*.

No.	RT	RI	Identified Compounds	Area Percentage			
				C. lemon	*C. aurantifolia*	*C. maxima*	*C. sinensis*
1	7.344	948	α-Pinene	0.83	0.72	0.43	0.29
2	7.772	955	Camphene	0.23	–	–	–
3	8.643	982	β-Pinene	11.15	7.63	0.49	–
4	9.086	991	β-Myrcene	0.87	0.79	1.24	1.05
5	10.126	998	p-Cymene	1.14	0.88	–	–
6	10.360	1018	D-Limonene	44.36	61.89	97.51	96.71
7	11.189	1063	γ-Terpinene	2.91	7.50	–	–
8	12.098	1079	α-terpinolene	1.53	0.93	–	–
9	12.471	1082	Linalool	1.69	2.14	0.09	1.53
10	12.615	1097	Nonanal	0.17	–	–	–
11	12.905	1098	Fenchol	0.39	0.14	–	–
12	13.856	1121	Camphor	–	–	–	0.24
13	14.579	1125	p-Mentha-1,5-dien-8-ol	1.02	0.33	–	–
14	14.878	1137	Terpinen-4-ol	2.79	2.03	–	–
15	15.307	1143	α-Terpineol	9.95	6.43	0.23	0.11
16	16.793	1174	β-Citral	13.51	2.52	–	–
17	16.904	1190	Carvone	3.53	0.68	–	–
18	17.198	1228	cis-Geraniol	0.48	–	–	–
19	17.815	1268	Perillaldehyde	–	0.32	–	–
20	20.895	1339	β-Bourbonene	0.13	–	–	–
21	21.845	1410	Caryophyllene	0.13	–	–	–
22	23.475	1515	Germacrene D	0.1	–	–	–
23	24.115	1518	cis-α-Bisabolene	0.64	–	–	–
24	25.439	1603	Germacrene B	0.11	–	–	–
Percentage of identified constituents				97.66	94.93	99.99	99.93
Percentage of identified hydrocarbons				64.13	80.34	99.67	98.05
Percentage of oxygenated hydrocarbons				33.53	14.59	0.324	1.88

2.2. Preparation and Characterization of Nano-Hexosomal Dispersions

The hexosomal system with its unique architecture is a promising nanoplatform that is capable of being loaded with hydrophobic and hydrophilic active molecules. In the current study, the hexosomal system was successfully prepared by the hot emulsification method. The melted lipids (glyceryl monooleate (GMO)/oleic acid together with volatile oil) were emulsified into an aqueous phase containing a stabilizer (Pluronic F127) at 70 °C. The presence of oleic acids changes the GMO effective packing parameter that induces the phase transition from the cubic phase (cubosomes) to the hexagonal phase (hexosomes). The prepared volatile oil-loaded hexosomal system showed a particle size (PS) in the nano-range (210.35 ± 3.18 nm) and acceptable polydispersity index (PDI) values (0.31 ± 0.0.6). Moreover, the prepared system demonstrated a negative and zeta potential (ZP) value (−16 ± 0.84 mV), as shown in Figure 1. This negative surface charge is probably attributed to the presence of a carboxylic end group in the fatty acid, oleic acid [23]. A representative transmission electron microscope (TEM) of the volatile oil-loaded hexosomal system is illustrated in Figure 2, where nearly hexagonal nonaggregated particles were observed.

2.3. Antifungal Activity of Different Citrus Oil and Nano-Hexosomal Formula of the Most Active One

Different citrus peel essential oils (*C. lemon*, *C. aurantifolia*, *C. maxima*, and *C. sinensis*) were assessed against different phytopathogenic fungi (*Rhizoctonia solani*, *Sclerotium rolfsii*, *Fusarium solani*, *Fusarium oxysporum*, *Fusarium semtectium*, *Botrytis cinerea*, and *Alternaria alternata*). All tested CEOs exerted strong antifungal activity in a dose-dependent manner where percentage inhibition increased with higher doses, as illustrated in Figure 3. *C. lemon* and *C. aurantifolia* essential oils with a concentration of 100 μL/mL showed complete inhibition of *S. rolfsii*, *F. solani*, and *F. oxysporum* mycelium growth, as shown in Figure 3.

The lowest IC_{50} was recorded with *F. oxysporum* (36.92, 41.72), *S. rolfsii* (37.59, 45.60), and *F. solani* (42.17, 50.79 µL/mL) in case of *C. lemon* and *C. aurantifolia*, respectively. On the other hand, *B. cinerea* was more tolerant to *C. lemon* with a high IC_{50} (78.60 µL/mL) compared to the other fungi. From the above results, it is evident that *C. lemon* and *C. aurantifolia* exert potent fungi toxic effects.

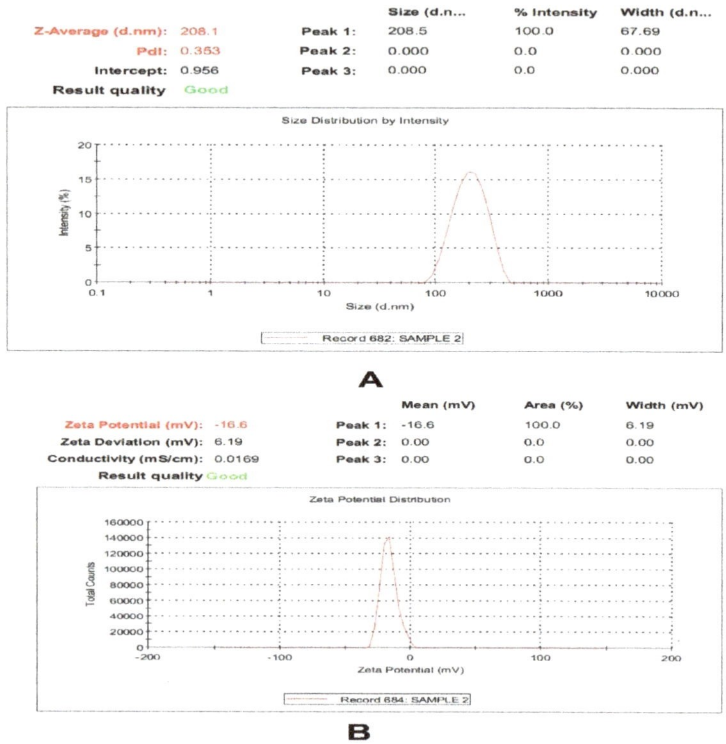

Figure 1. Particle size distribution and zeta potential of *Citrus lemon* essential nanohexosomes.

The results in Table 2 and Figure 3 indicate that the treatment with *C. maxima* and *C. sinensis* reduced the linear growth of all tested phytopathogenic fungi at concentrations of 100, 200, 300, and 400 µL/mL. They exerted complete inhibition of mycelium growth for *S. rolfsii, R. solani*, and *F. oxysporum* with a concentration of 400 µL/mL, as represented in Figure 3. *R. solani* was more resistant to *C. maxima* and *C. sinensis*, the IC_{50} of which was the lowest concentrations (191.1 and 180 µL/mL), respectively, compared to the other fungi. On the other hand, *F. solani* was more tolerant to *C. maxima* and *C. sinensis*, the IC_{50} of which was highest concentrations (395.9 and 457.9 µL/mL), respectively, compared to the other fungi. Their potent antifungal activity could be related to the high concentration of monoterpenes, especially limonene [20,24]. CEOs showed a strong fungi toxic effect through damage and loss of integrity, the rigidity of the cell wall and retraction of cytoplasm in the hyphae, and finally, death of the mycelium as reported in previous studies [25,26]. *C. lemon* essential oil is the only one that inhibited the mycelial growth of Alternaria alternata, with IC_{50} of 229.1 µL/mL. *C. lemon* essential oil is the most active oil with a high concentration of oxygenated compounds that could synergize the antifungal effect of monoterpene hydrocarbons [27]. The nano-hexosomal formula was prepared from the most active one (*C. lemon* essential oil), to boost its antifungal activity and allow the production of a natural potent pesticide drug against dangerous phytopathogenic fungi that affect many commercial crops and plants.

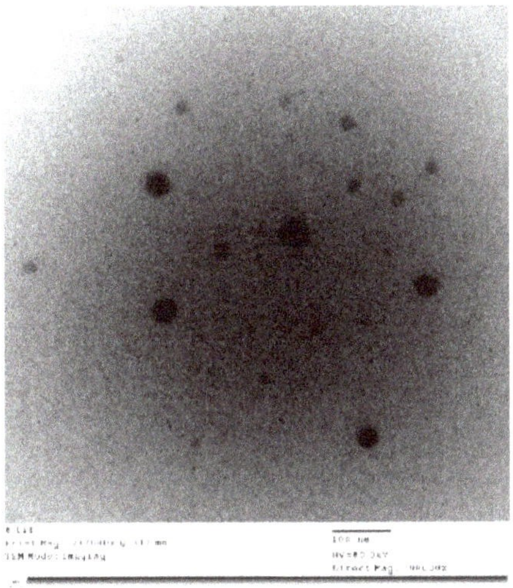

Figure 2. Transmission electron micrographs of different nano-hexosomal dispersions of *C. lemon*.

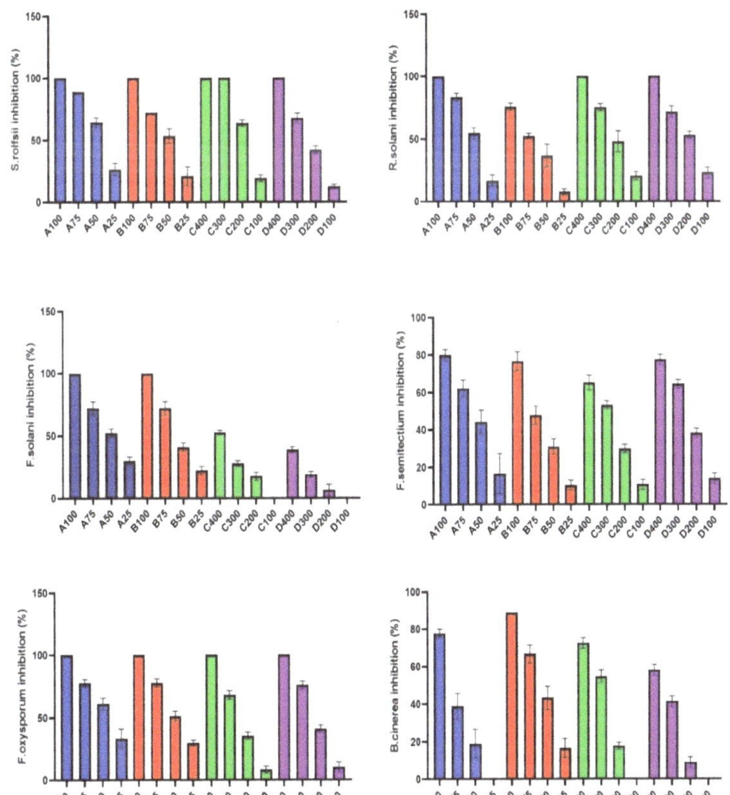

Figure 3. Mycelial growth % inhibition with different concentrations (μL/mL) of citrus essential oils (A: *C. lemon*, B: *C. aurantifolia*, C: *C. maxima*, D: *C. sinensis*) against different phytopathogenic fungi.

Table 2. IC_{50} of Citrus essential oils and hexosomal dispersion of *C. lemon* against different phytopathogenic fungi.

Plant Oil	IC_{50} (μL/mL)						
	Rhizoctonia solani	*Sclerotium rolfsii*	*Fusarium solani*	*Fusarium oxysporum*	*Fusarium semtectium*	*Botrytis cinerea*	*Alternarai alternata*
Citrus lemon	45.29	37.59	42.17	36.92	55.92	78.60	229.10
Citrus aurantifolia	66.52	45.60	50.79	41.72	70.03	53.56	0
Citrus maxima	191.10	159.00	395.90	232.70	290.70	294.80	0
Citrus sinensis	180.50	217.40	457.90	216.20	236.70	351.40	0
Citrus lemon nano-hexoosome	416.00	324.90	193.70	124.30	534.00	549.40	95.54

The treatment *potato dextrose agar* (PDA) medium with nano-hexosomes of *C. lemon* essential oil at different concentrations showed a potent antifungal effect against different phytopathogenic fungi under investigation, as illustrated in Figure 4. It was the most effective in inhibiting the mycelial growth of F. solani, which reached 100% inhibition at 600 μL/mL, while at 800 μL/mL, it completely inhibited the mycelial growth in the case of *S. rolfsii*, *F. oxysporum*, and A. alternata. Nano-hexosomes have a moderate effect on *R. solani*, *B. cinerea*, and *F. semitectium*, with IC_{50} of 416, 549.4, and 534 μL/mL, respectively. Nano-hexosomes of *C. lemon* essential oil showed potent activity against A. alternata, with IC_{50} of 95.54 μL/mL. Although CEOs did not inhibit A. alternata growth, only *C. lemon* weakly inhibited it, as shown in Figure 5. The current study proved that nano-hexosomes are among the most cost-effective drug-loaded lipid-based system that delivers essential oils in their bioactive form. CEO represent only 10% of the nano-hexosomes formula, so a low amount of oil is needed; thus, it is commercially useful to use this formula instead of chemical pesticides that show toxicity and harm to humans. Essential oils are composed of volatile constituents and undergo enzymatic reactions that decrease their activity and limit essential oil use [28]. Encapsulation of essential oil in a drug delivery system via nanotechnology overcomes the previously mentioned problems and improve its stability, bioavailability, and biological activities [11,28]. So, the CEOs nanohexosomes in our study potentiate the antifungal activity of the oil.

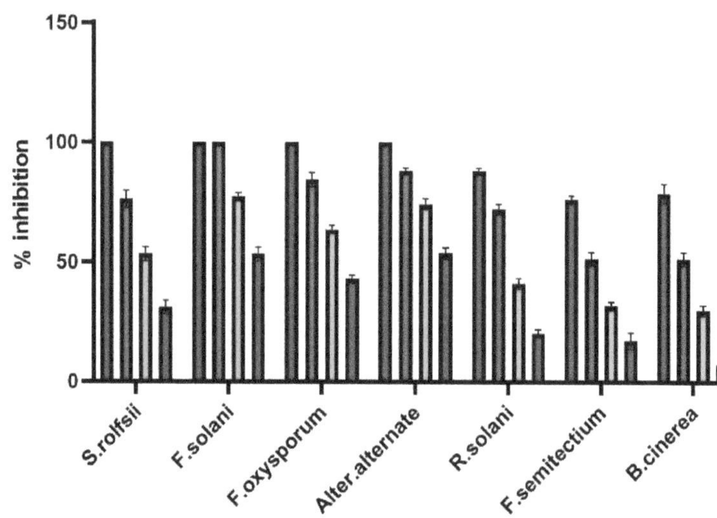

Figure 4. Mycelial growth % inhibition of *Citrus lemon* nanohexosomal formula.

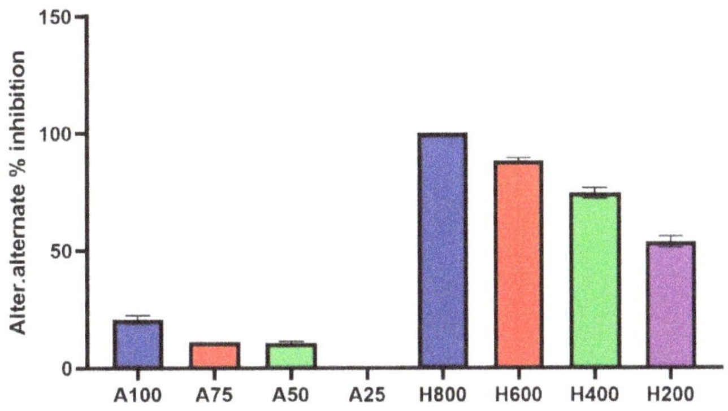

Figure 5. Mycelial growth % inhibition of *Citrus lemon* essential oil and its nano-hemosome formula.

3. Materials and Methods

3.1. Materials

Fruits of *Citrus lemon, Citrus aurantifolia, Citrus maxima,* and *Citrus sinensis* were collected from Horticulture Research Institute, Giza, Egypt. The plant's authenticity was generously validated by Mrs. Therese Labib, Botanical Specialist and consultant at Orman and Qubba Botanical Gardens, Giza, Egypt. Glyceryl monooleate (GMO), oleic acid, and Pluronic F127 (MW: 12,600 Da) were purchased from Sigma- Aldrich.

3.2. Methods

Preparation of the essential oil from fruits.

Peels of the fruits were washed, dried, and powdered mechanically. Their essential oils were extracted by hydro-distillation in a Clevenger's apparatus for 5 h according to the procedure described in the Egyptian Pharmacopeia [29]. The essential oils were dried with anhydrous sodium sulphate and stored in amber glass vials at 4 °C for use in further chemical and biological studies.

3.2.1. GC-MS Analysis and Quantification

Mass spectra were recorded using Shimadzu GCMS-QP2010 (Tokyo, Japan) equipped with a Rtx-5MS fused bonded column (30 m × 0.25 mm i.d. × 0.25 µm film thickness) (Restek, Bellefonte, PA, USA) equipped with a split–splitless injector. The capillary column was coupled to a quadrupole mass spectrometer (SSQ 7000; Thermo-Finnigan, Bremen, Germany). The initial column temperature was held at 45 °C for 2 min (isothermal) and programmed to 300 °C at a rate of 5 °C/min and kept constant at 300 °C for 5 min (isothermal). The injector temperature was 250 °C. The helium carrier gas flow rate was 1.41 mL/min. All the mass spectra were recorded applying the following conditions: (equipment current) filament emission current, 60 mA; ionization voltage, 70 eV; ion source, 200 °C. Diluted samples (1% *v*/*v*) were injected with split mode (split ratio 1: 15). The sample (1 µL) was injected automatically into the chromatograph using an AOC-20i auto-sampler. Volatile components were deconvoluted using AMDIS software (www.amdis.net) and identified by its mass spectrum matching to the database and with authentic standards (when available) [30].

3.2.2. Preparation of Hexosomal Dispersion

Hexosomal dispersions loaded with volatile oils were prepared using a hot emulsification method as described by Abdel-Bar et al. [31], with slight modifications. In brief, GMO (1 g), oleic acid (0.5 g), and volatile oil (1 g) were weighed accurately in a glass vial and allowed to melt at 70 °C. Pluronic F127 (0.5 g) was dissolved in deionized water (7 g) at the same temperature. The molten lipid phase was slowly added into the aqueous phase and homogenized for 5 min at 70 °C. The final milky dispersion was allowed to cool gradually to room temperature. It was kept in glass vials at 2–8 °C for further studies. The final concentrations of the hexosomes' components (GMO, oleic acid, volatile oil and Pluronic F127, and deionized water) were 10, 5, 10, 5, and 70% *w*/*w*, respectively.

3.2.3. Measurement of Particle Size, Polydispersity Index, and Zeta Potential

The average particle size (PS) and size distribution expressed as the polydispersity index (PDI) were estimated by dynamic light scattering (DLS) at 25 °C using a Zetasizer Nano ZS (Malvern Instruments, Malvern, UK) [32,33]. The zeta potential (ZP) was determined by the same instrument [34,35]. Prior to performing the determinations, all dispersions were appropriately diluted (100 times) using distilled water. Triplicate measurements were always obtained for each determined response [36,37].

3.2.4. Transmission Electron Microscopy (TEM)

The morphology of the volatile oil-loaded hexosomal system was envisioned via TEM (Joel JEM 1230, Tokyo, Japan). A copper grid was loaded with the diluted dispersion, which was subjected to negative staining with aqueous solution of phosphotungstic acid (2% *w*/*v*) for a duration of 5 min. Drying of the grid at ambient temperature for 10 min was then followed prior to visualization under a transmission electron microscope [35].

3.2.5. Fungal Strains

Cultures of plant pathogenic fungi (*Rhizoctonia solani*, *Sclerotium rolfsii*, *Fusarium solani*, *Fusarium oxysporum*, *Fusarium semtectium*, *Botrytis cinerea*, and *Alternaria alternata*) were provided by fungicide, Bactericide and Nematicide Department, Central Agricultural Pesticide Laboratory (CAPL). Each fungus was maintained on potato dextrose agar (PDA) and stored at 5 °C for further studies.

3.2.6. Antifungal Assay

The antifungal activity of CEOs and hexosomal dispersion of *C. lemon* was determined by the food poisoned technique [38,39]. Different concentrations of CEOs (μL/mL) were mixed with 50 mL of sterilized PDA medium and transferred equally into three petri dishes. The media was allowed to solidify. Then, a seven-day-old fungal culture disk with a 6 mm diameter was taken and inoculated to the center of the petri dishes containing plant extracts. PDA medium without plant extract served as a control. All dishes were incubated at 27 ± 2 °C and the radial growth of colonies was measured when the mycelia of the control had almost filled the petri dishes. Each test was performed in triplicate.

The fungal growth inhibition was calculated according to the treatment against the control using the following formula [40]:

$$\%\ \text{inhibition} = \text{C} - \text{T}/\text{C} \times 100 \tag{1}$$

where C is the average of three replicates of hyphal extension (mm) of the control and T is the average of three replicates of hyphal extension (mm) of plates treated with the tested material.

IC_{50} was calculated as the concentration of the tested compound that decreases the mycelial growth by half between the base and the maximum.

3.2.7. Statistical Data Interpretation

Data analysis and graphs were made using the GraphPad Prism Version 9 program. The data are expressed as an arithmetic mean, standard deviation, and 95% confidence interval for the IC_{50} parameter. P values less than or equal to 0.05 were considered statistically significant.

4. Conclusions

Results of the antifungal screening in this study support the recommendation of authors to use CEOs obtained from the peels of *C. lemon*, *C. aurantifolia*, *C. maxima*, and

C. sinensis as a first defense line against a wide spectrum of phytopathogenic fungi. The antifungal potential of *C. lemon* and *C. aurantifolia* could be attributed to the synergetic effect of their oxygenated hydrocarbons together with the limonene content. CEOs as natural fungicides are considered valuable sources for fungal control without the toxicity and the health hazards connected with synthetic fungicides. Besides, their use in crop management protocols aids in managing the residual peels generated during citrus fruit processing, which solves an environmental problem and compensates for economic losses. The introduction of the most active essential oil (*C. lemon*) into a hexosomal dispersion was performed to create a potential nano-fungicide against plant fungal pathogens. The nano-hexosomes of *C. lemon* essential oil exhibited powerful fungicidal activity that exceeded the oil itself specially against *A. alternata*. The presented results indicate that the nano-hexosomal dispersion helped to boost the antifungal properties of the oil and can be used as a natural nano-fungicide in plant pathogen control. Concisely, it is considered a potential carrier for enhancing the fungicidal activity of CEOs and can be used in integrated fungus management programs.

Author Contributions: M.S.S. and M.K.M. prepared essential oils, performed analysis and identification of essential oils, M.S.S. performed statistical analysis, M.S.S. and M.K.M. wrote the first draft of manuscript. A.M.A.-M. prepared nanohexosomes of essential oils and wrote his part. R.A.A.H. tested the antifungal activity of essential oils. A.M.A.A. and I.A.N. planned and funded the work and were responsible for co-writing and revision of the manuscript. All authors have read and agreed to the published version of the manuscript.

Funding: Taif University Researchers Supporting Project number (TURSP-2020/50), Taif University, Taif, Saudi Arabia.

Institutional Review Board Statement: Not applicable.

Informed Consent Statement: Not applicable.

Data Availability Statement: Data by authors are available upon request.

Acknowledgments: The authors would like to acknowledge the financial support offered by "Taif University Researchers Supporting Project number (TURSP-2020/50), Taif University, Taif, Saudi Arabia".

Conflicts of Interest: The authors declare no conflict of interest.

Sample Availability: No new compounds emerged in this study.

References

1. Shuping, D.S.S.; Eloff, J.N. The use of plants to protect plants and food against fungal pathogens: A review. *Afr. J. Tradit. Complement. Altern. Med.* **2017**, *14*, 120–127. [CrossRef] [PubMed]
2. Yang, J.; Hsiang, T.; Bhadauria, V.; Chen, X.-L.; Li, G. Plant Fungal Pathogenesis. *BioMed Res. Int.* **2017**, *2017*, 9724283. [CrossRef] [PubMed]
3. Suprapta, D.N. Potential of microbial antagonists as biocontrol agents against plant fungal pathogens. *J. ISSAAS* **2012**, *18*, 1–8.
4. Liu, Y.; Heying, E.; Tanumihardjo, S.A. History, global distribution, and nutritional importance of citrus fruits. *Compr. Rev. Food Sci. Food Saf.* **2012**, *11*, 530–545. [CrossRef]
5. Balu, A.M.; Budarin, V.; Shuttleworth, P.S.; Pfaltzgraff, L.A.; Waldron, K.; Luque, R.; Clark, J.H. Valorisation of orange peel residues: Waste to biochemicals and nanoporous materials. *ChemSusChem* **2012**, *5*, 1694. [CrossRef]
6. Sharma, K.; Mahato, N.; Cho, M.H.; Lee, Y.R. Converting citrus wastes into value-added products: Economic and environmently friendly approaches. *Nutrition* **2017**, *34*, 29–46. [CrossRef]
7. Mamma, D.; Christakopoulos, P. Biotransformation of citrus by-products into value added products. *Waste Biomass Valorization* **2014**, *5*, 529–549. [CrossRef]
8. Djilani, A.; Dicko, A. The therapeutic benefits of essential oils. *Nutr. Well-Being Health* **2012**, *7*, 155–179.
9. Lammari, N.; Louaer, O.; Meniai, A.H.; Elaissari, A. Encapsulation of essential oils via nanoprecipitation process: Overview, progress, challenges and prospects. *Pharmaceutics* **2020**, *12*, 431. [CrossRef]
10. Fisher, K.; Phillips, C. Potential antimicrobial uses of essential oils in food: Is citrus the answer? *Trends Food Sci. Technol.* **2008**, *19*, 156–164. [CrossRef]
11. Abdel-Kawy, M.A.; Michel, C.G.; Kirollos, F.N.; Hussien, R.A.A.; Al-Mahallawi, A.M.; Sedeek, M.S. Chemical composition and potentiation of insecticidal and fungicidal activities of *Citrus trifoliata* L. fruits essential oil against Spodoptera littoralis, Fusarium oxysporum and Fusarium solani via nano-cubosomes. *Nat. Prod. Res.* **2021**, *35*, 2438–2443. [CrossRef] [PubMed]

12. Prakash, B.; Kujur, A.; Yadav, A.; Kumar, A.; Singh, P.P.; Dubey, N.K. Nanoencapsulation: An efficient technology to boost the antimicrobial potential of plant essential oils in food system. *Food Control* **2018**, *89*, 1–11. [CrossRef]
13. Campolo, O.; Cherif, A.; Ricupero, M.; Siscaro, G.; Grissa-Lebdi, K.; Russo, A.; Cucci, L.M.; Di Pietro, P.; Satriano, C.; Desneux, N. Citrus peel essential oil nanoformulations to control the tomato borer, *Tuta absoluta*: Chemical properties and biological activity. *Sci. Rep.* **2017**, *7*, 1–10. [CrossRef] [PubMed]
14. Christofoli, M.; Costa, E.C.C.; Bicalho, K.U.; de Cássia Domingues, V.; Peixoto, M.F.; Alves, C.C.F.; Araújo, W.L.; de Melo Cazal, C. Insecticidal effect of nanoencapsulated essential oils from *Zanthoxylum rhoifolium* (Rutaceae) in Bemisia tabaci populations. *Ind. Crop. Prod.* **2015**, *70*, 301–308. [CrossRef]
15. De Oliveira, J.L.; Campos, E.V.R.; Bakshi, M.; Abhilash, P.C.; Fraceto, L.F. Application of nanotechnology for the encapsulation of botanical insecticides for sustainable agriculture: Prospects and promises. *Biotechnol. Adv.* **2014**, *32*, 1550–1561. [CrossRef] [PubMed]
16. Kaore, N.M.; Atul, A.R.; Khan, M.Z.; Ramnani, V.K. A rare case of human mycosis by Rhizoctonia solani. *Indian J. Med. Microbiol.* **2012**, *30*, 361. [CrossRef]
17. Zhang, N.; O'Donnell, K.; Sutton, D.A.; Nalim, F.A.; Summerbell, R.C.; Padhye, A.A.; Geiser, D.M. Members of the *Fusarium solani* species complex that cause infections in both humans and plants are common in the environment. *J. Clin. Microbiol.* **2006**, *44*, 2186–2190. [CrossRef]
18. Wang, C.-J.; Thanarut, C.; Sun, P.-L.; Chung, W.-H. Colonization of human opportunistic *Fusarium oxysporum* (HOFo) isolates in tomato and cucumber tissues assessed by a specific molecular marker. *PLoS ONE* **2020**, *15*, e0234517. [CrossRef]
19. Cody, D.T.; McCaffrey, T.V.; Roberts, G.; Kern, E.B. Effects of *Aspergillus fumigatus* and *Alternaria alternata* on human ciliated epithelium in vitro. *Laryngoscope* **1997**, *107*, 1511–1514. [CrossRef]
20. Singh, P.; Shukla, R.; Prakash, B.; Kumar, A.; Singh, S.; Mishra, P.K.; Dubey, N.K. Chemical profile, antifungal, antiaflatoxigenic and antioxidant activity of *Citrus maxima Burm.* and *Citrus sinensis* (L.) Osbeck essential oils and their cyclic monoterpene, DL-limonene. *Food Chem. Toxicol.* **2010**, *48*, 1734–1740. [CrossRef]
21. Ben Salha, G.; Herrera Díaz, R.; Lengliz, O.; Abderrabba, M.; Labidi, J. Effect of the chemical composition of free-terpene hydrocarbons essential oils on antifungal activity. *Molecules* **2019**, *24*, 3532. [CrossRef]
22. Haj Ammar, A.; Bouajila, J.; Lebrihi, A.; Mathieu, F.; Romdhane, M.; Zagrouba, F. Chemical composition and in vitro antimicrobial and antioxidant activities of *Citrus aurantium* L. flowers essential oil (*Neroli* oil). *Pak. J. Biol. Sci.* **2012**, *15*, 1034–1040. [CrossRef]
23. Al-Mahallawi, A.M.; Abdelbary, A.A.; El-Zahaby, S.A. Norfloxacin loaded nano-cubosomes for enhanced management of otitis externa: In vitro and in vivo evaluation. *Int. J. Pharm.* **2021**, *600*, 120490. [CrossRef] [PubMed]
24. Zohra, H.F.; Rachida, A.; Malika, M.; Benali, S.; Samir, A.A.; Meriem, B. Chemical composition and antifungal activity of essential oils of Algerian citrus. *Afr. J. Biotechnol.* **2015**, *14*, 1048–1055. [CrossRef]
25. Sharma, N.; Tripathi, A. Effects of *Citrus sinensis* (L.) Osbeck epicarp essential oil on growth and morphogenesis of *Aspergillus niger* (L.) Van Tieghem. *Microbiol. Res.* **2008**, *163*, 337–344. [CrossRef] [PubMed]
26. Viuda-Martos, M.; Ruiz-Navajas, Y.; Fernández-López, J.; Pérez-Álvarez, J. Antifungal activity of lemon (*Citrus lemon* L.), mandarin (*Citrus reticulata* L.), grapefruit (*Citrus paradisi* L.) and orange (*Citrus sinensis* L.) essential oils. *Food Control* **2008**, *19*, 1130–1138. [CrossRef]
27. Rostro-Alanis, M.d.J.; Báez-González, J.; Torres-Alvarez, C.; Parra-Saldívar, R.; Rodriguez-Rodriguez, J.; Castillo, S. Chemical composition and biological activities of oregano essential oil and its fractions obtained by vacuum distillation. *Molecules* **2019**, *24*, 1904. [CrossRef] [PubMed]
28. Cimino, C.; Maurel, O.M.; Musumeci, T.; Bonaccorso, A.; Drago, F.; Souto, E.M.B.; Pignatello, R.; Carbone, C. Essential oils: Pharmaceutical applications and encapsulation strategies into lipid-based delivery systems. *Pharmaceutics* **2021**, *13*, 327. [CrossRef]
29. *Egyptian Pharmacopoeia*; Central Administration and Pharmaceutical Affairs, Ministry of Health and Population: Cairo, Egypt, 2005.
30. Abdel-Kawy, M.A.; Michel, C.G.; Kirollos, F.N.; El, A.M. In vitro MAO-B Inhibitory Effects of Citrus trifoliata L. Fruits Extract, Self-Nano-Emulsifying Drug Delivery System and Isolated Hesperidin: Enzyme Assay and Molecular Docking Study. *Development* **2020**, *14*, 15.
31. Abdel-Bar, H.M.; Khater, S.E.; Ghorab, D.M.; Al-Mahallawi, A.M. Hexosomes as efficient platforms for possible fluoxetine hydrochloride repurposing with improved cytotoxicity against HepG2 cells. *ACS Omega* **2020**, *5*, 26697–26709. [CrossRef]
32. Khate, S.E.; El-Khouly, A.; Abdel-Ba, H.M.; Al-Mahallawi, A.M.; Ghorab, D.M. Fluoxetine hydrochloride loaded lipid polymer hybrid nanoparticles showed possible efficiency against SARS-CoV-2 infection. *Int. J. Pharm.* **2021**, *607*, 121023. [CrossRef]
33. Ghareeb, D.A.; Saleh, S.R.; Seadawy, M.G.; Nofal, M.S.; Abdulmalek, S.A.; Hassan, S.F.; Khedr, S.M.; AbdElwahab, M.G.; Sobhy, A.A.; Yassin, A.M. Nanoparticles of ZnO/Berberine complex contract COVID-19 and respiratory co-bacterial infection in addition to elimination of hydroxychloroquine toxicity. *J. Pharm. Investig.* **2021**, 1–23. [CrossRef]
34. Ahmed, M.A.; Al-Mahallawi, A.M.; El-Helaly, S.N.; Abd-Elsalam, W.H. The effect of the saturation degree of phospholipid on the formation of a novel self-assembled nano-micellar complex carrier with enhanced intestinal permeability. *Int. J. Pharm.* **2019**, *569*, 118567. [CrossRef]
35. Al-Mahallawi, A.M.; Fares, A.R.; Abd-Elsalam, W.H. Enhanced permeation of methotrexate via loading into ultra-permeable niosomal vesicles: Fabrication, statistical optimization, ex vivo studies, and in vivo skin deposition and tolerability. *AAPS PharmSciTech* **2019**, *20*, 1–10. [CrossRef]

36. Albash, R.; El-Nabarawi, M.A.; Refai, H.; Abdelbary, A.A. Tailoring of PEGylated bilosomes for promoting the transdermal delivery of olmesartan medoxomil: In-vitro characterization, ex-vivo permeation and in-vivo assessment. *Int. J. Nanomed.* **2019**, *14*, 6555. [CrossRef] [PubMed]
37. Saber, M.M.; Al-mahallawi, A.M.; Stork, B. Metformin dampens cisplatin cytotoxicity on leukemia cells after incorporation into cubosomal nanoformulation. *Biomed. Pharmacother.* **2021**, *143*, 112140. [CrossRef]
38. Ramaiah, A.K.; Garampalli, R.K.H. In vitro antifungal activity of some plant extracts against *Fusarium oxysporum* f. sp. lycopersici. *Asian J. Plant Sci. Res.* **2015**, *5*, 22–27.
39. Pawar, D.S.; Nasreen, S. HR-LCMS of phytoconstituents and antifungal activity of medicinal plants. *J. Med. Plants* **2018**, *6*, 173–176.
40. Gopalakrishnan, S.; Vadlamudi, S.; Bandikinda, P.; Sathya, A.; Vijayabharathi, R.; Rupela, O.; Kudapa, H.; Katta, K.; Varshney, R.K. Evaluation of Streptomyces strains isolated from herbal vermicompost for their plant growth-promotion traits in rice. *Microbiol. Res.* **2014**, *169*, 40–48. [CrossRef]

molecules

MDPI

Article

Phyto-Extract-Mediated Synthesis of Silver Nanoparticles Using Aqueous Extract of *Sanvitalia procumbens*, and Characterization, Optimization and Photocatalytic Degradation of Azo Dyes Orange G and Direct Blue-15

Madeeha Aslam [1], Fozia Fozia [2], Anadil Gul [3], Ijaz Ahmad [1,*], Riaz Ullah [4,*], Ahmed Bari [5], Ramzi A. Mothana [4] and Hidayat Hussain [6]

1 Department of Chemistry, Kohat University of Science & Technology, Kohat 26000, Pakistan; madeehaaslam06@gmail.com
2 Biochemistry Department, KMU Institute of Medical Sciences, Kohat 26000, Pakistan; drfoziazeb@yahoo.com
3 Beijing Key Laboratory for Green Catalysis and Separation, Department of Chemistry and Chemical Engineering, Beijing University of Technology, Beijing 100124, China; anadilgul97@gmail.com
4 Department of Pharmacognosy, College of Pharmacy, King Saud University, P.O. Box 2457, Riyadh 11451, Saudi Arabia; rmothana@ksu.edu.sa
5 Department of Pharmaceutical Chemistry, College of Pharmacy, King Saud University, P.O. Box 2457, Riyadh 11451, Saudi Arabia; abari@ksu.edu.sa
6 Department of Bioorganic Chemistry, Leibniz Institute of Plant Biochemistry, Weinberg 3, D-06120 Halle, Germany; Hidayat.Hussain@ipb-halle.de
* Correspondence: drijaz_chem@yahoo.com (I.A.); rullah@ksu.edu.sa (R.U.)

Citation: Aslam, M.; Fozia, F.; Gul, A.; Ahmad, I.; Ullah, R.; Bari, A.; Mothana, R.A.; Hussain, H. Phyto-Extract-Mediated Synthesis of Silver Nanoparticles Using Aqueous Extract of *Sanvitalia procumbens*, and Characterization, Optimization and Photocatalytic Degradation of Azo Dyes Orange G and Direct Blue-15. *Molecules* **2021**, *26*, 6144. https://doi.org/10.3390/molecules26206144

Academic Editors: Victoria Samanidou, Eleni Deliyanni and Dimitra Voutsa

Received: 19 September 2021
Accepted: 7 October 2021
Published: 12 October 2021

Abstract: Green synthesis of silver nanoparticles (AgNPs) employing an aqueous plant extract has emerged as a viable eco-friendly method. The aim of the study was to synthesize AgNPs by using plant extract of *Sanvitalia procumbens* (creeping zinnia) in which the phytochemicals present in plant extract act as a stabilizing and reducing agent. For the stability of the synthesized AgNPs, different parameters like $AgNO_3$ concentration, volume ratios of $AgNO_3$, temperature, pH, and contact time were studied. Further, AgNPs were characterized by UV–visible spectroscopy, FT-IR (Fourier Transform Infrared Spectroscopy), XRD (X-ray Diffraction), SEM (Scanning Electron Microscopy), and EDX (Energy Dispersive X-ray Spectrometer) analysis. FT-IR analysis showed that the plant extract contained essential functional groups like O–H stretching of carboxylic acid, N–H stretching of secondary amides, and C–N stretching of aromatic amines, and C–O indicates the vibration of alcohol, ester, and carboxylic acid that facilitated in the green synthesis of AgNPs. The crystalline nature of synthesized AgNPs was confirmed by XRD, while the elemental composition of AgNPs was detected by energy dispersive X-ray analysis (EDX). SEM studies showed the mean particle diameter of silver nanoparticles. The synthesized AgNPs were used for photocatalytic degradation of Orange G and Direct blue-15 (OG and DB-15), which were analyzed by UV-visible spectroscopy. Maximum degradation percentage of OG and DB-15 azo dyes was observed, without any significant silver leaching, thereby signifying notable photocatalytic properties of AgNPs.

Keywords: green synthesis; *Sanvitalia procumbens*; silver nanoparticles; azo dyes degradation

1. Introduction

Metal nanoparticles have a variety of features in various fields. Since the sizes, dimensions, and compositions of metallic nanomaterials are all tied to their optical, physical and chemical properties, nanoscale materials have already been used in a wide range of applications [1]. Nanostructured metallic nanoparticles with extraordinary tunable complex surface plasmon, optoelectronics, biomedical, and catalytic properties were created owing to their high surface area to volume ratio and limited effort to the surface functionalization [2]. Despite the fact that many metals occur in nature, only elements such as silver, palladium, platinum and gold are widely synthesized for their nanostructured aspect [3]. Silver nanoparticles (AgNPs), among the metals listed above, have drawn a great deal of interest because of their unique properties for use in the textile industries, agriculture, water detoxification, pharmaceutics, and as a catalyst in oxidation reactions and air filtration [4–7]. Chemical, physical, electrochemical, irradiative, photochemical, and biological methods are now available for the synthesis of silver nanoparticles [8,9]. Since most of these methods involve the use of hazardous chemicals and extreme reaction conditions, this often results in chemical toxicity and environmental contamination. The detrimental effects of chemical processes, as well as the growing focus on green chemistry, have increased the importance of biological methods for the synthesis of AgNPs [10,11]. Because of their versatility, organic design, and profitability, biomediated methods are considered a viable alternative to physical and chemical methods. Several groups have reported on the synthesis of silver nanoparticles by using different plants extract like *Andrographis paniculata* [12], *Cathranthus roseus* [13] and *Rhynchosia suaveolens* [14], which act as a reducing and stabilizing agent.

Recently, silver nanoparticles (AgNPs) have risen in prominence due to their wide range of applications in biomedicine such as anticancer [15], antibacterial [16], and antioxidant activity [17]. AgNPs toxicity has been linked to an increase in the synthesis of reactive oxygen species (ROS), which causes the release of silver ions to occur [18]. ROS development may induce oxidative stress, which is basically the reported mechanism of AgNP's toxicity, by lowering superoxide dismutase (SOD) and glutathione (GSH) levels and raising the lipid peroxidation levels in cells [19,20]. Nowadays, textile and synthetic dyes cause pollution that threaten living organisms and also contribute to changes in the climate. One of the possible strategies for achieving meaningful remediation is the catalytic degradation of synthetic pollutants by nanoparticles. Plant-mediated nanoparticles have proven to be effective photocatalytic agents in the fight against organic pollutants [21]. Numerous techniques like oxidation, adsorption, ion-exchange, chemical coagulation, and biological photocatalysis are available to overcome this issue [22]. AgNPs from *Phaseolus vulgaris* have recently been shown to degrade 4-nitrophenol, reactive red 141 dyes [23], and other fabricated AgNPs designed from *Solanum surattense,* which are utilized as an efficient catalyst towards Rhodamine B (RhB) [24]. Green synthesis of AgNPs has a wide range of applications for dye removal in a practical operating system that produces less sludges than other methods.

In the light of other discrepancies, different biological methods have reported on the synthesis of AgNPs and on the photocatalytic activities of other dyes. So, this work was carried out to synthesize AgNPs using *Sanvitalia procumbens* (*S. procumbens*) aqueous plant extract, and then to explore the photocatalytic applications of the synthesized AgNPs. The synthesis of AgNPs using aqueous extract of this plant and its photocatalytic degradation of azo dyes (Orange G and Direct blue-15) has not been reported in literature, and in this article we are reporting it for the first time. According to an ethnobotanical survey of weeds in the state of Mexico, *S. procumbens* (ornamental perpetually flowering plant) is one of the major plant families of commercial importance for medicinal, herbal, culinary, and ornamental uses [25]. Thus, the medicinally essential *S. procumbens* aqueous plant extract has since been used as a possible bioreducing agent for the green synthesis of AgNPs in such a cost-effective and environmentally sustainable manner. In light of the foregoing, the current study was conducted to produce low-cost, clear, and easily reusable AgNPs from *S. procumbens* that contains bioreductants and stabilizers such as alkamides, phenolics, proteins, and terpenoids, which are responsible for the reduction of Ag^{+} to Ag^{0} [25–28]. *Sanvitalia procumbens*-based silver nanoparticles can be highly reproducible. Large-scale utilization assessing the reproducible process can show the natural variability that occurs in the chemical composition of plant species that grow at the month of December, as well as assess how a plant can be used for this purpose so that the entire plant extract can be gainfully utilized in the process. Furthermore, the elemental composition, geometry, morphological, and optical study of synthesized AgNPs was carried out by using various techniques such as Energy Dispersive X-ray analysis (EDX), X-ray Diffractometer, Scanning Electron Microscope (SEM), Fourier Transform Infrared Spectroscopy (FT-IR), and UV–visible spectrophotometry. Moreover, the current study also assessed potential applications in the environment via green-synthesized AgNPs for the photocatalytic degradation of hazardous Orange G and Direct Blue-15 azo dyes by considering these dyes to be biodegradable.

2. Results and Discussion

Several studies have stressed and illustrated the possible use of shade-dried plants as a reducing and stabilizing source for nanoparticle synthesis. After integrating an aqueous plant extract of *S. procumbens* with the $AgNO_3$ solution at a ratio of 1:9 (0.01 mol/L solution of $AgNO_3$), the color change indicated that the synthesis of AgNPs occurred, where the plant extract was used as a reducing and stabilizing agent in this analysis. According to the literature, the initial indication of AgNPs was observed as a color transition from pale yellow to dark yellow and eventually colloidal brown, within a few minutes. The observed results were consistent with previous reports in which AgNPs were synthesized using various plant extracts, and aqueous silver nitrate solution (10^{-3} mol/L) altered the color from transparent to brown after extract addition [29,30].

2.1. *UV-Visible Spectroscopy*

On the nanoscale level, one of the most intriguing features of metal nanoparticles is their optical properties, which vary proportionally with their size and shape. The large-size AgNPs are obtained at the weak absorbance peak and shift towards higher wavelength, while maximum absorbance peak shifts towards a lower wavelength when the particle size becomes smaller. In the current study, the absence of an absorbance peak in the visible region was examined when the plant extract and $AgNO_3$ solution was observed in a UV-visible spectrophotometer. However, the synthesis of AgNPs was confirmed by a mixture of $AgNO_3$ and plant extract. The AgNPs formation in aqueous colloidal solution showed the sharp SPR (surface plasmon resonance) peak recorded at 438 nm as shown in Figure 1. This is because plant extracts contain plenty of phytochemicals that aid in the reduction of AgNPs and serve as a capping agent [31]. Due to the coupled oscillation of electrons in light wave resonance of metal nanoparticles, particularly AgNPs, a free electron to absorb the surface plasmon resonance phenomenon is provided [32]. The achieved results are comparable to those reported in the literature, wherein the *Datura stramonium* plant extract was used to reduce silver nanoparticles, and UV-visible absorbance spectra revealed that the SPR (surface plasmon resonance) band for silver occurred in the 400–450 nm range. The reproducibility of AgNPs synthesis was confirmed using the aqueous extract of a previously collected batch of plant, which provided silver nanoparticles when the optimized ratio of plant extract and $AgNO_3$ (1:9 mL) was used [33,34].

2.1.1. Effect of $AgNO_3$ Concentration

This parameter was measured by varying $AgNO_3$ concentrations under ideal circumstances. The size and forms of AgNPs were affected by varying the concentration of $AgNO_3$ solution. In this analysis, SPR (surface plasmon resonance) peak showed that at lower $AgNO_3$ concentrations, AgNPs with larger size distributions were present, implying the presence of AgNPs with broader size distributions. The SPR peak became sharper and narrower as the concentration of $AgNO_3$ rose, along with an increase in its intensity [35]. This suggests that AgNPs with a narrow particle size distribution were involved when the concentration of $AgNO_3$ solution was increased. However, if the concentration of $AgNO_3$ further increases to 0.02 mol/L, the intensity of such SPR peak weakens and broadens due to agglomerations induced by Van der Waals interactions between adjacent nanoparticles. It was concluded that with 0.01 mol/L $AgNO_3$ solution, a distinct peak was found at 438 nm, but as stated in the literature, the concentration rises from 0.01 mol/L, and the intensity peak of synthesized nanoparticles decrease, which was also mentioned in Figure 2 [36].

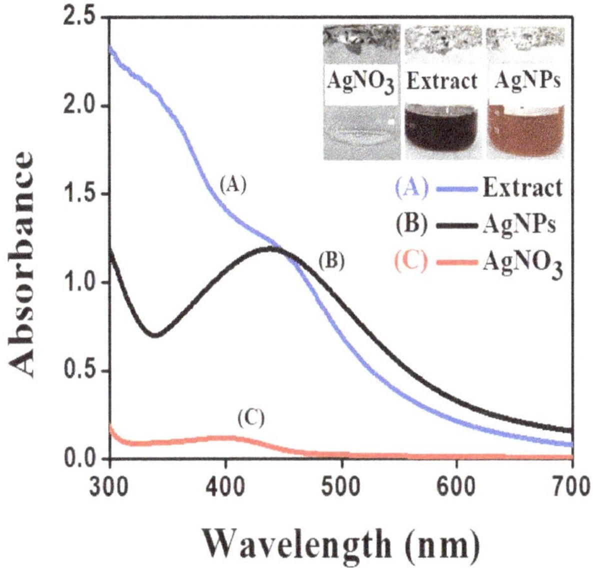

Figure 1. UV-Visible spectra of: (**A**) Pure extract of *S. procumbens*; (**B**) Synthesized AgNPs; and (**C**) 0.01 mol/L solution of $AgNO_3$.

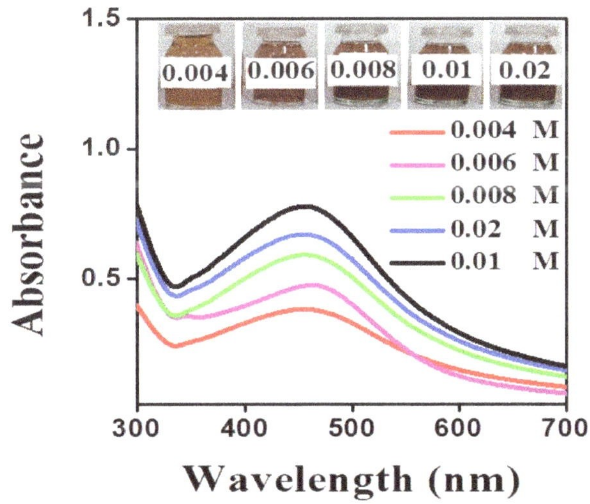

Figure 2. UV-visible spectrum of synthesized AgNPs from *S. procumbens* aqueous extract at various concentrations of $AgNO_3$.

2.1.2. Effect of $AgNO_3$ Volume to Plant Extract Ratios

To decide the impact of volume ratios on AgNPs, different extracts and $AgNO_3 v/v$ ratios were used. The transition in spectrum was observed at each ratio, as seen in Figure 3, where intensity increases as the ratios rises from 1:3 to 1:9; the peak becomes sharp as volume increases. The size and shape of AgNPs are also effected by increasing the volume of $AgNO_3$ solution, as mentioned in Figure 3, which showed that, with increasing the volume of $AgNO_3$ solution, the peak intensity also increased, indicating that small-size AgNPs were obtained. While at lower volume of $AgNO_3$ solution, non-uniform large-sized

AgNPs were synthesized [37]. As a result, from the given data, the maximum absorbance spectra were recorded at 438 nm for 1:9 v/v ratio, beyond this the peak becomes broader. The absorbance showed the quantity of AgNPs formed due to the reduction of silver ions.

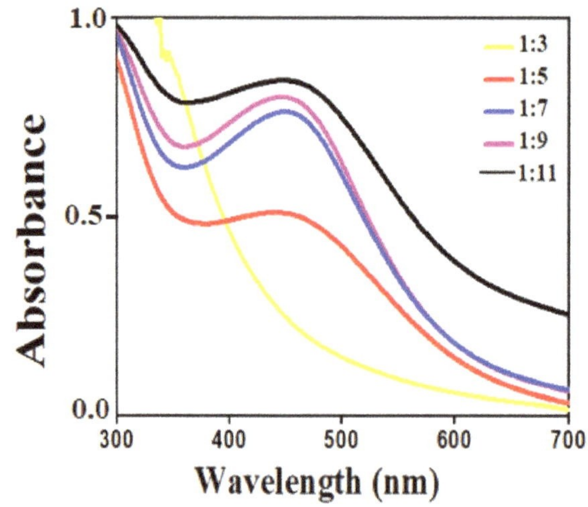

Figure 3. Effect of $AgNO_3$ volume to *S. procumbens* plant extract ratios.

2.1.3. Effect of Temperature on AgNPs

The production of AgNPs is heavily influenced by temperature. The reaction mixture of synthesized AgNPs was placed at various temperatures from 20–70 °C for 30 min. The color and intensity shift of the reaction mixture is illustrated in Figure 4. In general, silver nanoparticles aggregate together easily. The color changes from brown to black as shown in Figure 4, which is due to more aggregation within the particles at 40 °C. The structure and size of the particles formed during the aggregation were too large and lost the property of surface plasmon resonance. Hence, the synthesized AgNPs were stable up to 70 °C, but the more intense absorbance peak observed at 70 °C was due to the formation of small size AgNPs. This also implies that, at high temperature, the kinetic energy of the molecules increases and silver ions gets consumed faster, leaving less possibility for particle size growth. Thus, smaller particles of nearly uniform size distribution are formed at higher temperature. It was discovered that absorbance increased with increasing temperature [38]. Hence, the maximum intensity peak of 438 nm at 70 °C resulted.

2.1.4. Effect of Time on AgNPs

The impact of reaction time was studied by regularly reviewing the reaction mixture of an aqueous plant extract and $AgNO_3$ solution at different time intervals. The $AgNO_3$ solution reacts with the plant extract, resulting in a color shift within 1 h of the reaction. With the aging process, the color became more intense [39]. According to UV–visible spectrophotometric measurements taken at regular intervals of time, the broad surface plasmon resonance peak was obtained between 0 h and 2 h, which was due to the sluggish conversion of silver ion (Ag^+ to Ag^0). Since a significant amount of Ag^+ was transformed to Ag^0, an outstanding SPR (surface plasmon resonance) spectrum was found as the reaction time increased. The absorption peak of synthesized AgNPs was observed in the UV–vis spectra with strong SPR in the range of 438 nm at 8 h, indicating fast synthesis (Figure 5) [40].

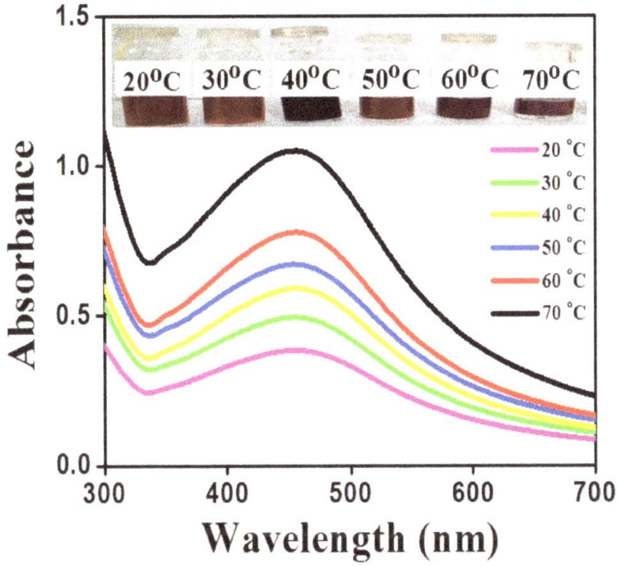

Figure 4. Absorption spectra of AgNPs achieved at different reaction temperatures using *S. procumbens* plant extract AgNPs.

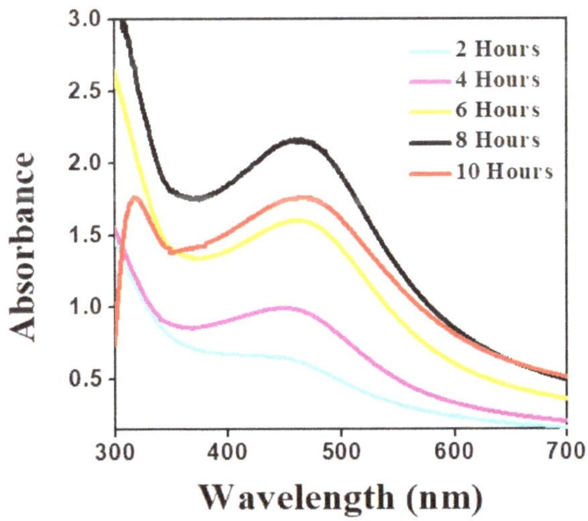

Figure 5. Absorption spectra of AgNPs obtained at different time intervals.

2.1.5. Effect of pH on AgNPs

The potential of the reaction pH is to comprehend the electrical charges of biomolecules, which may affect their reducing and capping capabilities, as well as the subsequent growth of the nanoparticles. The impact of pH on the absorption spectra of synthesized AgNPs via plant extract was shown in Figure 6. AgNPs were configured at various pH levels: 2, 4, 6, 8, 10, and 12. The absorption peak shifted towards lower wavelength, implying that the size of synthesized AgNPs was shrinking with the increasing pH of the reaction mixture. Furthermore, it was also noticed that the rate of reduction and the color of the solution turned colloidal brown faster at higher pH. As a result, the more favorable synthesis of AgNPs occurs at alkaline pH; this could be accredited to the ionization of the functional groups at higher pH. The slow rate of reduction observed in the acidic medium could be attributed to the electrostatic repulsion of anions present in the reaction mixture. Referring to the previously reported literature, the SPR (surface plasmon resonance) peak of AgNPs was not observed in the range of 400 to 500 nm at pH 2, 4, and 6 [41]. Whereas in the current study, the sharp and narrow SPR (surface plasmon resonance) peak was obtained within the range of 438 nm at pH 8, as shown in Figure 6, inferring that OH groups were responsible for a reduction of Ag^+ as supported by Davidovic et al. [42].

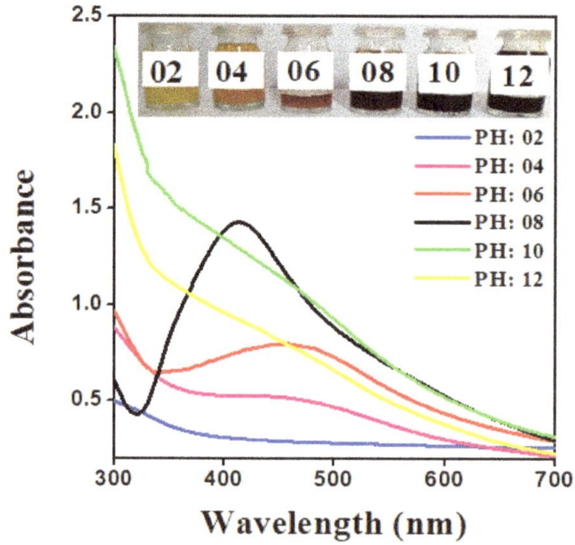

Figure 6. Normalized absorption spectra of AgNPs at different pH values of the reaction mixture employed S. *procumbens* plant extract.

2.1.6. Stability of AgNPs

The stability test of synthesized AgNPs was checked at 5, 10, and 15 days after synthesis. According to the results, there are no vibrational shifts at the surface plasmon resonance of AgNPs. The presence of a sharp peak indicated the development of mono-dispersed nanoparticles. As reported, from an aqueous extract of *Cassia roxburghii,* synthesized AgNPs were found to be stable for 2 months in a previous analysis [43].

2.2. Fourier Transform Infrared Spectroscopy (FT-IR) Analysis

The synthesis of AgNPs was achieved by biologically active functional groups and interaction sites present in an aqueous plant extract of *S. procumbens* that acts as a reducing, capping, and stabilizing agent, which were identified by using FT-IR (Bruker, Alpha-II) spectroscopy. The data suggests that during AgNPs synthesis, the phytochemicals present in a plant extract play an important role in the production of metal nanoparticles, which reduce the metal salt into its nanoparticles. It is tough to select an individual phytochemical entity, but in general, flavonoids, phenolic groups, organic acids, and proteins all are thought to be essential for the synthesis of nanoparticles [44]. In FT-IR spectroscopy, *S. procumbens* extract showed prominent bands of absorbance at around 3277, 2923, 2364, 1590, 1405, and 1033 cm^{-1} (Figure 7). The phenolic groups and proteins act as capping and stabilizing agents, which play an important role in the production of AgNPs; this was confirmed by a band at 3277 cm^{-1}, which was responsible for O–H stretching vibration of carboxylic acid [45]. The small band at 2923 cm^{-1} was due to **C**–H stretching of alkane, whereas the band at 2364 cm^{-1} represents the C–O vibration [46,47]. The band at 1590 cm^{-1} showed the N–H stretching vibration of amines present in proteins and 1405 cm^{-1} aligns to the C–N stretching of aromatic amine groups [48]. The 1033 cm^{-1} band was assigned C–O stretching vibration of alcohols, ester, or carboxylic acid. The reduction in bands at 3020, 2713, 2362, 1536, 1363, 1027, and 670 cm^{-1} is attributed to biomolecules present in the plant extract, which reduced Ag^{+} ions to Ag^{0}. We theorized that the synthesized AgNPs were capped by biomolecules that were involved in silver nanoparticles formation [49].

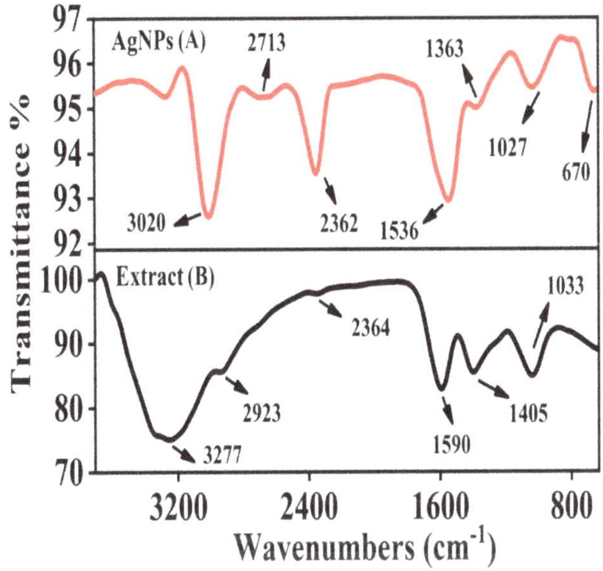

Figure 7. FTIR spectra of synthesized AgNPs (**A**) and *S. procumbens* plant extract (**B**).

2.3. X-ray Diffraction Analysis (XRD)

To investigate the crystalline nature and surface morphology of the synthesized AgNPs, the X-ray diffraction (XRD) analysis was performed in the range of 20–70 degrees at 2θ. The XRD pattern of the AgNPs synthesized by the reaction of silver salt solution with an aqueous plant extract of *S. procumbens* is shown in Figure 8. In XRD analysis, the crystal structure depends on the arrangement of atoms in a specific plane; the XRD peak showed higher intensity due to more atomic counts accumulated in its orientation. So, synthesized AgNPs, by using a plant extract of *S. procumbens*, were observed at high-intensity peaks at around 38°, 45°, and 64°. The mentioned peaks correspond to the *hkl* planes (111), (200), and (220) Bragg reflections, respectively, which were the exact peak positions as given for the face center cubic (FCC) lattice structure of AgNPs. This diffraction pattern was very similar to JCPDS file no. 04–0783 standard diffraction pattern [50]. The 38° peak was explained by assuming that the peak intensity of XRD becomes smaller along with the decrease in crystallite size. These facts revealed that the measuring content with the crystallites sizes of AgNPs were smaller than 30 nm [51]. The particle size of synthesized AgNPs was 26 nm, calculated by employing Scherrer's equation.

$$D = \frac{0.9\lambda}{\beta \cos\theta} \quad (1)$$

where D is crystallite size, λ corresponds to the X-rays wavelength that used 1.5406 °A, β is FWHM (Fill width half maximum), and θ corresponds to Bragg's angle.

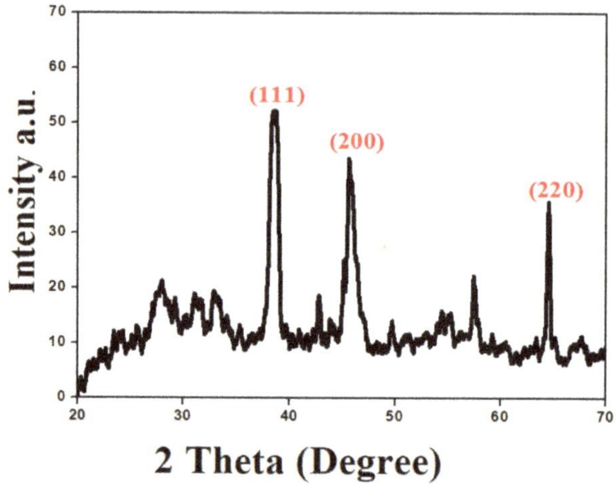

Figure 8. X-ray diffraction pattern of AgNPs synthesized from *S. procumbens*.

2.4. *Energy Dispersive X-ray (EDX) Analysis*

The elemental composition was determined by using energy dispersive X-ray (EDX) analysis. The EDX result of synthesized AgNPs revealed the existence of silver, as well as the residual carbon, oxygen, and chlorine peaks that were unwittingly inserted from the surface molecule of the plant extract. Green-synthesized AgNPs were responsible for the observed C, O, Cl, and Ag peaks. The peak of Ag at 3 eV confirms the formation of silver nanoparticles, where standard weight percent of silver was 90.88%. In this analysis, the green-synthesized AgNPs showed aggregated morphologies and a nonuniform size distribution (Figure 9). The particle size is dependent on how the aqueous extract is administered during the synthesis. Likewise, in accordance with these statements, the presence of Ag, Cl, and O were observed within the spectrum that appeared around 3 keV, which indicated the existence of elemental silver in the biosynthesized AgNPs [52].

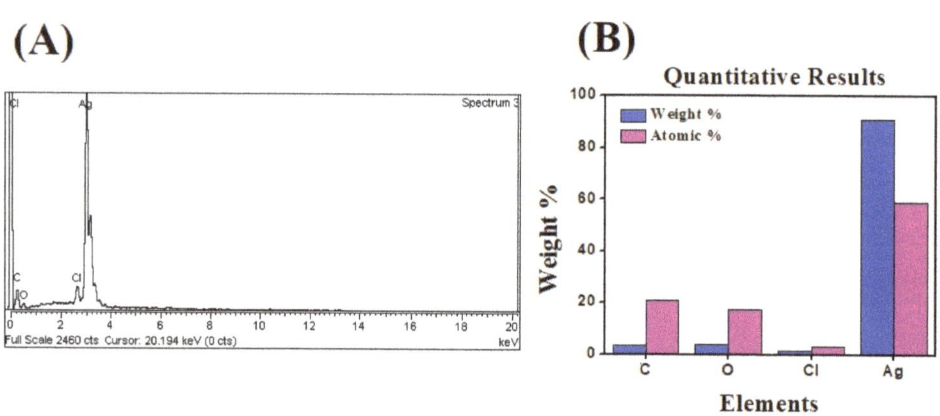

Figure 9. (**A**) Energy dispersive X-ray spectrum for AgNPs from *S. procumbens*; (**B**) Weight percentage of atoms present in AgNPs.

2.5. *Scanning Electron Microscopy (SEM)*

Researchers used scanning electron microscopy (SEM) to design the images of the sample and consider the morphology of the nanoparticles. The shape and size of synthesized AgNPs by using green synthesis were shown in a micrograph of a scanning electron microscope (SEM). According to the given micrograph, the scale of the synthesized AgNPs was up to 250 nm. The micrograph revealed that the particle sizes range from 10–80 nm and are spherical in shape. Most of them are synchronized in the form of large clusters, but few of them are dispersed as seen in Figure 10. So, the average particle size was found to be 46 nm, which is consistent with XRD results of 26 nm (crystallite size).

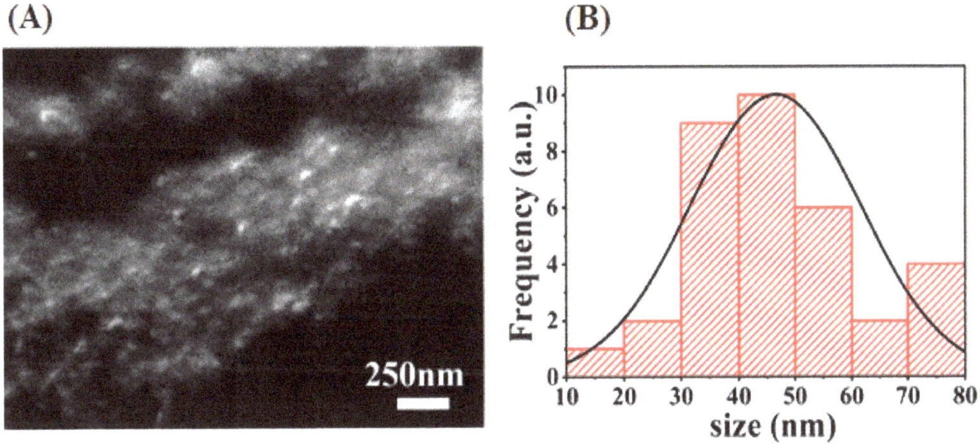

Figure 10. SEM image of AgNPs with 250 nm scale bar (**A**), and their corresponding size distribution histogram (**B**).

2.6. *Photocatalytic Degradation Analysis*

2.6.1. Photocatalytic Degradation of Orange G Azo Dye

To investigate the photocatalytic capability of stable AgNPs towards the degradation of various organic dye molecules, the commonly used Orange G dye in many dyeing industries was selected. Orange G is a well-known water-soluble dye that has always had widespread usage in industries such as leather, textiles, pharmaceuticals, paper, and printing [53]. To evaluate dye degradation in the presence of AgNPs as a catalyst, a UV–vis spectrophotometer was used. The high absorption peak of OG was centered around 483 nm; the reaction of degradation was carried out at room temperature. The photocatalytic degradation of dye was monitored by changing the absorbance peak. Degradation of OG did not proceed in the absence of synthesized AgNPs. Experiments were performed by using 0.05 g/100 mL of synthesized AgNPs, and the results were expressed in terms of dye degradation. Figure 11 showed the photocatalytic degradation of OG versus time. It was concluded that the maximum photocatalytic degradation of OG dye was ascertained by noting the time required for the OG absorption peak to reach baseline. Pertaining to this principle, the upper limit percentage of OG dye degradation was found to be 70.61% at about 180 min with *S. procumbens*-AgNPs [11].

2.6.2. Photocatalytic Degradation of Direct Blue-15 Azo Dye

Dye pollution has grown in tandem with the development of industries that use colorants, such as in clothing, leather, food, and agrochemicals. Since the majority of commonly used azo dyes are made from benzidine, these dyes contain a carcinogenic substance that is hazardous to marine biota. The azo dye DB-15 used in the textile had a variety of toxic effects on microalgae, Cladocera, and zebrafish embryos. So, the dumping of this colorant waterbody should be controlled to prevent environmental impacts [54]. As a result, deterioration of the azo dye DB-15 in wastewater is strongly recommended to alleviate their harmful impacts. A UV–visible spectrophotometer was used to monitor the extent of degradation by green-synthesized AgNPs. The absorption maxima for BD-15 were recorded at a wavelength of 602 nm. However, in the presence of 0.05 g/100 mL

AgNPs as a catalyst, the decrease in absorption peak of dye was observed as the time progressed, as shown in Figure 12. To determine the maximum photocatalytic degradation of DB-15, the amount of time required for the absorbance peak to reach baseline was measured. At 180 min, the maximum percentage of DB-15 dye degradation was found to be 70%. However, the synthesized catalyst showed more degradation efficiency against OG dye than DB-15, which might be due to the complex structure of DB-15 with many N atoms and its difficult conversion into oxidized nitrogen.

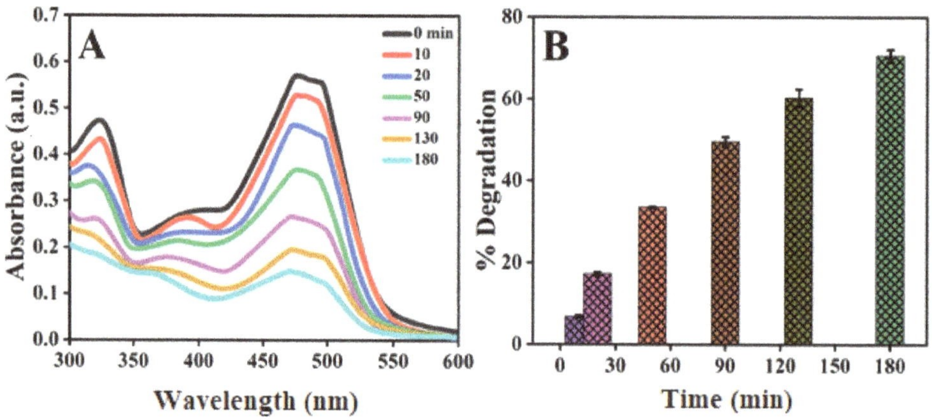

Figure 11. (**A**) Absorbance spectra of Orange G (15 ppm) dye at different time intervals, (**B**) Percentage degradation of Orange G dye at different time intervals.

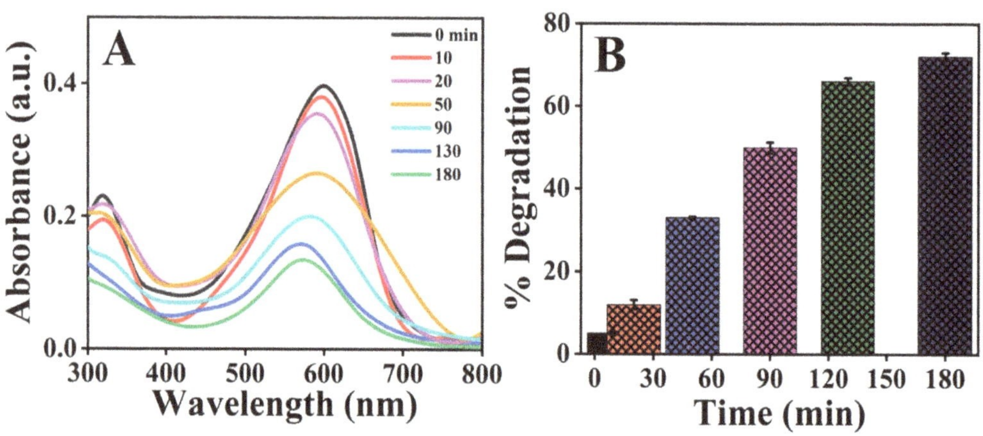

Figure 12. (**A**) Absorbance spectra of Direct Blue-15 (10 ppm) dye at different time intervals, (**B**) Percentage degradation of direct Blue-15 dye at different time intervals.

3. Materials and Methods

3.1. Plant Collection

The *S. procumbens* plant was harvested from the area of Kohat, Khyber Pakhtunkhwa, Pakistan and was identified by Dr. Nisar Ahmed, Department of Plants and Environmental Sciences, Kohat University of Science & Technology.

3.2. Preparation of *Sanvitalia procumbens* Aqueous Plant Extract

The plant of *S. procumbens* was washed thoroughly with tap water to remove debris and other contamination, followed by washing with distilled water. The plant was shade-dried at room temperature. Twenty grams of finely cut plant was heated at 70 °C for 30 min in 200 mL of distilled water. The extract was cooled down and then filtered via Whatman No. 1 filter paper, and the aqueous plant extract was stored in the refrigerator for further analysis.

3.3. Green Synthesis of AgNPs

At room temperature, 1 mL volume of aqueous plant extract was added to 9 mL of 0.01 mol/L $AgNO_3$ solution in a dropwise manner. After mixing the aqueous plant extract into silver ion solution, the reaction mixture was placed onto a rotatory orbital shaker operating at 200 rpm, then it was subjected to a water bath for 70 °C. Complete reduction of Ag^+ to Ag^0 ions in the reaction mixture was confirmed by the color change of $AgNO_3$; plant extract solution from pale yellow to dark brown implied the formation of AgNPs (further scanned by UV-visible spectrophotometry). Finally, the samples were centrifuged for 15 min at 6000 rpm. The organic moieties were removed after centrifugation, and the pellet was washed with distilled water and methanol. The powder form of nanoparticles was obtained, which were utilized for further characterization and photocatalytic studies.

3.4. Effects of Different Parameters on the Stability of AgNPs

Green synthesized AgNPs were produced by using *S. procumbens* aqueous plant extract and different parameters including $AgNO_3$ concentration, volume, pH, time, and temperature were studied to check the stability of AgNPs. Spectrophotometric study of AgNPs after 6–8 h of incubation in the dark at room temperature revealed the optimal conditions for efficient reduction of silver ions and eventual nucleation into silver nanoparticles. The impact of plant extract volume on green synthesis of AgNPs was explored through executing the reaction with various volumes of $AgNO_3$ solution (1, 3, 5, 7, 9, and 11 mL), whilst maintaining the volume of plant extract remains as constant. The reaction mixtures were deposited with various $AgNO_3$ concentrations, which were used to investigate the impact of $AgNO_3$ concentration on AgNPs synthesis (0.004, 0.006, 0.008, 0.01, and 0.02 mol/L). So, the optimal $AgNO_3$ concentration measured the absorbance of nanoparticles by UV-visible spectrophotometer. The aforementioned reaction mixture was subjected at different temperature conditions, namely 20 °C, 30 °C, 40 °C, 50 °C, 60 °C, and 70 °C, for synthesizing AgNPs. Following the synthesis, every single reaction tube was held at the specified temperature for 15 min. The pH effect was observed by exposing a reaction mixture of AgNPs at various pH level, and 0.1 mol/L NaOH solution was used to change the pH of the reaction solution to the optimal level.

3.5. Characterization of AgNPs

To calculate the reduction of silver nitrate into silver nanoparticles, UV-visible spectroscopy was used on a regular basis (Shimadzu-UV-1800). The AgNPs sample was diluted with distilled water and then examined under UV-visible spectrophotometer. The quartz cuvette cell filled with distilled water was used as a reference for recording the spectrograph at scanning speed of 300–700 nm [55]. The morphology of AgNPs was investigated by using SEM (SIGMA HV-Carl Zeiss, Oberkochen, Germany); this the sample of AgNPs was put onto a carbon grid, and the remaining sample was washed with blotting paper. Synthesized AgNPs were subsequently allowed to dry for 15 min under a mercury lamp before being imaged. A high-pressure mercury lamp was used to excite the storage molecules or proteins that exhibited autofluorescence [56]. At a 10 kV accelerating voltage, the characterization procedure of SEM was carried out. XRD was used to assess the crystalline nature and average grain size of AgNPs (XPERT-PRO). CuKa radiation was used to monitor X-ray diffraction in the 2θ range from 20–70 at 40 kV/40 mA current with dried AgNPs on a XRD grid [57]. The production of AgNPs was responsible for biologically active compounds that act as a reducing, capping, and stabilizing agents, which were identified by using a KBr pellet system in FT-IR (Bruker, Alpha-II) analysis. FT-IR (Bruker, Alpha-II) (Billerica, MA, USA) was performed by using a KBr pellet method to identify the functional groups of biological active compounds present in plant extract that act as reducing and capping agents for the formation of AgNPs. This spectral analysis was performed at 4 cm^{-1} resolution, within the range of 4000 to 500 cm^{-1} [58,59].

3.6. Photocatalytic Degradation Assay

The photocatalytic degradation studies were performed by mixing 100 mL of the reaction mixture into a 250 mL glass beaker, distinctly. Prior to being irradiated, the reaction sample (0.05 g/100 mL) was mixed with (15 ppm) OG and (10 ppm) DB-15 azo dye solution, respectively. Both solutions were constantly stirred for 30 min in the dark to ensure that equilibrium was reached. An absorption capacity of pure AgNPs for OG was evaluated at room temperature in the dark condition. Then the solution was irradiated by a Hg lamp with a high-pressure (max = 254 nm, 60 W, located 10 cm above the sample) sealed in to a rectangular steel box. The reactors were put on a magnetic stirrer to ensure that the solvent was evenly dispersed. After the irradiation times, aliquots of a previously utilized sample were removed at specific time intervals. Finally, the solution was centrifuged for 10 min at 4000 rpm to extract the suspended particles and then examined with a UV–visible spectrophotometer. The experiment was repeated twice, and the standard deviation did not reach 5%, indicating that the findings were relevant. The percentage degradation was calculated by using the following equation on the base of standard curve y = x0.0386 − 0.035 (R^2 = 0.9935) and y = x0.034 − 0.0428 (R^2 = 9.997) for OG and DB dyes, respectively.

$$\%\text{Degradation} = 100 \times (A_0 - A)/A_0 \quad (2)$$

where, 'A_0' is the initial absorbance of Orange G and DB-15 azo dyes, and 'A' shows the absorbance of dyes solution obtained at each time interval [60].

4. Conclusions

In the current study, the synthesis and optimization of AgNPs was demonstrated by using plant aqueous extract of *S. procumbens*. The synthesized AgNPs were successfully characterized by using different techniques like UV-Visible, FT-IR, SEM, XRD, and EDX. Furthermore, AgNPs exhibited potential photocatalytic degradation activity of Orange G and Direct blue-15; the synthesized catalyst showed more degradation efficiency against OG dye than DB-15. Overall, all these results suggest that AgNPs, as synthesized in this study, can be used as a potent photodegradation analyte.

Author Contributions: Conceptualization, F.F. and A.B.; Data curation, M.A.; I.A.; Formal analysis, I.A.; A.B.; R.U.; H.H.; R.A.M.; Funding acquisition, A.B., R.U.; R.A.M.; Investigation, M.A.; A.G.; Methodology, A.G.; Project administration, I.A.; R.A.M.; Resources, R.U.; A.B.; R.A.M.; H.H.; Software, M.A.; A.G.; F.F.; Supervision, I.A.; R.U.; Writing—original draft, M.A.; A.G.; F.F.; Writing—review & editing, R.A.M.; H.H., R.U. All authors have read and agreed to the published version of the manuscript.

Funding: Researchers Supporting Project, (RSP-2021/119) at King Saud University, Riyadh Saudi Arabia.

Institutional Review Board Statement: Not applicable.

Informed Consent Statement: Not applicable.

Data Availability Statement: Provided all data in the MS.

Acknowledgments: The authors of this study extend their appreciation to the Researchers Supporting Project, (RSP-2021/119) at King Saud University, Riyadh Saudi Arabia, for financial support.

Conflicts of Interest: Authors have declared that we have no conflict of interest.

References

1. Lee, S.H.; Sung, J.H.; Park, T.H. Nanomaterial-Based Biosensor as an Emerging Tool for Biomedical Applications. *Ann. Biomed. Eng.* **2012**, *40*, 1384–1397. [CrossRef] [PubMed]
2. Baruah, D.; Yadav, R.N.S.; Yadav, A.; Das, A.M. *Alpinia nigra* Fruits Mediated Synthesis of Silver Nanoparticles and Their Antimicrobial and Photocatalytic Activities. *J. Photochem. Photobiol. B* **2019**, *201*, 111649. [CrossRef]
3. Yoon, W.J.; Jung, K.Y.; Liu, J.; Duraisamy, T.; Revur, R.; Teixeira, F.L.; Sengupta, S.; Berger, P.R. Plasmon-Enhanced Optical Absorption and Photocurrent in Organic Bulk Heterojunction Photovoltaic Devices Using Self-Assembled Layer of Silver Nanoparticles. *Sol. Energy Mater. Sol. Cells.* **2010**, *94*, 128–132. [CrossRef]
4. Arvizo, R.R.; Bhattacharyya, S.; Kudgus, R.A.; Giri, K.; Bhattacharya, R.; Mukherjee, P. Intrinsic Therapeutic Applications of Noble Metal Nanoparticles: Past, Present and Future. *Chem. Soc. Rev.* **2012**, *41*, 2943–2970. [CrossRef] [PubMed]
5. Gan, L.; Zhang, S.; Zhang, Y.; He, S.; Tian, Y. Biosynthesis, Characterization and Antimicrobial Activity of Silver Nanoparticles by a Halotolerant *Bacillus endophyticus* SCU-L. *Prep. Biochem. Biotechnol.* **2018**, *48*, 582–588. [CrossRef]
6. Ni, Z.; Gu, X.; He, Y.; Wang, Z.; Zou, X.; Zhao, Y.; Sun, L. Synthesis of Silver Nanoparticle-Decorated Hydroxyapatite (HA@ Ag) Poriferous Nanocomposites and the Study of Their Antibacterial Activities. *RSC Adv.* **2018**, *8*, 41722–41730. [CrossRef]
7. Wu, Y.; Yang, Y.; Zhang, Z.; Wang, Z.; Zhao, Y.; Sun, L. A Facile Method to Prepare Size-Tunable Silver Nanoparticles and Its Antibacterial Mechanism. *Adv. Powder Technol.* **2018**, *29*, 407–415. [CrossRef]
8. Logeswari, P.; Silambarasan, S.; Abraham, J. Ecofriendly Synthesis of Silver Nanoparticles from Commercially Available Plant Powders and Their Antibacterial Properties. *Sci. Iran.* **2013**, *20*, 1049–1054.
9. Ahmed, S.; Saifullah; Ahmad, M.; Swami, B.L.; Ikram, S. Green Synthesis of Silver Nanoparticles Using *Azadirachta indica* Aqueous Leaf Extract. *J. Radiat. Res. Appl. Sci.* **2016**, *9*, 1–7. [CrossRef]
10. Thirunavoukkarasu, M.; Balaji, U.; Behera, S.; Panda, P.; Mishra, B. Biosynthesis of Silver Nanoparticle from leaf Extract of *Desmodium gangeticum* (L.) DC. and Its Biomedical Potential. *Spectrochim. Acta Mol. Biomol. Spectrosc.* **2013**, *116*, 424–427. [CrossRef]
11. Vidhu, V.; Philip, D. Catalytic Degradation of Organic Dyes Using Biosynthesized Silver Nanoparticles. *Micron* **2014**, *56*, 54–62. [CrossRef]
12. Kotakadi, V.S.; Gaddam, S.A.; Rao, Y.S.; Prasad, T.; Reddy, A.V.; Gopal, D.S. Biofabrication of Silver Nanoparticles Using *Andrographis paniculata*. *Eur. J. Med. Chem.* **2014**, *73*, 135–140. [CrossRef]
13. Kotakadi, V.S.; Rao, Y.S.; Gaddam, S.A.; Prasad, T.; Reddy, A.V.; Gopal, D.S. Simple and Rapid Biosynthesis of Stable Silver Nanoparticles using Dried Leaves of *Catharanthus roseus*. Linn. G. Donn and Its Anti-Microbial Activity. *Colloids Surf. B* **2013**, *105*, 194–198. [CrossRef]
14. Bethu, M.S.; Netala, V.R.; Domdi, L.; Tartte, V.; Janapala, V.R. Potential Anticancer Activity of Biogenic Silver Nanoparticles Using Leaf Extract of *Rhynchosia suaveolens*: An Insight into the Mechanism. *Artif. Cells Nanomed. Biotechnol.* **2018**, *46*, 104–114. [CrossRef]
15. Ratan, Z.A.; Haidere, M.F.; Nurunnabi, M.; Shahriar, S.M.; Ahammad, A.; Shim, Y.Y.; Reaney, M.J.T.; Cho, J.Y. Green Chemistry Synthesis of Silver Nanoparticles and Their Potential Anticancer Effects. *Cancers* **2020**, *12*, 855. [CrossRef] [PubMed]
16. Saravanan, M.; Arokiyaraj, S.; Lakshmi, T.; Pugazhendhi, A. Synthesis of Silver Nanoparticles from *Phenerochaete chrysosporium* (MTCC-787) and Their Antibacterial Activity against Human Pathogenic Bacteria. *Microb. Pathog.* **2018**, *117*, 68–72. [CrossRef]
17. Mohanta, Y.K.; Panda, S.K.; Jayabalan, R.; Sharma, N.; Bastia, A.K.; Mohanta, T.K. Antimicrobial, Antioxidant and Cytotoxic Activity of Silver Nanoparticles Synthesized by Leaf Extract of *Erythrina suberosa* (Roxb.). *Front. Mol. Biosci.* **2017**, *4*, 14. [CrossRef]
18. Kim, S.; Choi, J.E.; Choi, J.; Chung, K.H.; Park, K.; Yi, J.; Ryu, D.Y. Oxidative Stress-Dependent Toxicity of Silver Nanoparticles in Human Hepatoma Cells. *Toxicol. Vitro* **2009**, *23*, 1076–1084. [CrossRef]
19. Arora, S.; Jain, J.; Rajwade, J.; Paknikar, K. Cellular Responses Induced by Silver Nanoparticles: In Vitro Studies. *Toxicol. Lett.* **2008**, *179*, 93–100. [CrossRef] [PubMed]
20. Karuppaiya, P.; Satheeshkumar, E.; Tsay, H.S. Biogenic Synthesis of Silver Nanoparticles Using Rhizome Extract of *Dysosma pleiantha* and its Antiproliferative Effect against Breast and Human Gastric Cancer Cells. *Mol. Biol. Rep.* **2019**, *46*, 4725–4734. [CrossRef] [PubMed]
21. Arunachalam, R.; Dhanasingh, S.; Kalimuthu, B.; Uthirappan, M.; Rose, C.; Mandal, A.B. Phytosynthesis of Silver Nanoparticles Using *Coccinia grandis* Leaf Extract and Its Application in the Photocatalytic Degradation. *Colloids Surf. B* **2012**, *94*, 226–230. [CrossRef]
22. Lei, C.; Zhu, X.; Zhu, B.; Jiang, C.; Le, Y.; Yu, J. Superb Adsorption Capacity of Hierarchical Calcined Ni/Mg/Al Layered Double Hydroxides for Congo Red and Cr (VI) Ions. *J. Hazard. Mater.* **2017**, *321*, 801–811. [CrossRef]
23. Rani, P.; Kumar, V.; Singh, P.P.; Matharu, A.S.; Zhang, W.; Kim, K.H.; Singh, J.; Rawat, M. Highly Stable AgNPs Prepared via a Novel Green Approach for Catalytic and Photocatalytic Removal of Biological and Non-Biological Pollutants. *Environ. Int.* **2020**, *143*, 105924. [CrossRef]
24. Mani, M.; Chang, J.; Gandhi, A.D.; Vizhi, D.K.; Pavithra, S.; Mohanraj, K.; Mohanbabu, B.; Babu, B.; Balachandran, S.; Kumaresan, S. Environmental and Biomedical Applications of AgNPs Synthesized Using the Aqueous Extract of *Solanum surattense* leaf. *Inorg. Chem. Commun.* **2020**, *121*, 108228. [CrossRef]
25. Bal, U.; Touraev, A. Microspore Embryogenesis in Selected Medicinal and Ornamental Species of the *Advances*. In *Haploid Production in Higher Plants*; Springer: Dordrecht, The Netherlands, 2009; pp. 219–229.

26. Perez-Ochoa, M.L.; Chavez-Servia, J.L.; Vera-Guzman, A.M.; Aquino-Bolanos, E.N.; Carrillo-Rodriguez, J.C. Medicinal Plants Used by Indigenous Communities of Oaxaca, Mexico, to Treat Gastrointestinal Disorders. In *Pharmacognosy: Medicinal Plants*; BoD-Books on Demand GmbH: Norderstedt, Germany, 2016.
27. Rios, M.Y. Natural Alkamides: Pharmacology, Chemistry and Distribution. In *Drug Discovery: Research in Pharmacognosy*; BoD-Books on Demand GmbH: Norderstedt, Germany, 2012; pp. 107–144.
28. Wang, L.; Wang, T.; Guo, Q.S.; Huang, Y.; Xu, H.K. Comparative Study on Four Major Active Compounds of *Sanvitalia procumbens* and *Chrysanthemum morifolium* cv'Hangju'and'Gongju'. *China J. Chin. Mater. Med.* **2013**, *38*, 3442–3445.
29. Shankar, S.S.; Rai, A.; Ahmad, A.; Sastry, M. Rapid Synthesis of Au, Ag, and Bimetallic Au core-Ag Shell Nanoparticles Using Neem (*Azadirachta indica*) Leaf Broth. *J. Colloid Interface Sci.* **2004**, *275*, 496–502. [CrossRef]
30. Leela, A.; Vivekanandan, M. Tapping the Unexploited Plant Resources for the Synthesis of Silver Nanoparticles. *Afr. J. Biotechnol.* **2008**, *7*, 120–234.
31. Moldovan, B.; David, L.; Achim, M.; Clichici, S.; Filip, G.A. A Green Approach to Phytomediated Synthesis of Silver Nanoparticles Using *Sambucus nigra* L. Fruits Extract and Their Antioxidant Activity. *J. Mol. Liq.* **2016**, *221*, 271–278. [CrossRef]
32. Chandirika, J.U.; Annadurai, G. Biosynthesis and Characterization of silver Nanoparticles Using Leaf Extract *Abutilon indicum*. *Glob. J. Biotechnol. Biochem.* **2018**, *13*, 7–11.
33. Gomathi, M.; Rajkumar, P.; Prakasam, A.; Ravichandran, K. Green Synthesis of Silver Nanoparticles Using *Datura stramonium* Leaf Extract and Assessment of Their Antibacterial Activity. *Resour.-Effic. Technol.* **2017**, *3*, 280–284. [CrossRef]
34. Ganaie, S.U.; Abbasi, T.; Anuradha, J.; Abbasi, S.A. Biomimetic Synthesis of Silver Nanoparticles Using the Amphibious Weed *Ipomoea* and Their Application in Pollution Control. *J. King Saud Uni. Sci.* **2014**, *26*, 222–229. [CrossRef]
35. Ider, M.; Abderrafi, K.; Eddahbi, A.; Ouaskit, S.; Kassiba, A. Silver Metallic Nanoparticles with Surface Plasmon Resonance: Synthesis and Characterizations. *J. Clust. Sci.* **2017**, *28*, 1051–1069. [CrossRef]
36. Filippo, E.; Serra, A.; Buccolieri, A.; Manno, D. Green Synthesis of Silver Nanoparticles with Sucrose and Maltose: Morphological and Structural Characterization. *J. Non-Cryst. Solids.* **2010**, *356*, 344–350. [CrossRef]
37. Saware, K.; Venkataraman, A. Biosynthesis and Characterization of Stable Silver Nanoparticles Using *Ficus religiosa* Leaf Extract: A Mechanism Perspective. *J. Clust. Sci.* **2014**, *25*, 1157–1171. [CrossRef]
38. Oluwaniyi, O.O.; Adegoke, H.I.; Adesuji, E.T.; Alabi, A.B.; Bodede, S.O.; Labulo, A.H.; Oseghale, C.O. Biosynthesis of Silver Nanoparticles Using Aqueous Leaf Extract of *Thevetia peruviana Juss* and its Antimicrobial Activities. *Appl. Nanosci.* **2016**, *6*, 903–912. [CrossRef]
39. Balavandy, S.K.; Shameli, K.; Biak, D.R.B.A.; Abidin, Z.Z. Stirring Time Effect of Silver Nanoparticles Prepared in Glutathione Mediated by Green Method. *Chem. Cent. J.* **2014**, *8*, 11. [CrossRef]
40. Anandalakshmi, K.; Venugobal, J.; Ramasamy, V. Characterization of Silver Nanoparticles by Green Synthesis Method Using *Pedalium murex* Leaf Extract and Their Antibacterial Activity. *Appl. Nanosci.* **2016**, *6*, 399–408. [CrossRef]
41. Vanaja, M.; Rajeshkumar, S.; Paulkumar, K.; Gnanajobitha, G.; Malarkodi, C.; Annadurai, G. Kinetic Study on Green Synthesis of Silver Nanoparticles Using *Coleus aromaticus* Leaf Extract. *Adv. Appl. Sci. Res.* **2013**, *4*, 50–55.
42. Davidovic, S.; Vesna, L.; Ivana, V.; Jelena, P.; Anhrenkiel, S.; Suzana, D.; Nedeljkovic, J.M. Dextran Coated Silver Nanoparticles-Chemical Sensor for Selective Cysteine Detection. *Colloids Surf. B Biointerfaces* **2017**, *160*, 184–191. [CrossRef]
43. Balashanmugam, P.; Kalaichelvan, P.T. Biosynthesis Characterization of Silver Nanoparticles Using *Cassia roxburghii* DC. Aqueous Extract, and Coated on Cotton Cloth for Effective Antibacterial Activity. *Int. J. Nanomed.* **2015**, *10*, 87. [CrossRef]
44. Blume, Y.B.; Pirko, Y.V.; Burlaka, O.; Borova, M.; Danylenko, I.; Smertenko, P.; Yemets, A.I. Green Synthesis of Noble Metal Nanoparticles and CdS Semiconductor Nanocrystals Using Biological Materials. *Sci. Innov.* **2015**, *11*, 55–66. [CrossRef]
45. Firdhouse, M.J.; Lalitha, P. Green Synthesis of Silver Nanoparticles Using the Aqueous Extract of *Portulaca oleracea* (L.). *Asian J. Pharm. Clin. Res.* **2012**, *6*, 92–94.
46. Mallikarjuna, K.; Narasimha, G.; Dillip, G.; Praveen, B.; Shreedhar, B.; Lakshmi, C.S.B.V.S.; Ressb, D.; Raju, D.P. Green Synthesis of Silver Nanoparticles Using *Ocimum* Leaf Extract and Their Characterization. *Dig. J. Nanomat. Biostruct.* **2011**, *6*, 181–186.
47. Kumar, V.; Singh, D.K.; Mohan, S.; Hasan, S.H. Photo-Induced Biosynthesis of Silver Nanoparticles Using Aqueous Extract of *Erigeron bonariensis* and Its Catalytic Activity against Acridine Orange. *J. Photochem. Photobiol. B* **2016**, *155*, 39–50. [CrossRef]
48. Ahmed, M.J.; Murtaza, G.; Mehmood, A.; Bhatti, T.M. Green Synthesis of Silver Nanoparticles Using Leaves Extract of *Skimmia laureola*: Characterization and Antibacterial Activity. *Mater. Lett.* **2015**, *153*, 10–13. [CrossRef]
49. Kokila, T.; Ramesh, P.; Geetha, D. Biosynthesis of Silver Nanoparticles from Cavendish Banana Peel Extract and its Antibacterial and Free Radical Scavenging Assay: A Novel Biological Approach. *Appl. Nanosci.* **2015**, *5*, 911–920. [CrossRef]
50. Johnson, P.; Krishnan, V.; Loganathan, C.; Govindhan, K.; Raji, V.; Sakayanathan, P.; Sathishkumar, P.; Palvannan, T. Rapid Biosynthesis of *Bauhinia variegata* Flower Extract-Mediated Silver Nanoparticles: An Effective Antioxidant Scavenger and α-Amylase Inhibitor. *Artif. Cells. Nanomed. Biotechnol.* **2018**, *46*, 1488–1494. [CrossRef]
51. Ohtani, B.; Ogawa, Y.; Nishimoto, S.I. Photocatalytic Activity of Amorphous-Anatase Mixture of Titanium (IV) Oxide Particles Suspended in Aqueous Solutions. *J. Phys. Chem.* **1997**, *101*, 3746–3752. [CrossRef]
52. Hebeish, A.; El-Rafie, M.; El-Sheikh, M.; Seleem, A.A.; El-Naggar, M.E. Antimicrobial Wound Dressing and Anti-Inflammatory Efficacy of Silver Nanoparticles. *Int. J. Biol. Macromol.* **2014**, *65*, 509–515. [CrossRef]

53. El-Ghenymy, A.; Centellas, F.; Garrido, J.A.; Rodriguez, R.M.; Sires, I.; Cabot, P.L.; Brillas, E. Decolorization and Mineralization of Orange G Azo Dye Solutions by Anodic Oxidation with a Boron-Doped Diamond Anode in Divided and Undivided Tank Reactors. *Electrochim. Acta* **2014**, *130*, 568–576. [CrossRef]
54. Hernandez-Zamora, M.; Martinez-Jeronimo, F. Exposure to the Azo Dye Direct Blue 15 Produces Toxic Effects on Microalgae, Cladocerans, and Zebrafish Embryos. *Ecotoxicology* **2019**, *28*, 890–902. [CrossRef] [PubMed]
55. Raj, S.; Mali, S.C.; Trivedi, R. Green Synthesis and Characterization of Silver Nanoparticles Using *Enicostemma axillare* (Lam.) Leaf Extract. *Biochem. Biophys. Res. Commun.* **2018**, *503*, 2814–2819. [CrossRef]
56. Jung, K.L.; Wheeler, D.R.; Polsky, R.; Brozik, S.M.; Brozik, J.A.; Rudolph, A.R. Liquid-Cell Scanning Transmission Electron Microscopy and Fluorescence Correlation Spectroscopy of DNA-Directed Gold Nanoparticle Assemblies. *Micron* **2019**, *119*, 54–63.
57. Malaikozhundan, B.; Vaseeharan, B.; Vijayakumar, S.; Sudhakaran, R.; Gobi, N.; Shanthini, G. Antibacterial and Antibiofilm Assessment of *Momordica charantia* Fruit Extract Coated Silver Nanoparticle. *Biocatal. Agric. Biotech.* **2016**, *8*, 189–196. [CrossRef]
58. Malaikozhundan, B.; Vaseeharan, B.; Vijayakumar, S.; Pandiselvi, K.; Kalanjiam, M.A.R.; Murugan, K. Biological Therapeutics of *Pongamia pinnata* Coated Zinc Oxide Nanoparticles against Clinically Important Pathogenic Bacteria, Fungi and MCF-7 Breast Cancer Cells. *Microb. Pathog.* **2017**, *104*, 268–277. [CrossRef]
59. Gul, A.; Fozia; Shaheen, A.; Ahmad, I.; Khattak, B.; Ahmad, M.; Ullah, R.; Bari, A.; Ali, S.S.; Alobaid, A.; et al. Green Synthesis, Characterization, Enzyme Inhibition, Antimicrobial Potential, and Cytotoxic Activity of Plant Mediated Silver Nanoparticle Using Ricinus communis Leaf and Root Extracts. *Biomolecules* **2021**, *11*, 206. [CrossRef]
60. Ullah, R.; Sun, J.; Gul, A.; Bai, S. One-Step Hydrothermal Synthesis of TiO_2-Supported Clinoptilolite: An Integrated Photocatalytic Adsorbent for Removal of Crystal Violet Dye from Aqueous Media. *J. Colloid Interface Sci.* **2020**, *8*, 103852. [CrossRef]

molecules

MDPI

Article

Automatic On-Line Purge-and-Trap Sequential Injection Analysis for Trace Ammonium Determination in Untreated Estuarine and Seawater Samples

Athina Dimitriadou and Aristidis Anthemidis *

Laboratory of Analytical Chemistry, Department of Chemistry, Aristotle University, 54124 Thessaloniki, Greece; athina.dimitriadou@gmail.com

* Correspondence: anthemid@chem.auth.gr; Tel.: +30-2310-99-7826

Academic Editors: Victoria Samanidou, Eleni Deliyanni and Dimitra Voutsa
Received: 21 February 2020; Accepted: 27 March 2020; Published: 29 March 2020

Abstract: An innovative automatic purge-and-trap (P&T) system coupled with fluorimetric sequential injection (SI), for the on-line separation and preconcentration of volatile compounds, is presented. The truth of concept is demonstrated for the ammonium fluorimetric determination in environmental water samples with complex matrices without any pretreatment. The P&T flow system comprises a thermostated purge-vessel where ammonium is converted into gaseous ammonia and a trap-vessel for ammonia collection. This configuration results in matrix removal as well as analyte preconcentration, avoiding membrane-associated problems. All the main parameters affecting the efficiency of a P&T system were studied and optimized. The proposed method is characterized by a working range of 2.7–150.0 μg L^{-1} of NH_4^+, with a detection and quantification limit of 0.80 and 2.66 μg L^{-1}, respectively, for a 10-mL sample consumption. The accuracy of the method was assessed by recovery assays in seawater, estuarine, and lake water samples as well as by the analysis of standard reference material.

Keywords: automation; purge-and-trap; sequential injection analysis; fluorimetric; ammonium determination; environmental samples

1. Introduction

Automatic on-line analytical methods based on flow injection (FIA) and sequential injection (SIA) analysis have been proved to be versatile, fast, accurate, robust, and efficient analytical tools [1]. These methods have led to the development of less invasive chemical assays with low reagent consumption and waste generation, fulfilling the principles of Green Analytical Chemistry (GAC) [2]. A remarkable advantage of automatic on-line systems is their ability to integrate and handle a plethora of sample pretreatment techniques, including sample separation and preconcentration [3,4].

Ammonium is an analyte of high environmental interest due to its double role as a significant ecological nutrient in the nitrogen cycle of seawater and a common pollutant in alkaline aquatic ecosystems. In natural waters, the ratio of ammonia to ammonium is significantly dependent on pH level and temperature. At alkaline environments, toxic ammonia is the predominant form, which is an indicator of water quality. Various flow-based methods have been implemented for ammonium determination in water samples typically based on the spectrophotometric indophenol blue or fluorimetric o-phthaldialdehyde (OPA) techniques [5]. In particular, the OPA reaction was modified by replacing mercaptoethanol with sulfite, in order to avoid the interference from organic amine compounds resulting in higher sensitivity and selectivity [6]. Afterwards, the method was further modified for the determination of ammonium in seawater [7]. Regarding the direct analysis of ammonium in natural and seawater samples, it suffers from several interferences arising from complicated matrices and possible dissolved materials [8]. Hence, separation and preconcentration techniques are required to eliminate matrix problems and to enhance the sensitivity and selectivity of the method.

Several approaches for the on-line ammonia gas–liquid separation have been reported and are classified as: a) membrane-based, such as gas-diffusion (GD) [9–11] and pervaporation [12, 13], b) membranelles-gas-diffusion (MGD) [14–16], and c) headspace single drop (HS-SD)-based techniques [17,18]. Membrane-based methods make use of a hydrophobic membrane that allows the diffuse of gaseous ammonia through the membrane, from a donor to an acceptor liquid stream. Nevertheless, serious problems, such as clogging, cracking, and the deterioration of membrane in the GD units, are often observed due to the contact of the membrane with the sample solution. On the other hand, pervaporation units involve an air gap between the membrane and the donor stream in order to overcome the GD limitations. However, reduced diffusion effectiveness has been recorded due to water condensation on the surface of membrane, in pervaporation units [13,19]. Regarding both MGD and HS-SD techniques, they require careful and accurate handling of solution, as well as complicated equipment. Lately, a membranelles automated on-line dual-headspace lab-in-syringe (PA-D-HS-LIS) system for the gas separation of volatile compounds was presented for ammonia determination [20]. This system is based on decreased pressure under increased temperature into the headspace of sample solution for effective ammonia gas release. Then, the generated ammonia is transported to the second headspace of the extraction solution for its dissolution under increased pressure.

Purge-and-trap pretreatment is a well-established methodology for separation of volatile organic compounds (VOCs) in aqueous samples with complex matrices. In principle, an inert gas passes through the water to purge the volatile compounds, from the aqueous to gas phase, which are subsequently collected in an appropriate trap for further analysis [21]. Compared to GD and MGD, the P&T configuration can process higher sample volumes in the pretreatment steps, thus enhanced sensitivity can be achieved [22]. As far as we know, only a few P&T methods for ammonium determination in a manual operation, have been presented [23,24].

In this study, a novel automatic dynamic purge-and-trap platform based on sequential injection analysis (P&T-SIA) for the separation and preconcentration of volatile compounds, has been developed for the first time. The proposed system eliminates the serious problems arising from the use of membrane in the GD and pervaporation units. The effectiveness of the P&T-SIA system was demonstrated for ammonium determination. Gaseous ammonia is generated by adding an appropriate volume of NaOH solution to the sample. After its transportation, ammonia is dissolved in the acidic trapping solution and is determined by fluorimetric sequential injection analysis (SIA). All the main factors influencing the efficiency of the P&T system, such as purging flow rate, purging time, temperature, liquid depth, and the concentration of the trapping solution, have been commented and optimized. The applicability of the P&T-SIA system has been evaluated in seawater, estuarine, and lake water samples for ammonium determination.

2. Results and Discussion

2.1. Optimization of the Automatic Purge-and-Trap System

The aim of the present study is to develop and evaluate an effective automatic on-line purge-and-trap system, which is able to be connected with the miniSIA device for the fluorimetric determination of ammonium or other volatile compounds. For setting up the P&T-SIA system, several key issues should be considered. Some of them include the pH, the temperature, and the volume of the sample solution, the flow rate of the purge gas, the time of purging, as well as the type and the volume of trapping solution.

2.1.1. Sample Alkalization

In aqueous solutions, ammonia exists simultaneously as two forms: the free unionized gaseous ammonia (NH_3) and as the ionized ammonium ion (NH_4^+). Since the pK_a of ammonium ions is 9.26, the pH value of the sample should be higher than that value to ensure the formation of ammonia is favored. The chemical equilibrium in the water of the two nitrogenous forms (ammonia and ammonium ion) reveals that a 100% conversion of NH_4^+ to NH_3(g) requires a pH value higher than 11.5 at 20 °C. However, higher values than pH > 13 are recommended for trace ammonia removal from water due to its high solubility [23,24].

Gaseous NH_3 is produced into the purge vessel from the ammonium ion NH_4^+ by sample alkalization with NaOH. Thus, for 10.0 mL of sample volume, a volume of 1000 µL of 1.0 mol L^{-1} NaOH aqueous solution was employed for ensuring a pH higher than 13 in the sample solution.

2.1.2. Trapping Solution

Ammonium gas dissolution is higher in slightly acidic solution than in double-de-ionized water (DDW), as it is shown from preliminary experiments. On the other hand, when ammonium concentration is at mg L^{-1} levels, higher acid concentration is necessary for the best trapping efficiency [25]. Thus, considering the very low concentrations of ammonium (ppb levels) in the samples, the versatility and the requirements of an miniSIA fluorimetric detection system, a 0.001-mol L^{-1} hydrochloric acid solution was adopted as the trapping solution. It should be mentioned that low volumes of the trapping solution result in a high preconcentration ratio. Thus, for higher sensitivity, a volume of 300 µL of HCl solution into the trap-vessel was used throughout the study.

2.1.3. Temperature Effect

Increasing the temperature of the sample solution inside the purge-vessel (PV), causes the decrease of ammonia solubility, as arises from van't Hoff's equation [26]. Raising the temperature increases the kinetic energy of the molecules. Therefore, a higher kinetic energy causes more actuation between the molecules that break the intermolecular bonds and escape from the solution, increasing the generation of the gaseous ammonia [20]. The proposed purge-vessel involves a lab-made heating device to adjust the temperature of the sample solution during purging. The effect of the temperature on the fluorescence intensity has been studied in the range 20–80 °C and the obtained results are presented in Figure S1 (Supplementary Material). Temperatures above 80 °C were not applied in order to avoid the possible occurrence of boiling phenomena in the purge vessel and the production of water vapor, which then can be condensate in the tubing and in the trapping vessel. As the recorded signal increased throughout the studied range, for high sensitivity, 80 °C was adopted for further experiments.

2.1.4. Sample Volume

In principle, purging efficiency is affected by several factors, such as Henry's law constant, temperature, the flow rate of the purge gas, purging time, as well as the number and the size of the purge gas bubbles. Another significant parameter is the depth of the solution in the purge vessel, which is proportional to the sample volume. If all other parameters are constant, a higher solution depth results in more bubbles in the liquid phase, increasing the interfacial area and the remaining time of the bubbles in the solution.

In order to estimate the effect of the solution depth on the purging efficiency, five purge-vessels with the same internal diameter (i.d. = 16 mm) but at various lengths, were tested. Specifically, for a solution depth between 3.0 and 10.0 cm, an appropriate sample volume was added into the PV, which ranged from 6.0 to 20.0 mL, respectively. Each vessel contained a fixed amount of 5.0 µg NH_4^+. The obtained results are presented in Figure 1, showing that short depth resulted in lower efficiency, as is expected. The optimum depths vary between 4.0 and 7.0 cm. For higher depths, the effectiveness is decreased to some extent, as it is also referred previously [27].

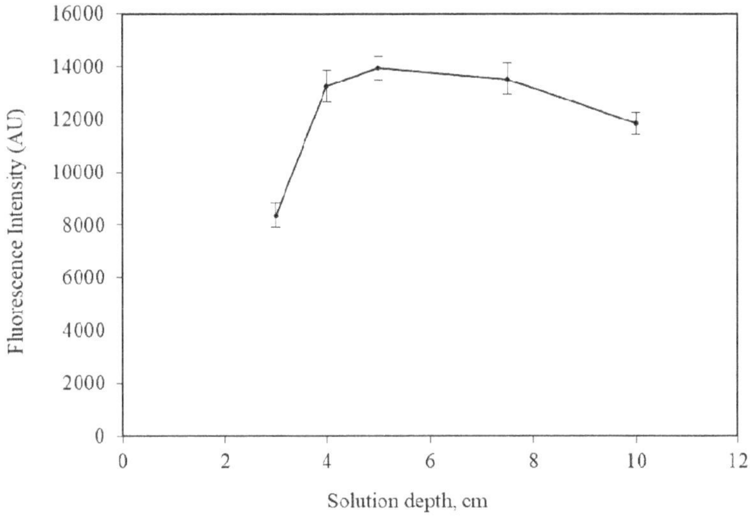

Figure 1. Effect of solution depth in purge-vessel on the intensity of 5.0 μg NH_4^+. The error bars were calculated based on standard deviation (± 1 s). NaOH volume = 1000 μL (1.0 mol L^{-1} NaOH); Trapping solution = 300 μL, 0.001 mol L^{-1} HCl; Purge-gas flow rate = 75 mL min^{-1}.

2.1.5. Effect of Purge-Gas Flow Rate and Purging Time

The effect of the flow rate of the purge-gas on the sensitivity of the method was examined in the range 10–100 mL/min and the recorded results presented diagrammatically in Figure S2 (Supplementary Material). As is shown, the intensity increased by increasing the flow rate of argon gas and leveled off for values higher than 75 mL/min. At flow rates higher than 90 mL/min, there was a possible transportation of small drops of the liquid through the tubing. Hence, the optimum flow rate of argon gas was set at 75 mL/min.

Typically, flow analysis methods are not based on thermodynamic equilibrium for their results. In this concept, the purging and trapping operation is not necessary to be completed at the time of measuring. Moreover, the time of purging positively affects the sensitivity of the method and can be considered as preconcentration time. The effect of purging time on sensitivity is shown in Figure S3 (Supplementary Material). It was examined from 1 to 15 min and it was found that the intensity increased by increasing the purging time, while it was leveled off after 10 min. Consequently, 10 min was adopted as the optimum purging time.

2.2. Interferences

An inherent feature of the purge-and-trap system is the effective separation of volatile compounds, such as ammonia from the matrix of the sample, eliminating the interferences.

Possible gases, such as CO_2, SO_2, and H_2S are transformed into non-volatile ions CO_3^{2-}, SO_3^{2-} and HS^- by alkalization with NaOH. Hence, they remain in the PV and do not interfere. The developed method is possibly affected by metal ions and volatile amines, such as primary amines. The interference studies were performed using a 50.0-μg L^{-1} NH_4^+ standard solution, and a relative deviation of less than 5% was considered as interference. Commonly encountered matrix components in natural water, such as metal ions K^+, Na^+, Ca^{2+}, and Mg^{2+}, were examined individually for each interferent and found to be tolerated at least up to 500 mg L^{-1}, while NaCl was tolerated up to 35 g L^{-1}. Other metals that could act as interferences are Zn^{2+}, Fe^{3+}, Al^{3+}, Cu^{2+}, and Ag^+, which is used as a disinfectant in water recycling systems. No interference was observed for these metals at concentration levels up to 1.0 mg L^{-1}. Primary amine (CH_3NH_2) can be tolerated at least up to 0.2 mg L^{-1}.

2.3. Figures of Merit

The figures of merit of the proposed method (P&T-SIA) under the optimum conditions for ammonium determination were calculated as follows. Under the optimum conditions, for a sample volume of 10.0 mL, the working linear range was between 2.7 and 150.0 μg L^{-1} of NH_4^+ with a correlation coefficient (r^2) of 0.9978. The calibration curve was characterized by the following regression equation: intensity = (258.7 ± 13.8) × [NH_4^+] + (47.9 ± 986.4), (7-standards; n = 5; [NH_4^+] expressed in μg L^{-1}). The detection and quantification limits were calculated based on the 3 and 10s criterion [28] and found to be 0.80 and 2.66 μg L^{-1}, respectively. The precision of the proposed method expressed as relative standard deviation (RSD%, n = 10), was calculated at 5.0, 20.0, and 100.0 μg L ammonium concentration levels and was found to be 4.8%, 4.2%, and 3.7% respectively. A complete analytical cycle lasts for 900s, resulting in a sampling frequency (f) of 4 h^{-1}.

For comparative purposes, the performance characteristics of the proposed method and other selected on–line separation methods based on gas diffusion, pervaporation, membranelles-GD, and HS-SD devises, reported in the literature for ammonium determination, are compiled in Table 1. As it is shown herein, the P&T-SIA method reveals satisfactory sensitivity with better detection limits than earlier works [10,12,15–17,29–31] for ammonium assays. In addition, it should be considered that the proposed system overcomes membrane-related problems and allows for the determination of untreated samples, minimizing the cost of analysis. According to the GAC requirements, the consumption of o-phthaldialdehyde and sodium sulfite is low enough (≤ 40 μL) and generated organic waste is limited. Regarding the time of analysis of the proposed methodology is rather higher than the other methods, even though it can be reduced to some extent by a strict synchronization of the analytical steps.

The accuracy of the presented method was estimated using a standard reference material SRM from NIST CertiPUR® NH_4Cl in H_2O 1000 ± 10 mg L^{-1} NH_4^+. After appropriate dilution, the analytical result for ammonium concentration was 935 ± 42 mg L^{-1} (standard deviation, n = 3) with a recovery of 93.5%.

The trueness of the P&T-SIA method was also investigated by analyzing two potable artificial water (PAW) and two hygiene artificial water (HAW) samples, and the obtained results were compared with that obtained by the certified method [32]. The recorded results and the t_{exp}, values are given in Table 2. The overall relative errors were 2.1% and 3.3% for PAW and HAW samples, respectively, indicating the efficient applicability of the developed method in similar water samples. Moreover, since all t_{exp} values are lower than the t_{crit}, 95% = 4.30, no statistically significant differences were found between the two studied methods.

Table 1. Comparative figures of merit of the proposed purge-and-trap platform based on sequential injection analysis (P&T-SIA) method and other on–line separation methods reported in the literature for ammonium determination.

Flow System	Separation Technique	Detection System	Linear Range (mg L^{-1})	LOD (μg L^{-1})	RSD (%)	F (h^{-1})	Sample Type	Ref.
MPFS	GD	Chemiluminescence	0.3–0.5	20	< 1.2	50	Tap, river, wastewater	[29]
MCFIA	GD	Spectrophotometry/BTB	0.050–1.0	27	< 1.5	20	Surface and tap water	[30]
SIA	GD	Spectrophotometry/BTB	0.10–1.0	27	< 2	20	Estuarine, river, well, marine water	[10]
FIA	PV	Spectrophotometry/CR - TB	0.2–20	100	< 3	11	Industrial effluents	[12]
FIA	PV	Spectrophotometry/BTB	0.05–50	30	1.9	10/8	Surface, urban sewage, industrial effluents	[31]
LIS	HS-SDME	Spectrophotometry/BTB	up to 0.425	30	< 8	17	River, coastal seawater	[17]
MSFIA	MGD	Spectrophotometry/BTB	10.0–50.0	2200	4.8	11	River, wastewater	[15]
SIA	MBL-VP	Conductivity	0.09–1.44	36	2.0	12	Canal water	[16]
LIS-SIA	PA-D-HS	Fluorimetry	0.15–10.0 *	0.05	3.6	8	Seawater, river, lake, ditch water	[20]
FIA	P&T (off-line)	Fluorimetry	0.18–7.2 *	0.13	4.4	4	Seawater	[24]
SIA	P&T	Fluorimetry	2.66–150 *	0.80	4.2	4	Estuarine, lake, seawater	**

* expressed as μg L^{-1}; ** present work; HS-SDME, headspace-single-drop microextraction; MBL-VP, membranelles-vaporization; MCFIA, multi-commuted FIA; MGD, membranelles gas-diffusion; MPFS, multi-pumping flow system; MSFIA, multisyringe FIA; PA-D-HS, pressure-assisted-dual-headspace; P&T, purge-and-trap; PV, pervaporation; BTB, bromothymol blue; CR, cresol red; TB, thymol blue; *f*, sampling frequency; LOD, limit of detection.

Table 2. Analytical results obtained by the P&T-SIA method and by the certified method for the determination of ammonium in potable (PAW) and hygiene (HAW) artificial water samples.

Sample	True Value (μg L^{-1})	Certified Method * (μg L^{-1})	P&T-SIA Method * (μg L^{-1})	Relative Error (%)	t_{exp}
PWA-1	25.0	24.2 ± 1.8	23.8 ± 1.5	1.7	0.831
PWA-2	50.0	48.0 ± 3.2	46.8 ± 2.5	2.5	0.462
Overall relative error				2.1	
HWA-1	50.0	51.6 ± 4.3	48.8 ± 2.4	5.4	2.021
HWA-2	100.0	97.8 ± 8.6	96.6 ± 5.2	1.2	0.400
Overall relative error				3.3	

* Mean value ± standard deviation (n = 3).

2.4. *Applications in Spiked Environmental Water Samples*

The proposed method has been applied to the analysis of collected (March 2019) environmental water samples namely estuarine, lake, and seawater. The recovery (R) values were estimated by analyzing the spiked samples with 20.0 μg L^{-1} NH_4^+ concentration level. The recorded results are presented in Table 3. The recoveries varied within the range 94.0%–102.0%, showing the applicability of the developed method for ammonium determination in natural water samples, even in samples with high salinity such as seawaters.

Table 3. Application of the proposed P&T-SIA method for ammonium determination in spiked natural water samples.

Sample Type	Added * (μg L^{-1})	Found * (μg L^{-1})	R (%)
Strymon estuarine water		N.D.	
	20.0	19.2 ± 0.9	96.0
	50.0	48.8 ± 1.5	97.6
Prespa lake water		35.2 ± 2.2	
	20.0	54.7 ± 2.7	97.5
	50.0	86.2 ± 2.8	102.0
Thermaikos gulf seawater		12.5 ± 0.5	
	20.0	31.3 ± 1.8	94.0
	50.0	60.2 ± 2.8	95.4
Toroneos gulf seawater		N.D.	
	20.0	20.3 ± 1.3	101.5
	50.0	47.3 ± 2.8	94.6

* Mean value ± standard deviation; N.D, not detected.

3. Materials and Methods

3.1. *The Automatic Purge-and-Trap Platform*

The developed purge-and-trap (P&T-SIA) system for ammonium separation/preconcentration and determination is presented schematically in Figure 2. It consists of a high precision bi-directional micro-syringe pump, SP2 (MicroCSP-3000, FIAlab Instruments, Bellevue, WA, USA), which is equipped with a 5000-μL glass barrel and a nine-position Teflon/Kel-F SV2 selection valve, directly connected at the top of the barrel. The purge-vessel (PV) is made of a borosilicate glass tubing (1.6 mm i.d., 100 mm length) with two push-fit poly tetra fluoro ethylene (PTFE) connectors at its two ends, as is shown in Figure 3. The lower connector is a male flangeless fitting end in order to be connected at the port-5 of SV2 (Figure 1). This connector enables the insertion of the purge gas at the bottom of the PV by four tiny holes, which are connected with a purge gas tubing. The PV is placed in a vertical position in the P&T manifold.

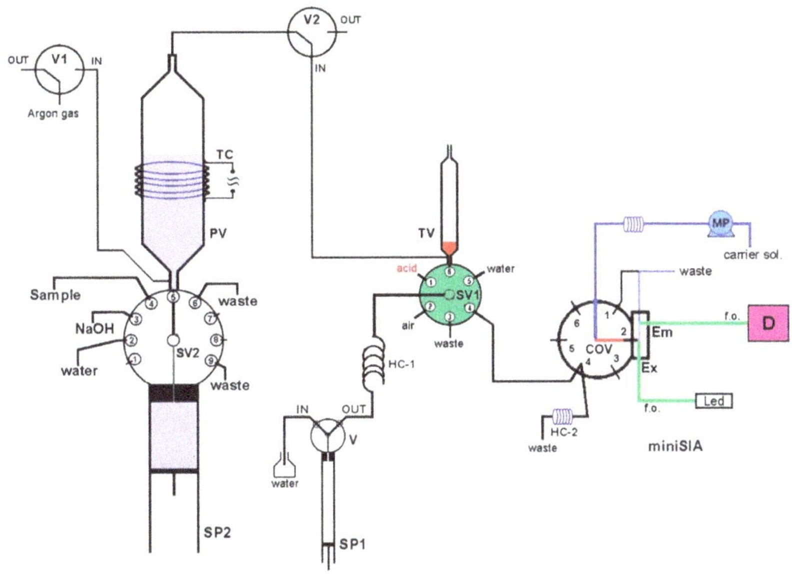

Figure 2. Schematic diagram of the P&T-SIA manifold for ammonium determination. PV, purge vessel; TV, trap vessel; SV1 and SV2, selection valves; COV, chem-on-valve unit with the flow-through cell; V, V1, V2, valves; SP1, SP2, syringe pumps; TC, thermostated coil; HC-1, HC-2, holding coils; MP, milliGUT pump; f.o., fiber optics; D, detector.

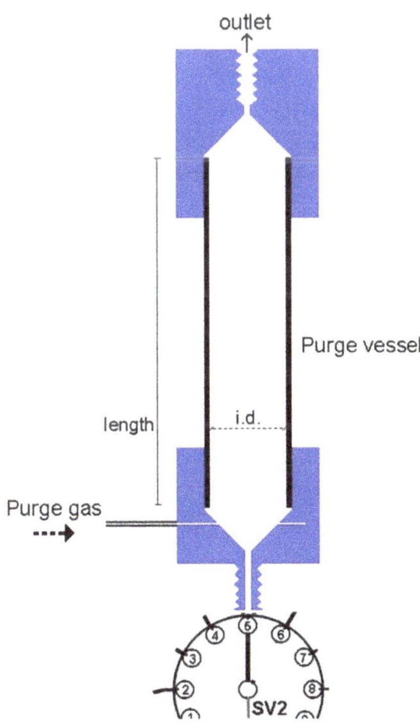

Figure 3. Schematic diagram of the purge-vessel. Internal diameter, i.d. = 1.6 mm; length = 100 mm.

In order to facilitate the gaseous ammonia release, the PV is heated at the appropriate temperature by a lab-made heating device, which consists of a thermostat and a chromium-nickel resistor coil (TC), electrically powered with a 12V/5A power supply. The trap-vessel (TV) is made of a plastic syringe (4.6 mm i.d., 70 mm length), located in the P&T manifold in a vertical position, by using a push-fit connector with a male fitting end. The TV is connected at the SV1 (port-6) of the FIAlab-3000® SI system.

3.2. Apparatus

A FIAlab®-3000 sequential injection (SI) system (Alitea FIAlab, Bellevue, WA, USA) was used for the handling of the P&T system. This system is equipped with a syringe pump, SP1 (Cavro, Sunnyvale, CA, USA) with a glass syringe barrel of 1000 μL capacity and a two-position (IN/OUT) valve (V), a six-port selection valve (SV1) and a holding coil (HC-1). Two additional three-port, two-position valves, V1 and V2 controlled by the FIAlab®-3000 SI system, were used for the purge gas handling. The Micro CSP-3000 syringe pump is interfaced with the FIAlab®-3000 system and controlled by a personal computer that runs the FIAlab software for windows v. 5.9.245 (http://www.flowinjection.com).

A miniSIA flow analyzer (https://www.globalfia.com) equipped with an acrylic Chem-on-Valve™ (COV) monolithic manifold was employed for ammonium SI fluorimetric determination. The miniSIA flow analyzer is equipped with a fluorescence flow cell which is located directly in the COV at position-2 (Figure 2). Two fiber optic cables (f.o.) are used for light emission (Em) and excitation (Ex). For the fluorometric detection of ammonium, an Ocean-Optics USB-4000 fluorescence spectrometer with a monochromator set at 425 nm emission wavelength, is used. In addition, a monochrome white LED is used as an excitation light source and has been set at 365 nm excitation wavelength. The recorded fluorescence intensity is given as intensity arbitrary units (AU). More details of the miniSIA flow analyzer are presented in previous work [33]. The quantification of ammonium in miniSIA flow analyzer is based on the OPA reaction, where the ammonia reacts with o-phthaldialdehyde in the presence of a strong reducing agent, sulfite, in order to produce the fluorescent isoindole derivative. The reaction is time and temperature depended.

For the control of miniSIA system and the data acquisition, a laptop running the FloZF 5.2 software (https://www.globalfia.com), is used. The integrated P&T-SIA system is controlled by the two discrete software programs namely FIAlab and FloZF, which are synchronized and activated simultaneously.

3.3. Chemicals and Samples

Analytical-grade chemicals were provided by Merck (Darmstadt, Germany, http://www.merck.de) and were used throughout the study. A Milli-Q system (Millipore, Bedford, MA, USA, http://www.millipore.com) was employed for ultra-pure quality water. All ammonium working standard solutions were prepared daily by the appropriate dilution of 5000 mg L^{-1} of NH_4^+ stock standard solution with water. A phosphate buffer at 0.1 mol L^{-1} concentration level was prepared by dissolving 13.4 g Na_2HPO_4 in 400 mL of water, adjusting the pH to 11.0 with 2.0 mol L^{-1} NaOH solution and diluting into a 500-mL volumetric flask. A solution of 6.0 mmol L^{-1} sodium sulfite was prepared by dissolving 75.6 mg of Na_2SO_3(s) in phosphate buffer and diluting to 100 mL. A solution of 15.0 mmol L^{-1} o-phthaldialdehyde (OPA) was prepared by dissolving 201.0 mg of solid $C_8H_6O_2$ in 20.0 mL methanol and diluting to 100.0 mL with water. Argon (99.997%, Grade 4.7) was used as a purge gas for triggering the release of the ammonia gas from the liquid mixture and transportation to the trap vessel. For ammonium determination by the certified method [32], the following solutions have been prepared properly: a 5% *m/v* sodium hypochlorite solution (NaClO); a 11.1% (*v/v*) phenol (C_6H_6O) solution in ethanol; a 20% (*m/v*) alkaline citrate solution; a 0.5% (*m/v*) sodium nitroprusside ($Na_2Fe(CN)_5NO$) aqueous solution; and a 1% (*m/v*) NaClO in 16% *m/v* citrates solution.

For the accuracy validation of the developed method, an ammonium standard reference solution traceable to SRM from the NIST (National Institute of Standard and Technology, Gaithersburg, MD, USA) NH_4Cl in H_2O 1000 mg L^{-1} NH_4^+ CertiPUR® has been analyzed, after dilution. Two potable artificial water samples (PAW) and two hygiene artificial water samples (HAW), at 25.0/50.0 μg L^{-1} and at 50.0/100.0 μg L^{-1} concentration levels of $NH_4{}^+$, respectively, were prepared according to the chemical composition of the recycled water in the International Space Station (ISS). The chemical constitution of them is given elsewhere [33].

The environmental water samples were collected from sampling sites that are located in northern Greece, during March 2019. The estuarine water was from the Strymon river, the lake water was from lake Prespa, and the seawater was from the Thermaikos gulf (Thessaloniki) and the Toroneos gulf (Chalkidiki). The collected samples were filtered through 0.45-μm membrane filter, acidified to ca. pH 2.0 with dilute nitric acid, and stored at 4 °C in acid-cleaned polyethylene bottles prior to analysis. The laboratory glassware was rinsed with water after decontaminated overnight in a 10% (*v*/*v*) nitric acid solution.

3.4. On-Line Analytical Procedure

The schematic diagram of the on-line P&T-SIA system for the fluorimetric determination of ammonium is given in Figure 2, while the operational sequence of the proposed method is summarized in Table 4.

At the start-up of the system, the argon purge gas flows through the PV and TV (V1 and V2 in IN position), while the thermostated heater is activated, in order to remove possible ammonia gas molecules from the two vessels. In steps 1-2, 300 μL of trapping solution (0.001 mol L^{-1} HCl) is delivered into the trap vessel, while V1 and V2 are in the "OUT" and "IN" position, respectively. Next (steps 4–7), appropriate volumes of 1.0-mol L^{-1} NaOH and sample/standard solutions are sequentially loaded into the PV by the operation of SP2 through ports 3 and 4 of SV2, respectively. The presence of NaOH creates an alkaline environment inside PV which results in the conversion of ammonium ion to ammonia (gas). Then, the operation of Purge-and-Trap starts (step 10). Specifically, the argon purge gas is flowing in the PV through the sample solution in the PV by switching the V1 in the "IN" position, for a purging time of 600 s. During this step, the sample solution is heated into PV at 80 °C. Consequently, ammonia-gas is released and transported by the argon gas into the TV in order to be dissoluted in the HCl solution (trapping solution).

In the following steps (11–13), the trapping solution is delivered and "parked" in the holding coil HC-2, by the operation of SP1. Then, a small segment of 40 μL is aspirated through port 4 of the COV, into miniSIA analyzer for fluorimetric quantification. The determination of ammonia is based on its reaction with o-phthaldialdehyde (OPA) and sodium sulfite, resulting in the formation of the isoindole derivative. The reaction is depended from the time and temperature. The operational protocol of the miniSIA system, the optimization procedure, as well as all performance characteristics have been presented in detail in our previous work [33]. In the last steps (14–29), a thorough cleaning of the system, including the purge-and-trap vessels and the entire tubing, is performed using ultra-pure water to eliminate any possible carryover phenomenon. Five replicate measurements have been made in all instances.

Table 4. Operational sequences of the P&T-SIA system for ammonia preconcentration and determination.

Step	Valve Position					Operation		Volume (μL)	Flow Rate (μL s^{-1})	Commentary
	V1	V2	SV1	SV2	V	SP1	SP2			
					Thermostated Heater: ON					PV is thermostated at 80 °C
1	OUT	IN	2	5	OUT	Aspirate	-	10	5	
2	OUT	IN	1	5	OUT	Aspirate	-	300	50	Trapping solution into TV
3	OUT	IN	6	5	OUT	Dispense	-	310	50	
4	OUT	IN	6	3	OUT	-	Aspirate	1000	100	Transportation of NaOH into PV
5	OUT	IN	6	5	OUT	-	Dispense	1000	100	
6	OUT	IN	6	4	OUT	-	Aspirate	5000	200	× 2-repeats Loading of Sample into PV
7	OUT	IN	6	5	OUT	-	Dispense	5000	200	
10	IN	IN	6	5	OUT	-	-	-	-	Start of Purge-and-Trap operation
					Purging Time: 600s					Purge Gas in PV, Transportation and Dilution of $NH_3(g)$ Separation/Preconcentration
11	OUT	OUT	2	5	OUT	Aspirate	-	10	5	
12	OUT	OUT	6	5	OUT	Aspirate	-	400	50	Delivery of trapping solution into HC-2
13	OUT	OUT	4	5	OUT	Dispense	-	410	40	
					Start miniSIA operation					Measurement/Quantification
14	OUT	OUT	4	5	OUT	-	Aspirate	5000	300	× 2-repeats
15	OUT	OUT	4	6	OUT	-	Dispense	5000	300	
18	OUT	OUT	4	5	OUT	-	Aspirate	1000	300	
19	OUT	OUT	4	6	OUT	-	Dispense	1000	300	
20	OUT	OUT	4	2	OUT	-	Aspirate	5000	300	Cleaning of PV
21	OUT	OUT	4	5	OUT	-	Dispense	5000	300	
22	OUT	OUT	4	5	OUT	-	Aspirate	5000	200	
23	OUT	OUT	4	6	OUT	-	Dispense	5000	200	
24	OUT	OUT	3	6	IN	Aspirate	-	500	100	
25	OUT	OUT	6	5	OUT	Dispense	-	500	100	
26	OUT	OUT	6	5	OUT	Aspirate	-	500	100	Cleaning of TV
27	OUT	OUT	3	5	OUT	Dispense	-	500	100	
28	OUT	IN	3	6	IN	Aspirate	-	1000	100	
29	OUT	IN	4	6	OUT	Dispense	-	1000	100	Cleaning of HC-2 of miniSIA

4. Conclusions

An automatic thermostated on-line purge-and-trap system based on the instrumentation of sequential injection analysis was developed for volatile compounds for the first time. The efficiency of the proposed P&T-SIA system coupled to a miniSIA flow analyzer has been successfully demonstrated for direct ammonium separation, preconcentration, and determination in environmental water samples. The presented flow method combines the advantages of membraneless purge-and-trap pretreatment and sequential injection systems. The simplicity, versatility, ease, and low costs of the operation are some distinct advantages that the P&T-SIA flow manifold exhibits. According to the obtained results for ammonium determination, it can be concluded that the newly developed method offers remarkable analytical features, such as good accuracy and reproducibility as well as high sensitivity. The applicability was examined in environmental water samples and specifically in seawater samples. The fluorimetric detection method is regarded as a green and environmentally friendly approach, minimizing the use of toxic reagents in μL levels and generating limited wastes. All these characteristics make the proposed P&T-SIA platform an attractive new tool for determining trace amounts of volatile compounds, suitable for monitoring water pollution.

Supplementary Materials: The following are available online: Figure S1: Effect of the temperature on the fluorescence intensity of 50.0 μg L^{-1} NH_4^+. Error bars were calculated based on standard deviation (± 1s). Sample volume = 10.0 mL; NaOH volume = 1000 μL of 1.0 mol L^{-1} NaOH; Trapping solution: 300 μL, 0.001 mol L^{-1} HCl; Purge-gas flow rate = 75 mL min^{-1}, Figure S2: Effect of purge-gas flow rate of the on the sensitivity of the method. Error bars were calculated based on standard deviation (± 1s). Other parameters as in Figure S1, Figure S3: Effect of purging time on the fluorescence intensity of 50.0 μg L^{-1} NH_4^+. Error bars were calculated based on standard deviation (± 1s). All other parameters as in Figure S1.

Author Contributions: Conceptualization, A.A. and A.D.; Data curation and A.A.; Investigation, A.D.; Supervision, A.A.; Validation, A.D.; Writing—original draft, A.D.; Writing—review & editing A.A. All authors have read and agreed to the published version of the manuscript.

Funding: This research received no external funding.

Conflicts of Interest: The authors declare no conflict of interest.

References

1. Clavijo, S.; Avivar, J.; Suárez, R.; Cerdà, V. Analytical strategies for coupling separation and flow-injection techniques. *Trends Anal. Chem.* **2015**, *67*, 26–33. [CrossRef]
2. Melchert, W.R.; Reis, B.F.; Rocha, F.R.P. Green chemistry and the evolution of flow analysis. A review. *Anal. Chim. Acta* **2012**, *714*, 8–19. [CrossRef] [PubMed]
3. Miró, M.; Hansen, E.H. On-line sample processing involving microextraction techniques as a front-end to atomic spectrometric detection for trace metal assays: A review. *Anal. Chim. Acta* **2013**, *782*, 1–11. [CrossRef] [PubMed]
4. Trojanowicz, M.; Kołacińska, K. Recent advances in flow injection analysis. *Analyst* **2016**, *14*, 2085–2139. [CrossRef] [PubMed]
5. O'Connor Šraj, L.; Almeida, M.I.G.S.; Swearer, S.E.; Kolev, S.D.; McKelvie, I.D. Analytical challenges and advantages of using flow-based methodologies for ammonia determination in estuarine and marine waters. *Trends Anal. Chem.* **2014**, *59*, 83–92. [CrossRef]
6. Genfa, Z.; Dasgupta, P.K. Fluorometric measurement of aqueous ammonium ion in a flow injection system. *Anal. Chem.* **1989**, *61*, 408–412. [CrossRef]
7. Kerouel, R.; Aminot, A. Fluorometric determination of ammonia in sea and estuarine waters by direct segmented flow analysis. *Mar. Chem.* **1997**, *57*, 265–275. [CrossRef]
8. Lin, K.; Zhu, Y.; Zhang, Y.; Lin, H. Determination of ammonia nitrogen in natural waters: Recent advances and applications. *Trends Environ. Anal. Chem.* **2019**, *24*, e00073. [CrossRef]
9. Kolev, S.D.; Fernandes, P.R.L.V.; Satinsky, D.; Solich, P. Highly sensitive gas-diffusion sequential injection analysis based on flow manipulation. *Talanta* **2009**, *79*, 1021–1025. [CrossRef]

10. Segundo, R.A.; Mesquita, R.B.R.; Ferreira, M.T.S.O.B.; Teixeira, C.F.C.P.; Bordalo, A.A.; Rangel, A.O.S.S. Development of a sequential injection gas diffusion system for the determination of ammonium in transitional and coastal waters. *Anal. Methods* **2011**, *3*, 2049–2055. [CrossRef]
11. Timofeeva, I.I.; Bulatov, A.V.; Moskvin, A.L.; Kolev, S.D. A gas-diffusion flow injection method coupled with online solid–liquid extraction for the determination of ammonium in solid samples. *Talanta* **2015**, *142*, 140–144. [CrossRef] [PubMed]
12. Wang, L.; Cardwell, T.J.; Cattrall, R.W.; Luque de Castro, M.D.; Kolev, S.D. Pervaporation-flow injection determination of ammonia in the presence of surfactants. *Anal. Chim. Acta* **2000**, *416*, 177–184. [CrossRef]
13. Wang, L.; Cardwell, T.J.; Cattrall, R.W.; Luque de Castro, M.D.; Kolev, S.D. Determination of ammonia in beers by pervaporation flow injection analysis and spectrophotometric detection. *Talanta* **2003**, *60*, 1269–1275. [CrossRef]
14. Mornane, P.; van den Haaka, J.; Cardwell, T.J.; Cattrall, R.W.; Dasgupta, P.K.; Kolev, S.D. Thin layer distillation for matrix isolation in flow analysis. *Talanta* **2007**, *72*, 741–746. [CrossRef]
15. Almeida, M.I.G.S.; Estela, J.M.; Segundo, M.A.; Cerdà, V. A membraneless gas-diffusion unit – multisyringe flow injection spectrophotometric method for ammonium determination in untreated environmental samples. *Talanta* **2011**, *84*, 1244–1252. [CrossRef]
16. Alahmad, W.; Pluangklang, T.; Mantim, T.; Cerdà, V.; Wilairat, P.; Ratanawimarnwong, N.; Nacapricha, D. Development of flow systems incorporating membraneless vaporization units and flow-through contactless conductivity detector for determination of dissolved ammonium and sulfide in canal water. *Talanta* **2018**, *177*, 34–40. [CrossRef]
17. Sramkova, I.; Horstkotte, B.; Sklenarova, H.; Solich, P.; Kolev, S.D. A novel approach to lab-in-syringe head-space single-drop microextraction and on-drop sensing of ammonia. *Anal. Chim. Acta* **2016**, *934*, 132–144. [CrossRef]
18. Timofeeva, I.; Khubaibullin, I.; Kamencev, M.; Moskvin, A.; Bulatov, A. Automated procedure for determination of ammonia in concrete with headspace single-drop micro-extraction by stepwise injection spectrophotometric analysis. *Talanta* **2015**, *133*, 34–37. [CrossRef]
19. Luque de Castro, M.D.; Gamiz-Gracia, L. Analytical pervaporation: An advantageous alternative to headspace and purge-and-trap techniques. *Chromatographia* **2000**, *52*, 265–272. [CrossRef]
20. Giakisikli, G.; Anthemidis, A.N. Automatic pressure-assisted dual-headspace gas-liquid microextraction. Lab-in-syringe platform for membraneless gas separation of ammonia coupled with fluorimetric sequential injection analysis. *Anal. Chim. Acta* **2018**, *1033*, 73–80. [CrossRef]
21. Martínez, E.; Lacorte, S.; Llobeta, I.; Viana, P.; Barceló, D. Multicomponent analysis of volatile organic compounds in water by automated purge and trap coupled to gas chromatography–mass spectrometry. *J. Chromatogr. A* **2002**, *959*, 181–190. [CrossRef]
22. Zhu, Y.; Chen, J.; Yuan, D.; Yang, Z.; Shi, X.; Li, H.; Jin, H.; Ran, L. Development of analytical methods for ammonium determination in seawater over the last two decades. *Trends Anal. Chem.* **2019**, *119*, 115627. [CrossRef]
23. Wang, P.-Y.; Wu, J.-Y.; Chen, H.-J.; Lin, T.-Y.; Wu, C.-H. Purge-and-trap ion chromatography for the determination of trace ammonium ion in high-salinity water samples. *J. Chromatogr. A* **2008**, *1188*, 69–74. [CrossRef] [PubMed]
24. Zhu, Y.; Yuan, D.; Lin, H.; Zhou, T. Determination of ammonium in seawater by purge-and-trap and flow injection with fluorescence detection. *Anal. Lett.* **2016**, *49*, 665–675. [CrossRef]
25. Lin, T.-Y.; Pan, Y.-T.; Lee, H.-Y.; Wang, P.-Y.; Wu, C.-H. Markedly enhanced purge-and-trap performance and efficiency for the determination of ammonium ion in high-salinity water samples. *J. Chin. Chem. Soc.* **2012**, *59*, 718–726. [CrossRef]
26. Sander, R. Compilation of Henry's law constants (version 4.0) for water as solvent. *Atmos. Chem. Phys.* **2015**, *15*, 4399–4981. [CrossRef]
27. Saridara, C.; Brukh, R.; Mitra, S. Development of continuous on-line purge and trap analysis. *J. Sep. Sci.* **2006**, *29*, 446–452. [CrossRef]
28. Inczedy, J.; Lengyel, T.; Ure, A.M. *International Union of Pure and Applied Chemistry, Compendium of Analytical Nomenclature (definitive rules 1997)*, 3rd ed.; Blackwell Science: Oxford, UK, 1998.

29. Marques, K.L.; Pires, C.K.; Santos, J.L.M.; Zagatto, E.A.G.; Lima, J.L.F.C. A multi-pumping flow system for chemiluminescent determination of ammonium in natural waters. *Int. J. Environ. Anal. Chem.* **2007**, *87*, 77–85. [CrossRef]
30. Oliveira, S.M.; Lopes, T.I.M.S.; Tóth, I.V.; Rangel, A.O.S.S. A multi-commuted flow injection system with a multi-channel propulsion unit placed before detection: Spectrophotometric determination of ammonium. *Anal. Chim. Acta* **2007**, *600*, 29–34. [CrossRef]
31. Hong, L.; Sun, X.; Wang, L. Determination of ammonia in water using flow injection analysis with automatic pervaporation enrichment. *Anal. Lett.* **2009**, *42*, 2364–2377. [CrossRef]
32. Federation, Water Environmental, and American Public Health Association. *Standard Methods for the Examination of Water and Wastewater*; American Public Health Association (APHA): Washington, DC, USA, 1999.
33. Giakisikli, G.; Trikas, E.; Petala, M.; Karapantsios, T.; Zachariadis, G.; Anthemidis, A. An integrated sequential injection analysis system for ammonium determination in recycled hygiene and potable water samples for future use in manned space missions. *Microchem. J.* **2017**, *133*, 490–495. [CrossRef]

Sample Availability: Samples of the compounds are not available from the authors.

molecules

MDPI

Review

Review of Fluorescence Spectroscopy in Environmental Quality Applications

Despoina-Eleni Zacharioudaki, Ioannis Fitilis and Melina Kotti *

Department of Electronic Engineering, Hellenic Mediterranean University, Romanou 3, 73133 Chania, Greece; despoinazax@gmail.com (D.-E.Z.); fitilis@hmu.gr (I.F.)
* Correspondence: kotti@hmu.gr; Tel.: +30-28210-23007; Fax: +30-28210-23010

Abstract: Fluorescence spectroscopy is an optical spectroscopic method that has been applied for the assessment of environmental quality extensively during the last 20 years. Most of the earlier works have used conventional light sources in spectrofluorometers to assess quality. Many recent works have used laser sources of light for the same purpose. The improvement of the energy sources and of the higher resolution spectrometers has led to a tremendous increase in applications. The motivation for the present review study is the increasing use of laser sources in environmental applications. The review is divided in two parts. The fundamental principles of fluorescence spectroscopy are described in the first part. The environmental applications are described in the second part.

Keywords: conventional lamps; fluorescence; laser sources; laser induced fluorescence; environmental quality

Citation: Zacharioudaki, D.-E.; Fitilis, I.; Kotti, M. Review of Fluorescence Spectroscopy in Environmental Quality Applications. *Molecules* **2022**, 27, 4801. https://doi.org/10.3390/molecules27154801

Academic Editors: Victoria Samanidou, Eleni Deliyanni and Dimitra Voutsa

Received: 21 June 2022
Accepted: 23 July 2022
Published: 27 July 2022

1. Introduction

Optical spectroscopy deals with the study of interactions between matter and light such as absorption, emission and scattering, among others. Fluorescence spectroscopy (or fluorometry) is based on the emission of photons from a substance after excitation from light absorption. The molecules, due to their vibrational energy levels, emit light of lower energy (longer wavelength) than the absorbed light. This is called Stokes' shift and happens due to an energy loss in non-radiative decay. These processes are shown in the so-called Jablonski diagram in Figure 1a. When a molecule is in the ground level (S_0) and absorbs a photon of sufficient energy, an electron is promoted to a higher energy singlet level (S_1 or S_2) equal to the energy of the absorbed photon. If the electron, following a relaxation pathway, returns to its original state emitting a photon, it is called photoluminescence. The two types of photoluminescence are fluorescence and phosphorescence. Fluorescence refers to when the electron from a singlet excited level decays radiatively to the ground singlet state, within a characteristic decay time of the order of 10^{-10} s to 10^{-7} s. Phosphorescence occurs when the electron, after a transition from a singlet excited state to a triplet state, returns radiatively to the ground singlet state. Since the last process involves change of the electron spin, the phosphorescence time could be from 10^{-6} s up to seconds.

For a compound, its emission spectrum and absorption spectrum (excitation spectrum) are usually almost mirror images as shown in Figure 1b for the radiative transition from S_0 to S_1 and vice versa. This symmetry is attributed to the same vibrational levels structures that are involved in each of these processes.

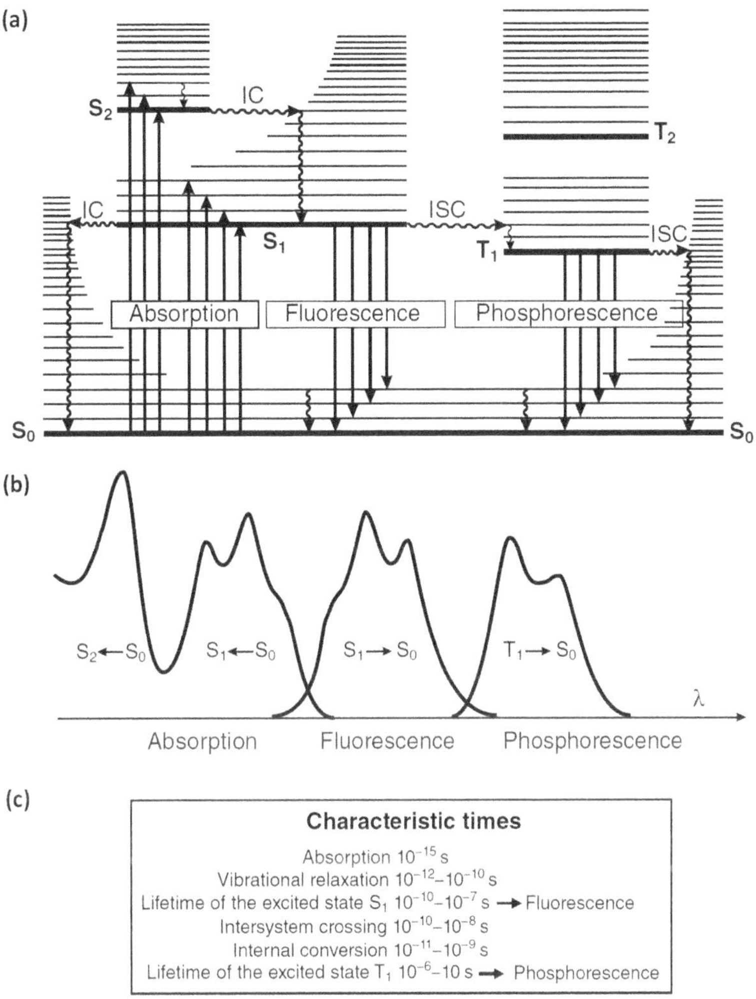

Figure 1. (**a**) Perrin–Jablonski diagram with the possible radiative transitions (straight arrows) of absorption, fluorescence and phosphorescence, as well as the non-radiative transitions (wavy arrows) of vibrational relaxation, internal conversion (IC) and intersystem crossing (ISC). (**b**) Illustration of the spectra for the radiative transitions between electronic states that shown in (**a**). (**c**) Characteristic times for each transition. Reprinted from [1]. 2012, John Wiley and Sons.

1.1. Light Sources

Fluorescence excitation can be induced by a lamp, a light-emitting diode (LED) or a laser source. When a laser is used for the excitation, the fluorescence is called laser induced fluorescence (LIF). As lamps are conventional sources of light, the fluorescence is considered as conventional in that case. For multispectral light sources, a spectral filter or monochromator is used for selecting the excitation wavelength.

Typically, the commercial spectrofluorometers utilize arc-lamps as light sources for the excitation of compounds. The main sources used for ultraviolet (UV) and visible light are high-pressure and low-pressure lamps of Xenon (Xe) and mercury (Hg). The line spectrum emission of low-pressure Hg-Ar and Hg lamps are used mainly for calibration purposes. The high-pressure Xe and Hg lamps provide broadband light emissions from 250 nm and from 350 nm, respectively, up to IR. The intensities of high-pressure Hg lamps are higher compared with Xe lamps. There are also high-pressure Xe-Hg lamps with higher intensities than Xe lamps. Deuterium lamps are also used, providing continuum emission in UV, from 160 nm to 400 nm. The tungsten-halogen lamps are less used in fluorescence spectroscopy than the previously mentioned ones due to their weak light emission below 400 nm [2].

In the last two decades, the light-emitting diodes (LEDs) began to appear as light sources in spectrofluorometers. The LED is an electroluminencence device that produces photon emission by the recombination of electrons with electron holes at semiconductor junction. They are inexpensive light sources with bright illumination and the development of LEDs emitted up to deep UV range make them very attractive for fluorescence excitation [3].

The laser sources are categorized in four types according to the active medium: solid state, gas, liquid and semiconductors/diode lasers. Among the solid state lasers, the Neodynium-doped Yttrium/Aluminium Garnet (Nd: YAG) is the most common type used in fluorescence spectroscopy [4]. The fundamental wavelength of the laser is in the infrared (IR) at 1064 nm but with the use of nonlinear crystal the second or higher harmonics could be produced, providing laser wavelength at 532 nm, 355 nm, 266 nm or even 213 nm.

The most common gas lasers include argon ion, HeNe, nitrogen and excimers [5]. Argon ion lasers emit radiation at 488 nm (blue) and 514 nm (green); hence, they are called also blue-green lasers. They are relatively large lasers and external cooling is required. Helium-neon (HeNe) lasers operate at a single wavelength of light, most often at 632 nm. They do not produce high-power light (from few to tens mW), but they are more stable and are often used for metrology applications. HeNe lasers are more compact than argon ion lasers and do not require external cooling. Another type of gas lasers used in spectroscopy are nitrogen lasers, which operate in the UV range, as they emit radiation at 337 nm. They have been used for air pollution monitoring. Excimer lasers are another type of gas lasers for UV emission. The laser medium are short-live dimeric molecules generated in excited state of inner nobble gas atoms themselves (excimer) such as argon, krypton, xenon or by their combination with halogens (exciplex) such as fluorine or chlorine gas atoms.

The liquid lasers consist of a dye as lasing material. An organic dye dissolved in water or other solvent (at a typical concentration of the order of 1 part in 10000), is radiated by an intense light source [6]. After the absorption of light in one wavelength, the dye emits light over a broad range of visible wavelengths. Different dyes can produce a variety of wavelength emissions. Coumarin and Rhodamine are some of the most used dyes [7,8]. By changing the cavity length, or through other techniques, the output wavelength can be easily tuned in a wide range of tens of nm. Their drawbacks are that the dye undergoes photodecomposition and needs to be renewed by circulation, either in a container cell or as stream of jet in open air. In addition, the dyes are hazardous toxic materials and great attention when handling them is needed.

The last category of lasers is the diode (or semiconductor) lasers that work on a somewhat different principle. Light is emitted by flowing electrical current through the semiconductor due to the energy gap in the diode's junction. They are compact lasers that emit wavelengths typically from visible to IR, but recently there are devices emitting in the UV region [9]. Their output has a wide range of power from low to moderate. They have small size, are lightweight and less expensive than other lasers [10].

Generally, the choice of laser type depends on the wavelength range that is needed in each research. The application of laser depends on the properties (monochromaticity, directionality, spatial and temporal characteristics) of the laser beam.

1.2. Detectors

The fluorescence light is detected and quantified either by a photomultiplier tube (PMT) or avalanche photodiodes (APD) or by a charge-coupled device (CCD). The majority of commercial fluorometers use photomultiplier tubes (PMT) to detect the low light intensity of fluorescence [2]. A PMT is a vacuum glass tube that consists of a photocathode where incident photon induces electron emission, a series of dynodes for electron multiplication and an anode. The material that the photocathode is made from determines the spectral range of the photomultiplier. In order to cover the whole UV-visible range, two types of photocathodes are required.

The APD is another low-intensity light detector used in fluorescence spectroscopy. In a photodiode an electron-hole pair generated when a photon is captured in the diode junction area. In APD, a high reverse bias voltage creates a strong electric field where the electron generated from photon is accelerated to produce secondary electrons by impact ionization. The resulting electron avalanche produces measurable electrical signal even from few photons [11].

Another type of detector used in fluorometers is the CCD with which the whole fluorescence spectrum can be analyzed at once without the need for wavelength scanning with a monochromator. It is mainly used when the emitted light is of higher intensity, such as in LIF. The CCD consists of an array of semiconductive photosensitive elements (pixels) where photons induce an electrical charge which is accumulatively stored in each of them. Then, the read out of each element charge is made in series by shifting the charge of each element to their neighbor toward the charge measurement circuit where it is also digitized.

Typical schemes of a conventional and a laser induced fluorescence system are shown in Figure 2.

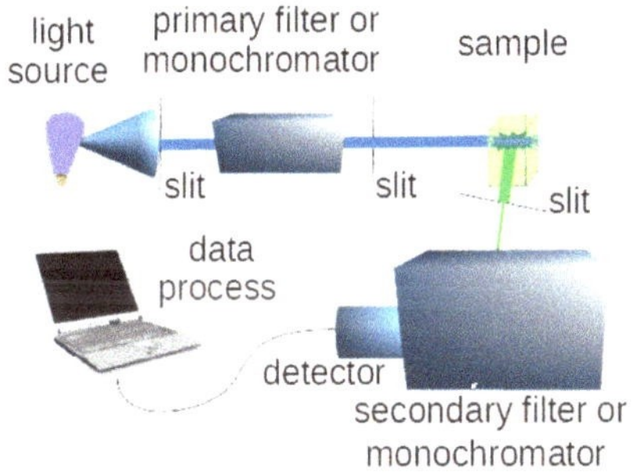

(a)

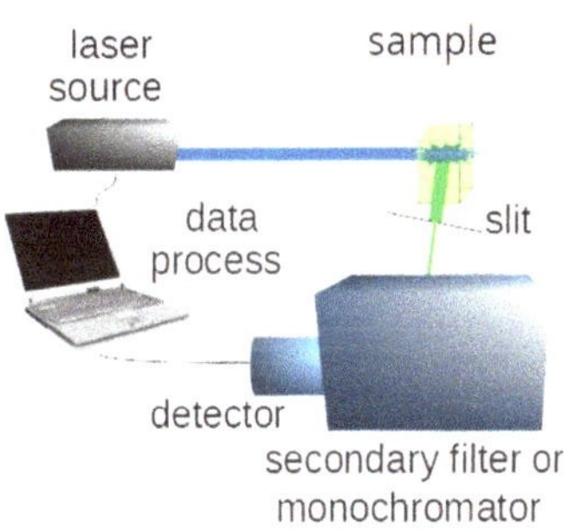

(b)

Figure 2. Schemes of a conventional (**a**) and a laser-induced (**b**) fluorescence spectroscopy system.

1.3. Fluorescent Compounds

The compounds that absorb visible and/or UV light are called chromophores. The compounds that emit light are called fluorophores. Compounds with fluorescent characteristics are those with several aromatic (fused) rings and/or with conjugated double bonds. There are two types of fluorophores depending on the groups: those with electron-donating groups, such as -OH, -NH_2 and -OCH_3 that increase fluorescence, and those with electron-withdrawing groups, such as COOH and -N=N- that reduce fluorescence. There are some exceptions, for example fluorophores such as tryptophan and tyrosine, that showed lower quantum yields than expected [12]. Compounds such as halogens, oxygen and acrylamide are also known to reduce fluorescence and will be described in the next section. Heterocyclic compounds do not fluoresce significantly unless they are attached to an aromatic ring.

The major characteristics of a fluorophore include the fluorescence lifetime (τ_F) and the quantum yield (Φ). Fluorescence lifetime is defined as the average time that the molecule remains in the excited state before it returns to its ground state, of the order of ns. The quantum yield is defined as the ratio of the number of the emitted photons to the number of the excited molecules. The higher the quantum yield, the higher is the fluorescence. It is almost equal to unity when the non-radiative decay rate is much smaller than the radiative (fluorescence) decay rate.

The intensity of fluorescence (I_F) is proportional to the molecule's concentration in diluted solutions ($\varepsilon \cdot b \cdot C < 0.05$), as described from the following equation:

$$I_F = k \cdot I_o \cdot \Phi \cdot (\varepsilon \cdot b \cdot C) \quad (1)$$

where:

k: a constant dependent on the instrument;
I_o: the intensity of the incident light;
Φ: the quantum yield;
ε: the molar absorptivity;
b: the path length and
C: the molecule's concentration.

The optical properties of standard organic fluorophores are founded in many databases and include: the wavelengths of maximum absorption and emission and their bandwidths (full width at half maximum), extinction coefficient, photoluminescence quantum yield and fluorescence lifetime [13].

1.4. Factors That Affect Fluorescence

The factors that affect fluorescence include solvents, temperature, pH and ionic strength as well as the presence of other substances. The decrease in fluorescence intensity caused by any factor is called quenching, while the substances that may induce this are called quenchers. The decrease in the fluorescence intensity happens either by intramolecular or by intermolecular interactions.

One of the most frequently present quenchers is the molecular oxygen which quenches almost all known fluorophores. Other quenchers are acrylamide, amines and halogens. Table 1 summarizes examples for quenchers of some typical fluorophores. There are two mechanisms of quenching: static and dynamic. Static quenching happens when a non-fluorescent complex is formed between the fluorophore and the quencher. Dynamic quenching, otherwise called collisional quenching, happens when the quencher interferes with the fluorophore during the lifetime of the excited state. The excited molecule is then deactivated either by contacting other molecules or by intermolecular interactions (collisions). The Stern–Volmer equation describes the fluorescence quenching as follows [2]:

$$I_o/I = 1 + K_q \cdot \tau_F \cdot [Q] \quad (2)$$

where:

I_o: the intensity without quencher;
I: the intensity with quencher;
K_q: the quencher rate coefficient;
τ_F: the fluorescence lifetime and
Q: the quencher's concentration.

Table 1. Examples of typical fluorophores and their quenchers.

Typical Fluorophores	Quenchers
Polycyclic aromatic hydrocarbons (PAHs)	Nitrocompounds, nitromethane [14,15]
Anthracene	Diethylaniline [16]
Tyrosine	Disulfides [17], phosphates [18]
Tryptophan, indole	Acrylamide [19,20], cations [21], anions [22]
Aromatic hydrocarbons	Aromatic and aliphatic amines, pyridinium salts [23]
Majority of known fluorophores	Oxygen [24,25]

The fluorescence of a compound can be affected from the solvent used for the fluorophore solution. The solvent effect can be observed as a shift of the spectral maxima (solvatochromic shift), as a change of the intensity of the spectral line or band and as a change of the shape and width of the band [26]. By increasing the solvent polarity, shifts of the emission spectrum to longer wavelengths (red shifts) are usually observed.

The pH parameter affects fluorescence as the structure of a molecule can be altered by altering the pH. For example, a compound can become a spherical or linear shape in different pH values [27]. The ionization of a molecule after pH modification may alter the molecular orbital of the excitable electrons. Furthermore, H^+ ions compete with metal ions in complexation with dissolved organic matter (DOM) and thus, there are metals that increase the fluorescence in water samples [28].

Additionally, the effect of temperature is very important. The electrons return to the ground state by radiationless processes. By decreasing the temperature from 45 °C to 10 °C, it was found that the intensity of DOM has been increased by 48%. [29]. Finally, the ionic strength affects fluorescence by changing the conformity and by charge transfer [30].

1.5. Types of Laser Induced Fluorescence

Laser induced fluorescence (LIF) is classified as steady-state (or continuous wave, CW) when only the spectral information of the emission is recorded, or as time-resolved LIF, where information of the fluorescence lifetime is derived [1]. For CW LIF, where the time integrated fluorescence is recorded, both CW and pulsed laser sources can be used. In time-resolved LIF, for the fluorescence analysis in time domain, a pulsed laser is necessary with pulse duration below ns, while in frequency-domain techniques, an optical modulation of CW laser can be used. Most types of lasers can operate in both CW and pulsed mode, while there are few types that in principle cannot be run in CW mode.

The LIF spectroscopy can also be categorized according to its spectral operation, as excitation LIF when selecting the excitation wavelength or as emission LIF when the emitted light is spectrally analyzed. In excitation LIF spectroscopy, the excitation wavelength is varied using a tunable laser and each time the total emitted light is detected using a filter in front of the detector to remove any scattered laser light. In emission LIF spectroscopy, a fixed wavelength laser is used to excite the sample and the emission spectrum is measured by using a monochromator before the detector to scan the wavelength or by using a spectrometer [31].

The high intensity light, achieved from focusing the pulsed laser beam, is able to induce fluorophore excitation by the simultaneous absorption of two or more photons of lower energy (longer wavelength) [32]. Multiphoton excitation has a rather small probability, as it depends non-linearly to the light intensity (photon fluence) and becomes efficient only in the region of the focal spot. For the two-photon absorption, there is quadratic dependence on laser intensity. The excitation photons originate from one laser beam or by different laser sources. The two-photon absorption spectrum may differ in shape from the one-photon spectrum, as there are different selection rules applied for them that allow or prohibit specific transitions [33]. The two-photon absorption laser induced fluorescence (TALIF) spectroscopy could be complementary to the conventional LIF capable of revealing energy states not accessible from one-photon transition, to excites in UV range using laser of double wavelengths that are more easily available to clearly separate the excitation laser wavelength from the detected emitted light.

LIF spectroscopy is often combined with the laser-induced breakdown spectroscopy (LIBS) as an emerging analytical technique for the elemental analysis of various samples [34]. In LIBS, the sample is irradiated by a high-power pulsed laser to generate localized plasma and breakdown the material into excited ionic and atomic species, whose emission is then spectroscopically analyzed. By utilizing a second laser beam tuned to selectively excite the plasma species, a great enhancement of their emission capability is achieved [35].

1.6. Fluorescence Recording

The fluorescence signals could be recorded in several ways. In emission LIF spectroscopy, the intensity for each wavelength of the emitted light is recorded for a fixed excitation wavelength to derive the fluorescence emission spectrum. In excitation LIF spectroscopy, the intensity of the total light emitted is recorded whilst scanning the excitation wavelength to derive the fluorescence excitation spectrum.

There are other sophisticated methods of conventional fluorescence spectroscopy that also apply in LIF, such as the synchronous fluorescence spectroscopy (SFS), the total synchronous fluorescence spectroscopy (TSFS) and the excitation-emission matrix (EEM). In SFS, the emission spectrum is recorded while both the excitation and emission wavelengths are scanned simultaneously but maintaining a constant difference (offset) between them $\Delta\lambda$ [36]. Synchronous spectra provide more information compared to a single scan. In TSFS, the output result is a contour map that contains numerous synchronous spectra at different offsets. Finally, the EEM method that was introduced in 1977 [37,38] results in a three-dimensional (3D) plot of fluorescence excitation wavelength versus emission wavelength and intensity. It provides full detailed information that can be used to identify several fluorophores present in complex mixtures. They are easy distinguished because the maximum fluorescence intensity for each of them is resulted only from one pair of $\lambda_{ex}/\lambda_{em}$ in the matrix.

Furthermore, for quantitative measurements, the data from recording spectra need to be quantified. For this purpose, many standard fluorescence indices have been developed and are defined as the ratios of emission intensity at two different points or areas [39]. The data obtained from EEM are often reduced with statistical methods, with parallel factor analysis (PARAFAC) be the most popular discriminant analysis.

1.7. Interferences in Fluorescence Measurement

Scattering is a major problem encountered in fluorescence measurements. There are two types of scattering: the Raman (inelastic) and the Rayleigh (elastic) scattering. Rayleigh scattering occurs when there are molecules of smaller size than the wavelength of the excitation light, while the scattered light has the same wavelength. It is easily filtered out from the longer wavelengths of emission. The Rayleigh scatter in EEM appears as a visible diagonal line at the emission wavelength equal to the excitation wavelength, while for the second order of Rayleigh scatter, the bright line appears at the emission wavelength equal to double the excitation wavelength. Raman scatter happens because the light is absorbed and re-emitted with loss in photon energy. The loss is due to the vibrational states and the scattered light having a higher wavelength than the excitation light. Water has scattering properties from the vibration of O-H bonds, and in aqueous samples the Raman line appears in EEM as a diagonal line at excitation wavelengths from 260 to 350 nm and at emission wavelengths from 280 to 400 nm.

The inner filtering effect (IFE) is another phenomenon that affects the recorded fluorescence. The excitation IFE occurs because a part of the incident light is absorbed from the fluorophore before reaching the sample area form where the fluorescence emission is collected in. Therefore, when increasing the fluorophore concentration, the fluorescence intensity increases up to a certain point and then starts to decrease. Moreover, the emission of IFE is caused by the self-absorption of the emitted light by the fluorophore before it reaches the detector. This results in a reduction of the recorded spectrum, where the absorption overlaps with the emission which changes the shape of the recorded spectrum in the low wavelength portion. Additionally, IFE could be induced due to presence of other chromophores that absorb the excitation or the emission light. In most environmental samples, the main reason for the IFE is the naturally dissolved humic material [40]. The IFE problem is compensated by reducing the path length in the sample for the excitation or/and emission light and by diluting the water samples to have an absorbance of less than 0.1 at 254 nm [41].

2. Applications of Fluorescence in Environmental Samples

The detection of target pollutants in liquid and solid environmental samples has been extensively implemented by chromatographic methods coupled with fluorescence detectors. Fluorescence detectors are of high selectivity as there are few compounds that fluoresce. They are also of higher sensitivity (10–1000 times) compared to diode array UV detectors. A tremendous number of publications are dedicated to the determination of pollutants such as polycyclic aromatic hydrocarbons (PAHs) in water [42–44], in soils [45,46], in sediments [47,48], in pesticides in water [49,50], in soil [51,52], in pharmaceuticals in water [53,54], in soils [55,56], in sediments [57,58] and in metals in water [59,60] by chromatographic techniques coupled with fluorescence detectors. In all publications, a preconcentration of the samples is required, as the pollutants exist in very low concentrations. The preconcentration can be achieved by liquid–liquid extraction, solid phase extraction or solid phase microextraction for aqueous samples. For the solid samples such as soils and sediments, solid-liquid extraction and fractionation according to pH have usually been applied. Besides, fluorescence has been used for the direct characterization of quality of environmental samples without preconcentration. The applications of fluorescence in environmental samples without preconcentration are described as follows.

2.1. Applications of Conventional Fluorescence in Various Types of Water

Without any sample preconcentration and/or pretreatment (except filtration if necessary), conventional fluorescence has revealed useful information about the water and wastewater quality. The majority of studies have focused on the dissolved organic matter (DOM), which is the major water constituent, while less have focused on oil pollution.

The first significant studies were those of Coble [61,62] that have characterized marine and terrestrial DOM in seawater by EEM fluorescence spectroscopy. The peaks were classified into five categories and named as A, B, C, T and M. Both peaks A and C were of humic-like fluorescence, with peak A irradiated by UV excitation and peak C by visible excitation. Peaks B and T were of tyrosine-like and tryptophan-like fluorescence, respectively. Peak M was specific for marine humic-like fluorescence.

The distinguishment between surface water from eastern (Atlantic and modified polar water) and western (Canada-basin polar water) Arctic sectors was detected by using EEM fluorescence spectroscopy [63]. In this work, the Eurasian polar water showed higher visible DOM fluorescence signals than the water from the Canada basin.

The characterization and monitoring of wastewater in surface waters has been achieved by EEM fluorescence spectroscopy [64]. The peaks T (living and dead cellular material and their exudates) and C (microbially reprocessed organic matter) from wastewater samples presented much higher intensity compared with those from natural waters. Furthermore, peak T fluorescence was highly reduced after the biological treatment process, while peak C was almost completely removed after the chlorination and reverse osmosis processes.

Another application of EEM coupled with principal component analysis and second derivative analysis has characterized wastewater samples after each treatment process in a municipal wastewater plant [65]. Figure 3 shows an example of a wastewater sample after anaerobic treatment.

Another work [66] has used EEMs to distinguish the origin and the distribution of DOM in different water samples such as oligotrophic oceanic waters, reef waters, river waters and groundwater. The fluorophores that were identified within the samples were: humic-like A, humic-like C, marine humic-like M, tryptophan-like T1 and T2 and tyrosine-like B1 and B2. Some unknown peaks (U1 and U2) have also been identified.

Huang [67] identified six fluorophores by EEC-PARAFAC in the eutrophicated lake Taihu during autumn. They were named A, B, C, N, M and T. The results showed that red shift happened with increasing fluorescence intensity for peaks A (UV humic-like) and C (terrestrial humic-like) while peak M had the reverse red shift. From another survey in the same lake, only four fluorophores were detected during one month [68] by EEC-PARAFAC.

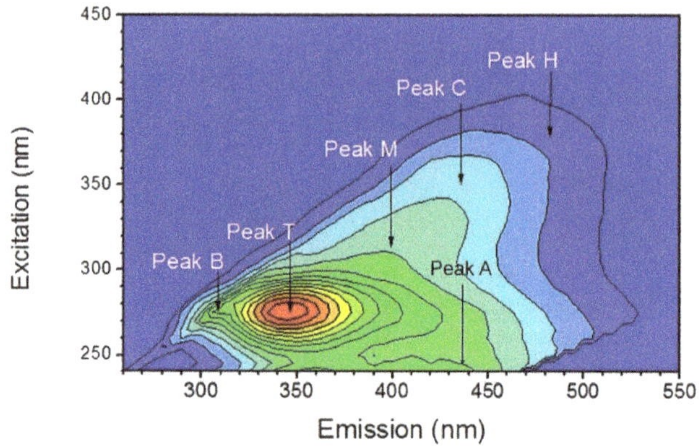

Figure 3. EEM feature and the position of peaks A, C, B, M, H and T in a wastewater sample after anaerobic treatment. Adapted with permission from Ref. [65]. 2018, Elesevier.

Goslan [69] studied the seasonal change of DOM and its effect on the treatment processes by EEM and by synchronous fluorescence. At least two fluorophores were identified in non-fractionated and fractionated (acids, bases, neutrals) water samples. The fractionation of samples was performed by resin according to their hydrophobicity.

In rainwater samples [70], three fluorophores were identified by EEM: humic-like C, tyrosine-like B and tryptophan-like T. In addition, the results show that humic-like fluorescence was strongly influenced by terrestrial/anthropogenic sources, while tyrosine and tryptophane-like fluorescence was not influenced from the meteorological variables.

Synchronized fluorescence by a conventional spectrofluorometer at $\Delta\lambda$ = 30 nm between emission and excitation wavelengths was applied in an estuary located in northeast Brazil [71]. Four peaks were identified at wavelengths 278–280 nm (peak I), 350 nm (peak II), 385 nm (peak III) and 458–460 nm (peak IV). Peak I was due to microbial production while peaks II, III and IV were related to humic and fulvic acids.

Generally, the main fluorophores and their positions as identified by most researchers [62,72] in various types of water are summarized in Table 2.

Other researchers [73] have studied the oil dispersion in seawater by recording the emission spectra. They found that the intensity at 445 nm is indicative of higher molecular weight PAHs without having to measure the oil concentration.

The concentrations of three PAHs were predicted by EEM-PARAFAC in urban run-offs from asphalt paved roads [74]. The PAHs that were studied were: phenanthrene, benzo (k) fluoranthene and benzo (a) pyrene.

Table 2. Main fluorophores and their positions.

Name	Letter	Ex (nm)/Em (nm)
Humic-like	C	340–360/420–480
Fulvic-like	A	240–260/380–460
Tyrosine-like	B (B1, B2)	265–285/290–310
Tryptophan-like	T (T1, T2)	265–285/290–340
Microbial-like (Marine-like)	M	310–330/390–410
Humic-like	H	370–390/480–500

2.2. Applications of LIF in Various Types of Water

Laser induced fluorescence has also been used for the characterization of water quality without any preconcentration. The majority of these studies have used solid state lasers.

Uebel [75] applied the LIF method in order to detect water pollutants and their possible interactions with phytoplankton and break-down products (yellow-substances) in situ. They used a frequency doubled dye laser as an excitation source with a pulse energy of 10 mJ and a pulse duration of 10 ns. The substances were excited in the range from 265 nm to 400 nm while the fluorescence signal was recorded in the range of 310 nm–750 nm. Different fluorescence spectra appeared during ageing and dying of the phytoplankton.

For the detection of PAHs and oils in groundwater, Baumann [76] chose a nitrogen laser and time-resolved LIF based on the different decay times of humic substances and PAHs. They calculated the concentrations of 16 PAHs and found a good correlation with the results obtained by HPLC coupled with a fluorescence detector.

In a work of Sivaprakasam and Killinger [77], two different LIF instruments were tested for the determination of DOC and quinine sulfate in natural water samples, bottled distilled and bottled drinking water samples, where it was found that plastic-related compounds were leached into the water from the containers. One instrument utilizes a UV tunable dye laser (200–285 nm, 0.2–5 μJ, 10 Hz) with a spectrometer and CCD detector, while the other one is a portable system utilizing a fixed wavelength microchip laser (at 266 nm, 1 μJ, 8 KHz) with a gated PMT detector and bandpass interference filters. Although both systems use laser pulses of the same energy, the higher repetition rate in the latter system provides a much greater signal-to-noise ratio (SNR) measurements due to the high pulse averaging and the higher total energy output. In conjunction with the detection system, it was found to have a 10 times higher sensitivity than the first system, but with much slower processing time to obtain a full emission spectrum. It was found to have up to 100 times higher sensitivity from the best commercial portable spectrofluorometer, while it was tested in situ for the determination of plastics and DOC in seawater samples [78]. Additionally, they upgraded it to have two interchangeable microchip lasers operating at 266 nm and 355 nm, achieving the tracking of DOC in the clean ocean water by continuously operating the portable LIF system for five days [79].

The same portable LIF unit was tested to measure the fluorescence from tap water and sea water after being treated by reverse osmosis [80] and in ground and drinking water samples [81]. It was found that the deeper UV laser showed more distinct spectra with quantitative features and gave better separation of the LIF from the Raman peak allowing the detection of unique spectral features. Most of the LIF systems utilize laser sources in deep UV.

Ghervase et al. [82] have chosen the fourth harmonic output of an Nd:YAG laser for the excitation of both microbial and humic-like substances due to the high energy of the excitation photons (266 nm). They detected the impact from human and chicken waste in rivers and found that chicken waste had a specific fluorescence signature.

LIF and excitation-emission matrix (EEM) fluorescence were applied in river samples from the lower basin of the Arges River. They detected urban sewage contamination by picking up fluorescence signals, such as that from tyrosine and the presence of folded and unfolded tryptophan residues [83].

Recently, Du et al. [84] have used a UV laser at 266 nm in order to detect the presence of three aromatic amino acids in seawater (tryptophan, tyrosine, and phenylalanine) in situ. The peaks of tryptophan, tyrosine and phenylalanine were detected at 350, 300 and 280 nm, respectively, as shown in Figure 4, and their concentrations were quantified.

In a recent work [85] a small-sized spectrofluorometer operated at 278 nm was used for the identification of oil products in seawater. They found that the spectral features were changed depending on the state of the oil product.

A combination of LIBS and LIF was used in order to detect trace amounts of heavy metals in water samples. Specifically, lead was detected by using a Q-switched Nd:YAG laser to produce plasma, at 1064 nm or 532 nm [86]. In another work, lead was detected by micro-LIBS with LIF by using a 170 μJ laser pulse for ablation and a 10 μJ laser pulse for re-excitement [87]. Other metals that were detected with LIBS were cadmium after enrichment with a resin [88] and chromium [89,90]. In addition, LIBS portable compact systems were employed for in situ seawater analysis for deep sea [91–93].

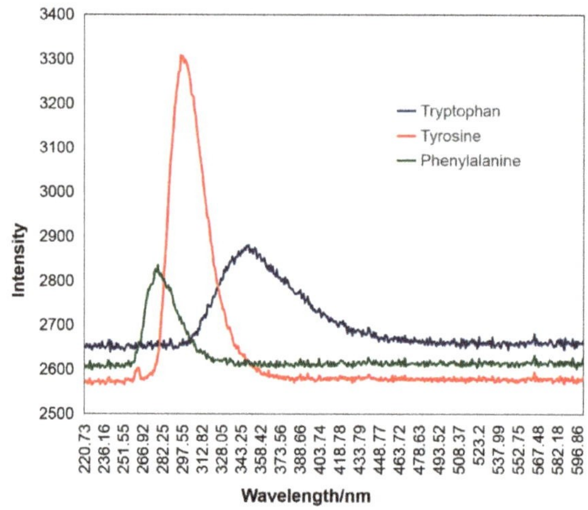

Figure 4. LIF spectra of the three aromatic amino acids as presented by [84].

A totally different application of LIF was used for the classification of viruses. In a recent study [94], a laser at 266 nm with a pulse repetition of 10 kHz and a power of 25 mW has been used for virological analysis of environmental samples. Although the high repetition rate of high energy pulses, the SNR was not always adequate to discriminate the virus in small concentrations. However the LIF method could significantly reduce the time and operational cost of virus analysis.

2.3. Applications of Conventional Fluorescence in Soils and Sediments

Conventional fluorescence has been applied in soils and sediments after isolation of the target components such as humic substances, oils, etc.

Surface marine sediments were extracted by proper solvent and examined for the presence of bulk PAH levels by a portable fluorescence apparatus [95]. The results showed good correlation with the lab results.

EEM and synchronous scans were obtained by [96] to examine the concentrations of PAHs in soils. By using a fluorescence fingerprints library with several EEM and SFS maps for various dilutions of Romanian crude oil in methanol, they confirmed the identity of the soil pollutant. They created a calibration to estimate the pollutant concentrations.

EEM–PARAFAC was used in a study [97] conducted in lake bottom sediments of selected lobelia lakes in order to assess their properties and their origin. The optical properties of HA extracted from the sediments were compared to the parameters that describe their structural and chemical properties. Four components were identified: two protein-like (C2 and C4), one humic-like (C1), and one fulvic-like (C2). The more dominant component was C2 and the less dominant one was C4. Each of the components revealed different information. The results showed that the organic matter (OM) present in the bottom sediments from sampled lakes had autochthonic origin and consisted of mostly labile organic compounds.

Five fluorescent components were identified by EEM-PARAFAC in soils and sediments from two different estuaries in Spain and were found in both FA and HA fractions with different abundance [98].

In surface and deep sediments from Toulon Bay in France, EEM-PARAFAC identified two components [99]: a "fresh" particulate OM in surface sediments, which produces protein like high molecular weight-DOM and low molecular weight-DOM and a "buried" particulate OM in the deeper sediment layer. In porewaters, three components were identified: C1, C2—both humic-like—and C3 of protein-like fluorescence.

From another study [100] conducted in a forested watershed, three fluorescent compounds were identified from soil/sediment HS: a terrestrial humic-like (C1), a microbial humic-like or fulvic-like (C2) and a protein-like component (C3) as shown in Figure 5. Component C3 was probably related to an autochthonous organic input to the reservoir sediments and/or phenolic compounds. They used EEM-PARAFAC spectroscopy.

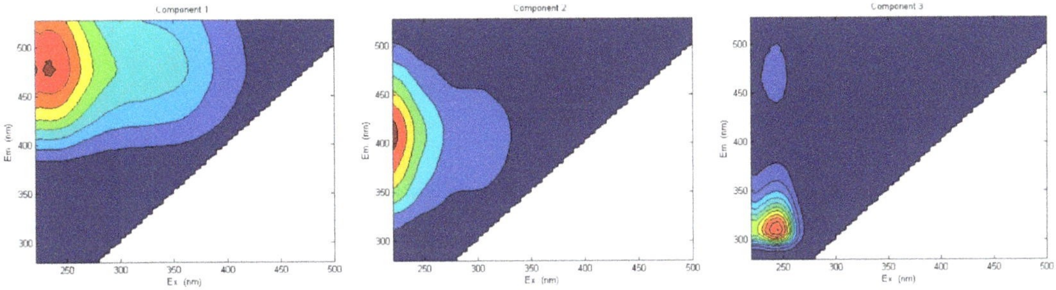

Figure 5. The three fluorescent components identified by EEM-PARAFAC. Adapted with permission from Ref. [100]. 2017, Springer Nature.

EEM-PARAFAC has also distinguished soils that were treated with mineral fertilizers and organic manure from those treated with only mineral fertilizer and those without fertilization [101].

Synchronous fluorescence at $\Delta\lambda = 20$ nm has been applied in different types of soil samples and revealed the presence of five main peaks at 360, 470, 488, 502 and 512 nm. They assessed the soil humification degree in different types of soils [102].

2.4. Applications of LIF in Soils and Sediments

LIF has been applied to soil samples without any chemical and or/physical pretreatment and extraction procedures.

The influence of a sewage-sludge addition on the OM of a Brazilian oxysol was studied by an argon laser emitted at 351 nm with an output power of 400 mW [103]. The application of LIF was performed on unfractionated soil samples, fractionated with chemical methods and fractionated by physical methods (particle size). The results indicate that every particle size fraction showed a different shape, revealing differences in organic compounds bounded to them. It was also used to obtain LIF spectra of pelletized whole soils from different origins and various depths [104]. A similar apparatus, utilizing a CW laser of 20 mW power at 405 nm for the excitation, was used for the characterization of OM in pelletized soil samples [105].

The LIBS spectra of soils were obtained using a Q-switched Nd:YAG laser at 1064 nm [106]. The humification degree (HD) of OM was evaluated for the first time by LIBS. The results show high correlation with those determined by LIF using a CW diode laser at 405 nm.

The results from the previous studies were compared in a review by Senesi [107] and it was found that there is a high correlation between (HD) LIBS and (H) LIF index values, and that LIBS can be used to evaluate the humidification degree of soil organic matter (SOM).

LIF was also applied in dry and hydrated crusts by an Nd:YAG laser operating at 532 nm with 7 ns pulses [108]. The microphytobenthos (MPB) showed three peaks (two main at 570 and 650 nm, and a secondary at 720 nm) as is shown in Figure 6. Furthermore, the only slight difference between bare soil, either dry or hydrated, was the intensity of the peaks.

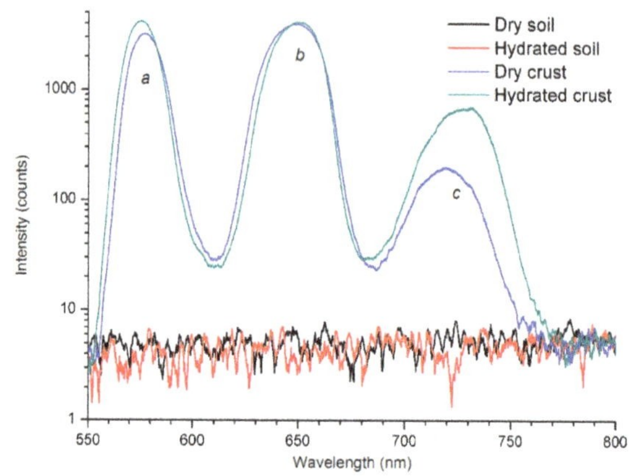

Figure 6. LIF spectra obtained from soils and crust samples [108].

3. Remote Sensing

LIF has the great advantage of being expended to the outdoor environment. It is a technique that can also be applied as remote sensing and, hence, it is called laser remote sensing system LIDAR (light detection and ranging). It makes it possible to analyze compounds from long distances and it can operate under full sunlight.

There are many studies about LIDAR systems for the measurements of hydrographic parameters and substances in water. Nunes [109] has used a LIDAR using the 2nd harmonic of Nd: YAG laser system with a pulse energy of 200 mJ, a duration of 10 ns and a repetition rate of 10 Hz to measure chlorophyll a (685 nm) and DOM (from 540 to 620 nm) in deep sea water. Babichenko [110] has used a LIDAR Nd: YAG laser at 355 nm and a four-channel detector (355, 403, 450 and 680 nm) in order to measure chlorophyll a and other pigments in the subsurface water layer.

Other researchers have used the laser beam of an excimer (308 nm) and of a tunable dye laser (367 and 460 nm) to measure dissolved organic matter (DOM), chlorophyll a and other pigments in the Black Sea [111,112] and in the Danube Delta [113].

The disadvantages of excimer lasers are their high purchase and maintenance costs and the continuous need for gas supply compared with the solid state Nd:YAG lasers [114].

To the best of our knowledge, there are not any applications of LIDAR for monitoring soil quality. The only recent application had the aim to evaluate the soil surface and furrow cross-sectional area after a trailing shoe sweep [115].

4. Conclusions

The present study reveals the benefits of the application of fluorescence spectroscopy to environmental samples in order to assess their quality. Fluorescence can be achieved by the excitation of the compounds either by lamps or by lasers. In the first occasion, it is considered as conventional fluorescence, while in the second it is called laser induced fluorescence (LIF). Both of them have been applied mainly for the characterization of DOM in various types of water, soil and sediment samples. In a lesser extent, they have been applied for the detection of pollutants such as oil products in these samples.

In conventional fluorescence spectroscopy, the most common lamps are Xe and Hg. The most use recording technique in conventional fluorescence is EEM combined with PARAFAC analysis. It has been applied mainly for the characterization of DOM in various types of water. For the majority of the studies, the results from EEM showed the presence of peaks that have been categorized in five categories. The application of conventional spectroscopy in soil and sediment samples after pretreatment for isolation has led to the categorization of the same fluorophores.

In LIF spectroscopy, the common light source is the solid state laser Nd:YAG, as it was found to be more suitable for the detection of DOC and pollutants, followed by dye lasers that offer excitation tunability for the recording the fluorescence intensity versus emission wavelength. LIF has been applied less than conventional spectroscopy, especially in soil samples, although no sample pretreatment was necessary. Furthermore, LIF has been used for remote sensing measurements of various water systems by excimer and dye lasers.

LIF has proven to be a powerful tool for assessing the environmental quality and its use is continuously extending. Lasers have the advantages of high sensitivity and selectivity, as well as being versatile sources. Furthermore, the problem of quenching is minimized because laser sources have very high intensities compared with the potential quenching. The disadvantage is the restriction in excitation tunability that makes it difficult to record the EEM in order to gain more information about the complexity of the environmental samples.

Funding: This research is financed by the Project "Strengthening and optimizing the operation of MODY services and academic and research units of the Hellenic Mediterranean University", funded by the Public Investment Program of the Greek Ministry of Education and Religious Affairs.

Conflicts of Interest: The authors declare no conflict of interest.

References

1. Valeur, B.; Berberan-Santos, M.N. *Molecular Fluorescence: Principles and Applications*, 2nd ed.; John Wiley and Sons: Weinheim, Germany, 2012.
2. Lakowicz, J.R. *Principles of Fluorescence Spectroscopy*, 3rd ed.; Springer: Baltimore, MD, USA, 2006.
3. Amano, H.; Collazo, R.; De Santi, C.; Einfeldt, S.; Funato, M.; Glaab, J.; Zhang, Y. The 2020 UV emitter roadmap. *J. Phys. D Appl. Phys.* **2020**, *53*, 503001. [CrossRef]
4. Abramczyk, H. *Chapter 4: Lasers in Introduction to Laser Spectroscopy*; Elsevier: Amsterdam, The Netherlands, 2005; pp. 59–106.
5. Telle, H.H.; Urena, A.G. *Laser Spectroscopy and Laser Imaging: An Introduction*; CRC: Boca Raton, FL, USA, 2017. [CrossRef]
6. Eichler, H.J.; Eichler, J.; Lux, O. *Lasers*; Springer International Publishing: Berlin/Heidelberg, Germany, 2018.
7. Pavlopoulos, T.G. Scaling of dye lasers with improved laser dyes. *Prog. Quantum Electron.* **2002**, *26*, 193–224. [CrossRef]
8. Murray, J.R. *Lasers for Spectroscopy. Laser Spectroscopy and Its Applications*; CRC Press: Boca Raton, FL, USA, 2017. [CrossRef]
9. Iwaya, M.; Tanaka, S.; Omori, T.; Yamada, K.; Hasegawa, R.; Shimokawa, M.; Miyake, H. Recent development of UV-B laser diodes. *Jpn. J. Appl. Phys.* **2022**, *61*, 040501. [CrossRef]
10. Hitz, C.B.; Ewing, J.; Hecht, J. Semiconductor Lasers. In *Introduction to Laser Technology*; John Wiley & Sons, Ltd.: Hoboken, NJ, USA, 2012. [CrossRef]
11. Donati, S. *Photodetectors: Devices, Circuits and Applications*, 2nd ed.; Wiley: Hoboken, NJ, USA, 2021.
12. Chen, R.F. Fluorescence Quantum Yields of Tryptophan and Tyrosine. *Anal. Lett.* **1967**, *1*, 35–42. [CrossRef]
13. Joung, J.F.; Han, M.; Jeong, M.; Park, S. Experimental database of optical properties of organic compounds. *Sci. Data* **2020**, *7*, 1–6. [CrossRef]
14. Dreeskamp, H.; Koch, E.; Zander, M. On the fluorescence quenching of polycyclic aromatic hydrocarbons by nitromethane. *Z. Nat.* **1975**, *30*, 1311–1314. [CrossRef]
15. Pandey, A.; Yadav, A.; Bhawna; Pandey, S. Fluorescence quenching of polycyclic aromatic hydrocarbons within deep eutectic solvents and their aqueous mixtures. *J. Lumin.* **2017**, *183*, 494–506. [CrossRef]
16. Guo, X.; Xu, H.; Guo, R. Fluorescence Quenching of Anthracene by *N,N*-Diethylaniline in the Sodium Dodecyl Sulfate/Benzyl Alcohol/Water System. *J. Colloid Interface Sci.* **2001**, *240*, 559–565. [CrossRef]
17. Arian, S.; Benjamini, M.; Feitelson, J.; Stein, G. Interaction between tyrosine and divalent sulfur in fluorescence quenching and in the photochemistry of ribonuclease. *Photochem. Photobiol.* **1970**, *12*, 481–487. [CrossRef]
18. Chen, R.F.; Cohen, P.F. Quenching of tyrosine fluorescence in proteins by phosphate. *Arch. Biochem. Biophys.* **1966**, *114*, 514–522. [CrossRef]
19. Tallmadge, D.H.; Huebner, J.S.; Borkman, R.F. Acrylamide Quenching of Tryptophan Photochemistry and Photophysics. *Photochem. Photobiol.* **1989**, *49*, 81–386. [CrossRef] [PubMed]
20. Froehlich, P.M.; Nelson, K. Fluorescence quenching of indoles by amides. *J. Phys. Chem.* **1978**, *82*, 2401–2403. [CrossRef]
21. Wang, G.; Wang, A.-J.; Hu, K.-S. Tryptophan fluorescence quenching by alkaline earth metal cations in deionized bacteriorhodopsin. *J. Photochem. Photobiol. B Biol.* **2000**, *59*, 38–41. [CrossRef]
22. Idress, M.; Ayaz, M.; Bibi, R.; Khan, M.N. Fluorescence Quenching of the Probes L-Tryptophan and Indole by Anions in Aqueous System. *Anal. Sci.* **2020**, *36*, 183–185. [CrossRef]
23. Davis, G.A. Quenching of aromatic hydrocarbons by alkylpyridinium halides. *J. Chem. Soc. Chem. Commun.* **1973**, *19*, 728–729. [CrossRef]

24. Willkinson, F. Quenching of electronically excited states by molecular oxygen in fluid solution. *Pure Appl. Chem.* **1997**, *69*, 851–858. [CrossRef]
25. Ware, W.R. Oxygen Quenching of Fluorescence in Solution: An Experimental Study of the Diffusion Process. *J. Phys. Chem.* **1962**, *66*, 455–458. [CrossRef]
26. Gemeda, F.T. A Review on Effect of Solvents on Fluorescent Spectra. *Chem. Sci. Int. J.* **2017**, *18*, 1–12. [CrossRef]
27. Ghosh, K.; Schnitzer, M. Macromolecular structures of humic substances. *Soil Sci.* **1980**, *129*, 266–276. [CrossRef]
28. Patel-Sorrentino, N.; Mounier, S.; Lukas, Y.; Benaim, J.Y. Effects of UV-visible irradiation on natural organic matter from the Amazon basin. *Sci. Total Environ.* **2004**, *321*, 231–239. [CrossRef]
29. Baker, A. Thermal fluorescence quenching properties of dissolved organic matter. *Water Res.* **2005**, *39*, 4405–4412. [CrossRef] [PubMed]
30. Mikalauskaite, K.; Ziaunys, M.; Sneideris, T.; Smirnovas, V. Effect of Ionic Strength on Thioflavin-T Affinity to Amyloid Fibrils and Its Fluorescence Intensity. *Int. J. Mol. Sci.* **2020**, *21*, 8916. [CrossRef] [PubMed]
31. Donges, A.; Noll, R. Laser-Induced Fluorescence. In *Laser Measurement Technology*; Springer Series in Optical Sciences; Springer: Berlin/Heidelberg, Germany, 2015; Volume 188. [CrossRef]
32. He, G.; Tan, L.; Zheng, Q.; Prasad, P. Multiphoton Absorbing Materials: Molecular Designs, Characterizations, and Applications. *Chem. Rev.* **2008**, *108*, 1245–1330. [CrossRef]
33. Fitilis, I.; Fakis, M.; Polyzos, I.; Giannetas, V.; Persephonis, P.; Mikroyannidis, J. Strong two photon absorption and photophysical properties of symmetrical chromophores with electron accepting edge substituents. *J. Phys. Chem. A* **2008**, *112*, 4742–4748. [CrossRef]
34. Noll, R. Combination of LIBS and LIF. In *Laser-Induced Breakdown Spectroscopy*; Springer: Berlin/Heidelberg, Germany, 2012. [CrossRef]
35. Anabitarte, F.; Cobo, A.; Lopez-Higuera, J.M. Laser-induced breakdown spectroscopy: Fundamentals, applications, and challenges. *ISRN Spectrosc.* **2012**, *2012*, 12. [CrossRef]
36. Lloyd, J.B.F. Multicomponent analysis by synchronous luminescence spectrometry. *Nature* **1971**, *231*, 64–67.
37. Johnson, D.W.; Callis, J.B.; Christian, G.D. Rapid scanning fluorescence spectroscopy. *Anal. Chem.* **1977**, *49*, 747–757. [CrossRef]
38. Warner, I.M.; Christian, G.D.; Davidson, E.R.; Callis, J.B. Analysis of multicomponent fluorescence data. *Anal. Chem.* **1977**, *49*, 2155–2159. [CrossRef]
39. Gabor, R.; Baker, A.; McKnight, D.M.; Miller, M. Fluorescence indices and their interpretation. In *Aquatic Organic Matter Fluorescence*; Coble, P., Lead, J., Baker, A., Reynolds, D., Spencer, R., Eds.; Cambridge Environmental Chemistry Series; Cambridge University Press: Cambridge, UK, 2014; pp. 303–338. [CrossRef]
40. Ohno, T. Fluorescence inner-filtering correction for determining the humification index of dissolved organic matter. *Environ. Sci. Technol.* **2002**, *36*, 742–746. [CrossRef]
41. Kubista, M.; Sjoback, R.; Eriksson, S.; Albinsson, B. Experimental correction for the inner-filter effect in fluorescence spectra. *Analyst* **1994**, *119*, 417–419. [CrossRef]
42. Brouwer, E.R.; Hermans, A.N.; Lingeman, H.; Brinkman, U.A.T. Determination of polycyclic aromatic hydrocarbons in surface water by column liquid chromatography with fluorescence detection, using on-line micelle-mediated sample preparation. *J. Chromatogr. A* **1994**, *669*, 45–57. [CrossRef]
43. Gasperi, J.; Garnaud, S.; Rocher, V.; Moilleron, R. Priority pollutants in surface waters and settleable particles within a densely urbanised area: Case study of Paris (France). *Sci. Total Environ.* **2009**, *407*, 2900–2908. [CrossRef]
44. Kotti, M.; Zacharioudaki, D.-E.; Kokinou, E.; Stavroulakis, G. Characterization of water quality of Almiros river (Northeastern Crete, Greece): Physicochemical parameters, polycyclic aromatic hydrocarbons and anionic detergents. *Modeling Earth Syst. Environ.* **2018**, *4*, 1285–1296. [CrossRef]
45. Song, Y.; Wilke, B.; Song, X.; Gong, P.; Zhou, Q.; Yang, G. Polycyclic aromatic hydrocarbons (PAHs), polychlorinated biphenyls (PCBs) and heavy metals (HMs) as well as their genotoxicity in soil after long-term wastewater irrigation. *Chemosphere* **2006**, *65*, 1859–1868. [CrossRef] [PubMed]
46. Chrysikou, L.; Gemenetzis, P.; Kouras, A.; Manoli, E.; Terzi, E.; Samara, C. Distribution of persistent organic pollutants, polycyclic aromatic hydrocarbons and trace elements in soil and vegetation following a large scale landfill fire in northern Greece. *Environ. Intern.* **2008**, *34*, 210–225. [CrossRef] [PubMed]
47. Vane, C.; Harrison, I.; Kim, A. Polycyclic aromatic hydrocarbons (PAHs) and polychlorinated biphenyls (PCBs) in sediments from the Mersey Estuary, U.K. *Sci. Total Environ.* **2007**, *374*, 112–126. [CrossRef]
48. Bihari, N.; Fafandel, M.; Hamer, B.; Kralj-Bilen, B. PAH content, toxicity and genotoxicity of coastal marine sediments from the Rovinj area, Northern Adriatic, Croatia. *Sci. Total Environ.* **2006**, *366*, 602–611. [CrossRef]
49. Fua, L.; Liub, X.; Hub, J.; Zhaob, X.; Wangc, H.; Wanga, X. Application of dispersive liquid–liquid microextraction for the analysis of triazophos and carbaryl pesticides in water and fruit juice samples. *Anal. Chim. Acta* **2009**, *632*, 289–295. [CrossRef]
50. Le Bot, B.; Colliaux, K.; Pelle, D.; Briens, C.; Seux, R.; Clement, M. Optimization and performance evaluation of the analysis of glyphosate and AMPA in water by HPLC with fluorescence detection. *Chromatographia* **2005**, *56*, 161–164. [CrossRef]
51. Murillo Pulgarin, J.A.; Garcia Bermejo, L.F. Determination of the pesticide napropamide in soil, pepper, and tomato by micelle-stabilized room-temperature phosphorescence. *J. Agric. Food Chem.* **2002**, *50*, 1002–1008. [CrossRef]

52. Salinas-Castillo, A.; Fernandez-Sanchez, J.F.; Segura-Carretero, A.; Fernandez-Gutierrez, A. Simple determination of herbicide napropamide in water and soil samples by room-temperature phosphorescence. *Pest Manag. Sci.* **2005**, *61*, 816–820. [CrossRef] [PubMed]
53. Peng, X.; Tan, J.; Tang, C.; Yu, Y.; Wang, Z. Multiresidue determination of fluoroquinolone, sulfonamide, trimethoprim, and chloramphenicol antibiotics in urban waters in China. *Environ. Toxicol. Chem.* **2008**, *27*, 73–79. [CrossRef] [PubMed]
54. Seifrtova, M.; Pena, A.; Lino, C.M.; Solich, P. Determination of fluoroquinolone antibiotics in hospital and municipal wastewaters in Coimbra by high performance liquid chromatography using a monolithic column and fluorescence detection. *Anal. Bioanal. Chem.* **2008**, *391*, 799–805. [CrossRef]
55. Blackwell, P.A.; Holten, H.C.; Mab, H.P.; Halling-Sorensen, B.; Boxall, A.B.A.; Kaya, P. Ultrasonic extraction of veterinary antibiotics from soils and pig slurry with SPE clean-up and LC–UV and fluorescence detection. *Talanta* **2004**, *64*, 1058–1064. [CrossRef]
56. Uslu, M.O.; Yediler, A.; Balcıoğlu, I.A.; Schulte-Hostede, S. Analysis and Sorption Behavior of Fluoroquinolones in Solid Matrices. *Water Air Soil Pollut.* **2008**, *190*, 55–63. [CrossRef]
57. Robinson, B.J.; Hellou, J. Biodegradation of endocrine disrupting compounds in harbour seawater and sediments. *Sci. Total Environ.* **2009**, *407*, 5713–5718. [CrossRef]
58. Sorensen, L.K.; Hansen, H. Determination of oxolinic acid in marine sediments by HPLC with fluorescence detection. *J. Liq. Chromatogr. Relat. Technol.* **2001**, *24*, 2469–2476. [CrossRef]
59. Wie, Q.; Yang, J.; Zhang, Y.; Chemg, G.; Du, B. Determination of antimony (III) in environmental water samples in microemulsion system by the fluorescence quenching method. *Talanta* **2002**, *58*, 419–426.
60. San Vicente, B.; Costa-Fernandez, J.M.; Jin, W.L.; Pereiro, R.; Sanz-Medel, A. Determination of trace levels of mercury in water samples based on room temperature phosphorescence energy transfer. *Anal. Chim. Acta* **2002**, *455*, 179–186.
61. Coble, P.G.; Green, S.A.; Blough, N.V.; Gagosian, R.B. Characterization of disove organic matter in the Black Sea by fluorescence spectroscopy. *Nature* **1990**, *348*, 432–435. [CrossRef]
62. Coble, P.G. Characterization of marine and terrestial DOM in seawater using excitation-emission matrix spectroscopy. *Mar. Chem.* **1996**, *51*, 325–346. [CrossRef]
63. Goncalves-Araujo, R.; Granskog, M.A.; Bracher, A.; Azetsu-Scott, K.; Dodd, P.A.; Stedmon, C.A. Using fluorescent dissolved organic matter to trace and distinguish the origin of Arctic waters. *Sci. Rep.* **2016**, *6*, 33978. [CrossRef] [PubMed]
64. Carstea, E.M.; Bridgeman, J.; Baker, A.; Reynolds, D.M. Fluorescence spectroscopy for wastewater monitoring: A review. *Water Res.* **2016**, *95*, 205–219. [CrossRef] [PubMed]
65. Guo, X.; Yu, H.; Yan, Z.; Gao, H.; Zhang, Y. Tracking variations of fluorescent dissolved organic matter during wastewater treatment by accumulative fluorescence emission spectroscopy combined with principal component, second derivative and canonical correlation analysis. *Chemosphere* **2018**, *194*, 463–470. [CrossRef] [PubMed]
66. Tedetti, M.; Cuet, P.; Guigue, C.; Goutx, M. Characterization of dissolved organic matter in a coral reef ecosystem subjected to anthropogenic pressures (La Réunion Island, Indian Ocean) using multi-dimensional fluorescence spectroscopy. *Sci. Total Environ.* **2011**, *409*, 2198–2210. [CrossRef]
67. Huang, C.-C.; Li, Y.-M.; Yang, H.; Sun, D.-Y.; Xu, L.-J.; Chen, X. Study of influencing factors to chromophoric dissolved organic matter absorption properties from fluorescence features in Taihu lake in autumn. *J. Limnol.* **2013**, *72*, 326–335. [CrossRef]
68. Qiu, Y.; Shi, H.; Jing, H.; Liu, R.; Cai, Q.; Takemura, M.; Haraguchi, S. Characterization and variations of dissolved organic matter in the Lake Taihu area of China. *Water Sci. Technol. Water Supply* **2012**, *12*, 439–450. [CrossRef]
69. Goslan, E.H.; Voros, S.; Banks, J.; Wilson, D.; Hillis, P.; Campbell, A.T.; Parsons, S.A. A Model for Predicting Dissolved Organic Carbon Distribution in a Reservoir Water using Fluorescence Spectroscopy. *Water Res.* **2004**, *38*, 783–791. [CrossRef]
70. Muller, C.L.; Baker, A.; Hutchinson, R.; Fairchild, I.; Kidd, C. Analysis of rainwater dissolved organic carbon compounds using fluorescence spectrophotometry. *Atmos. Environ.* **2008**, *42*, 8036–8045. [CrossRef]
71. Costa, A.S.; Passos, E.D.A.; Garcia, C.A.B.; Alves, J.D.P.H. Characterization of Dissolved Organic Matter in the Piauí River Estuary, Northeast Brazil. *J. Braz. Chem. Soc.* **2011**, *22*, 2139–2147. [CrossRef]
72. Carstea, E.M. Fluorescence Spectroscopy as a Potential Tool for In-Situ Monitoring of Dissolved Organic Matter in Surface Water Systems. In *Water Pollution*; Balkis, N., Ed.; IntechOpen: London, UK, 2012. [CrossRef]
73. Kepkay, P.E.; Yeung, C.W.; Bugden, J.B.C.; Li, Z.; Lee, K. Ultraviolet Fluorescence Spectroscopy (UVFS): A new means of determining the effect of chemical dispersants on oil spills. In Proceedings of the International Oil spill Conference, Savannah, GA, USA, 4–8 May 2008; pp. 639–642.
74. Nahorniak, M.L.; Booksh, K.S. Excitation-emission matrix fluorescence spectroscopy in conjuction with multiway analysis for PAH detection in complex matrices. *Analyst* **2006**, *131*, 1308–1315. [CrossRef]
75. Uebel, U.; Kubitz, J.; Anders, A. Laser Induced Fluorescence Spectroscopy of Phytoplankton and Chemicals with Regard to an in situ Detection in Waters. *J. Plant Physiol.* **1996**, *148*, 586–592. [CrossRef]
76. Baumann, T.; Haaszio, S.; Niessner, R. Applications of a Laser Induced Fluorescence Spectroscopy Sensor in aquatic systems. *Water Res.* **2000**, *34*, 1318–1326. [CrossRef]
77. Sivaprakasam, V.; Killinger, D.K. Tunable ultraviolet laser-induced fluorescence detection of trace plastics and dissolved organic compounds in water. *Appl. Opt.* **2003**, *42*, 6739–6746. [CrossRef]

78. Sivaprakasam, V.; Shannon, R.F.; Luo, C.; Coble, P.G.; Boehme, J.R.; Killinger, D.K. Development and initial Calibration of a portable Laser Induced Fluorescence system used for in situ measurements of trace plastics and organics in seawater and the Gulf of Mexico. *Appl. Opt.* **2003**, *42*, 6747. [CrossRef] [PubMed]
79. Sivaprakasam, V.; Killinger, D.K. How Water Glows: Water Monitoring with Laser Fluorescence. *Opt. Photonics News* **2006**, *17*, 34–39. [CrossRef]
80. Sharikova, A.V.; Killinger, D.K. Laser and UV-LED-induced fluorescence detection of dissolved organic compounds in water. In Proceedings of the SPIE 7666, Sensors, and Command, Control, Communications and Intelligence (C3I) Technologies for Homeland Security and Homeland Defence IX, 76661L, Orlando, FL, USA, 5–9 April 2010; SPIE Bellingham: Washington, DC, USA, 2010. [CrossRef]
81. Sharikova, A.V.; Killinger, D.K. LIF detection of trace species in water Using different UV laser wavelengths. *Int. J. High Speed Electron. Syst.* **2007**, *17*, 689–695. [CrossRef]
82. Gervase, L.; Carstea, E.M.; Pavelescu, G.; Savastru, D. Laser Induced Fluorescence Efficiency in water quality assessment. *Rom. Rep. Phys.* **2010**, *62*, 652–659.
83. Carstea, E.M.; Pavelescu, G.; Baker, A.; Roman, C.; Ioja, C.; Savastru, D. Pollution analysis on the Arges River using fluorescence spectroscopy. *J. Optoelectron. Adv. Mater.* **2008**, *10*, 1489–1494.
84. Du, R.; Yang, D.; Jiang, G.; Song, Y.; Yin, X. An Approach for In Situ Rapid Detection of Deep-Sea Aromatic Amino Acids Using Laser-Induced Fluorescence. *Sensors* **2020**, *20*, 1330. [CrossRef]
85. Bukin, O.; Proschenko, D.; Alexey, C.; Korovetskiy, D.; Bukin, I.; Yurchik, V.; Sokolova, I.; Nadezhkin, A. New Solutions of Laser-Induced Fluorescence for Oil Pollution Monitoring at Sea. *Photonics* **2020**, *7*, 36. [CrossRef]
86. Kang, J.; Li, R.; Wang, Y.; Chen, Y.; Yang, Y. Ultrasensitive detection of trace amounts of lead in water by LIBS-LIF using a wood-slice substrate as a water absorber. *J. Anal. At. Spectrom.* **2017**, *32*, 2292–2299. [CrossRef]
87. Lui, S.; Godwal, Y.; Taschuk, M.; Tsui, Y.; Fedosejevs, R. Detection of Lead in Water Using Lasr-Induced Breakdown Spectroscopy and Laser-Induced Fluorescence. *Anal. Chem.* **2008**, *80*, 1995–2000. [CrossRef]
88. Tian, H.; Jiao, L.; Dong, D. Rapid determination of trace cadmium in drinking water using laser-induced breakdown spectroscopy coupled with chelating resin enrichment. *Sci. Rep.* **2019**, *9*, 10443. [CrossRef]
89. Yao, M.; Lin, J.; Liu, M.; Xu, Y. Detection of chromium in wastewater from refuse incineration power plant near Poyang Lake by laser induced breakdown spectroscopy. *Appl. Opt.* **2012**, *51*, 1552–1557. [CrossRef]
90. Koch, S.; Garen, W.; Muller, M.; Neu, W. Detection of chromium in liquids by laser induced breakdown spectroscopy (LIBS). *Appl. Phys. A Mater. Sci. Process* **2004**, *79*, 1071–1073. [CrossRef]
91. Xue, B.; Tian, Y.; Lu, Y.; Li, Y.; Zheng, R. Characteristics of the secondary breakdown of DP-LIBS in bulk water with different axial focusing arrangements and laser energies. *Spectrochim. Acta B At. Spectrosc.* **2019**, *151*, 20–25. [CrossRef]
92. Li, Q.; Tian, Y.; Xue, B.; Li, N.; Ye, W.; Lu, Y.; Zheng, R. Improvement in the analytical performance of underwater LIBS signals by exploiting the plasma image information. *J. Anal. At. Spectrom.* **2020**, *35*, 366. [CrossRef]
93. Guo, J.; Lu, Y.; Cheng, K.; Song, J.; Ye, W.; Li, N.; Zheng, R. Development of a compact underwater laser induced breakdown spectroscopy (LIBS) system and preliminary results in sea trials. *Appl. Opt.* **2017**, *56*, 8196. [CrossRef]
94. Gabbarini, V.; Rossi, R.; Ciparisse, J.-F.; Malizia, A.; Divizia, A.; De Filippis, P.; Anselmi, M.; Carestia, M.; Palombi, L.; Divizia, M.; et al. Laser-induced fuorescence (LIF) as a smart method for fast environmental virological analyses: Validation on Picornaviruses. *Sci. Rep.* **2019**, *9*, 12598. [CrossRef]
95. NAVFAC (Naval Facilities Engineering Command). *Tech Data Sheet, Near-Real Time UV Fluorescence Technique for Characterization of PAHs in Marine Sediment, NFESC TDS-2075-ENV*; NAVFAC: Washington, DC, USA, 10 June 2000; pp. 20374–25065. Available online: https://frtr.gov/pdf/uvfluorescence_2.pdf (accessed on 10 June 2022).
96. Cristescu, L.; Pavelescu, G.; Cârstea, E.M.; Săvăstru, D. Evaluation of petroleum contaminants in soil by Fluorescence Spectroscopy. *Environ. Eng. Manag. J.* **2009**, *8*, 1269–1273. [CrossRef]
97. Mielnik, L.; Kowalczuk, P. Optical characteristic of humic acids from lake sediments by excitation-emission matrix fluorescence with PARAFAC model. *J. Soils Sediments* **2018**, *18*, 2851–2862. [CrossRef]
98. Santı'n, C.; Yamashita, Y.; Otero, X.L.; A'lvarez, M.A.; Jaffe, R. Characterizing humic substances from estuarine soils and sediments by excitation-emission matrix spectroscopy and parallel factor analysis. *Biogeochemistry* **2009**, *96*, 131–147. [CrossRef]
99. Dang, D.H.; Lenoble, V.; Durrieu, G.; Mullot, J.-U.; Mounier, S.; Garnier, C. Sedimentary dynamics of coastal organic matter: An assessment of the porewater size/reactivity model by spectroscopic techniques. *Estuar. Coast. Shelf Sci.* **2014**, *151*, 100–111. [CrossRef]
100. Derrien, M.; Lee, Y.K.; Ark, J.-E.; Li, P.; Chen, M.; Lee, S.H.; Lee, S.H.; Lee, J.-B.; Hur, J. Spectroscopic and molecular characterization of humic substances (HS) from soils and sediments in a watershed: Comparative study of HS chemical fractions and the origins. *Environ. Sci. Pollut. Res.* **2017**, *24*, 16933–16945. [CrossRef]
101. Wei, D.; Li, Y.; Cai, S.-S.; Jin, L.; Li, Y.-M.; Wang, W.; Bai, Y.; Hu, Y.; Clarke, N. Fluorescence Characteristics of Humic Acid in Chinese Black Soil under Long-Term Fertilization. *Adv. Polym. Technol.* **2019**, *2019*, 1–8. [CrossRef]
102. Fasurova, N.; Pospisilova, L. Characterization of soil humic substances by Ultraviolet-Visible and Synchronous Fluorescence Spectroscopy. *J. Cent. Eur. Agric.* **2010**, *11*, 351–358.
103. Perez, M.G.; Milori, D.M.B.P.; Martin-Neto, L.; Colnago, L.A.; de Camargo, O.A.; Berton, R.; Bettiol, W. Laser-Induced Fluorescence of organic matter from a Brazilian oxisol under sewage-sludge applications. *Sci. Agric.* **2006**, *63*, 269–275. [CrossRef]

104. Milori, D.M.B.P.; Galeti, H.V.A.; Martin-Neto, L.; Dieckow, J.; Gonzalez-Perez, M.; Bayer, C.; Salton, J. Organic matter study of whole soil samples using laser-induced fluorescence spectroscopy. *Soil Sci. Soc. Am. J.* **2006**, *70*, 57–63. [CrossRef]
105. Segnini, A.; Posadas, A.; Quiroz, R.; Milori, D.M.B.P.; Saab, S.C.; Martin-Neto, L.; Vaz, C.M.P. Spectroscopic assessment of soil organic matter in wetlands from high Andes. *Soil Sci. Soc. Am. J.* **2010**, *74*, 2246–2253. [CrossRef]
106. Ferreira, E.C.; Ferreira, E.J.; Villas-Boas, P.R.; Senesi, G.S.; Carvalho, C.M.; Romano, R.A.; Martin-Neto, L.; Milori, D.M.B.P. Novel estimation of the humification degree of soil organic matter by laser induced breakdown spectroscopy. *Spectrochim. Acta B* **2014**, *99*, 76–81. [CrossRef]
107. Senesi, G.S.; Martin-Neto, L.; Villas-Boas, P.R.; Nicolodelli, G.; Milori, D.M.B.P. Laser-based spectroscopic methods to evaluate the humification degree of soil organic matter in whole soils: A review. *J Soils Sediments* **2018**, *18*, 1292–1302. [CrossRef]
108. da Silva, J.M.; Utkin, A.B. Application of Laser-Induced Fluorescence in Functional Studies of Photosynthetic Biofilms. *Processes* **2018**, *6*, 227. [CrossRef]
109. Nunes, R.A.; Tabares, R.H.; Bunkin, A.; Persh, S. Compact lidar for remote sensing of water pollution. *Trans. Ecol. Environ.* **1997**, *14*, 401–408.
110. Babichenko, S. Laser Remote Sensing of the European Marine Environment: LIF Technology and Applications. In *Remote Sensing of the European Seas;* Barale, V., Gade, M., Eds.; Springer Science+Business Media B.V.: Berlin/Heidelberg, Germany, 2008; pp. 189–204.
111. Pavelescu, G.; Vasilescu, J.-G.; Strechie, C.; Babichenko, S.; Lisin, A.; Onciu, T.; Belegante, L. Water pollution analysis from LIDAR investigations on the Romanian Black Sea coast. In Proceedings of the SPIE—Volume 6743 Remote Sensing of the Ocean, Sea Ice, and Large Water Regions 2007, Florence, Italy, 17–18 September 2007. [CrossRef]
112. Vasilescu, J.; Belegante, L.; Sliwinski, C.; Cristescu, C.P. Using Active Remote Sesnsing to assess the seawater quality. *UPB Sci. Bull.* **2008**, *70*, 41–48.
113. Vasilescu, J.; Onciu, T.; Jugaru, L.; Belegante, L. Remote estimation of fluorescence marine components distribution. *Rom. Rep. Phys.* **2009**, *61*, 721–729.
114. Utkin, A.; Lavrov, A.; Vilar, R.; Babichenko, S.; Shchemelyov, S.; Sobolev, I.; Bastos, L.; Deurloo, R.; Palenzuela, J.T.; Yarovenko, N.; et al. Optical methods for water pollution monitoring, Spatial and Organizational Dynamics Discussion Papers. In *CIEO-Research Centre for Spatial and Organizational Dynamics;* University of Algarve: Faro, Portugal, 2011.
115. Foldager, F.F.; Pedersen, J.M.; Skov, E.H.; Evgrafova, A.; Green, O. LiDAR-Based 3D Scans of Soil Surfaces and Furrows in Two Soil Types. *Sensors* **2019**, *19*, 661. [CrossRef]

molecules

MDPI

Review

Recent Advances in the Extraction of Polycyclic Aromatic Hydrocarbons from Environmental Samples

Natalia Manousi * and **George A. Zachariadis** *

Laboratory of Analytical Chemistry, Department of Chemistry, Aristotle University of Thessaloniki, 54124 Thessaloniki, Greece

* Correspondence: nmanousi@chem.auth.gr (N.M.); zacharia@chem.auth.gr (G.A.Z.); Tel.: +2310-997-707 (N.M.)

Academic Editors: Victoria Samanidou and Eleni Deliyanni
Received: 15 April 2020; Accepted: 4 May 2020; Published: 7 May 2020

Abstract: Polycyclic aromatic hydrocarbons (PAHs) comprise a group of chemical compounds consisting of two or more fused benzene rings. PAHs exhibit hydrophobicity and low water solubility, while some of their members are toxic substances resistant to degradation. Due to their low levels in environmental matrices, a preconcentration step is usually required for their determination. Nowadays, there is a wide variety of sample preparation techniques, including micro-extraction techniques (e.g., solid-phase microextraction and liquid phase microextraction) and miniaturized extraction techniques (e.g., dispersive solid-phase extraction, magnetic solid-phase extraction, stir bar sorptive extraction, fabric phase sorptive extraction etc.). Compared to the conventional sample preparation techniques, these novel techniques show some benefits, including reduced organic solvent consumption, while they are time and cost efficient. A plethora of adsorbents, such as metal-organic frameworks, carbon-based materials and molecularly imprinted polymers, have been successfully coupled with a wide variety of extraction techniques. This review focuses on the recent advances in the extraction techniques of PAHs from environmental matrices, utilizing novel sample preparation approaches and adsorbents.

Keywords: PAHs; sample preparation; environmental samples; extraction; MSPE; SPME; FPSE; SBSE; DSPE; PT-SPE.

1. Introduction

Polycyclic aromatic hydrocarbons (PAHs) are a large group of chemical compounds composed of two or more fused benzene rings [1]. PAHs are hydrophobic compounds with low water solubility, and their solubility in water and volatility decrease with an increase in their molecular weight [2]. PAHs consisting of up to four rings are known as light PAHs, while PAHs that are made of more than four rings are known as heavy PAHs. Heavy PAHs are more stable and more toxic than the light PAHs. These chemical compounds are widespread environmental contaminants that are considered byproducts of the incomplete combustion of organic materials, such as coal, gas, garbage, meat, oil, tobacco and wood, during natural or anthropogenic processes [1,3]. PAHs are toxic substances, which are resistant to degradation and exposure to them may increase the risk of cancer [4]. As a result, PAHs are considered by US Environmental Protection Agency (EPA) and the European Environmental Agency to be priority pollutants [1]. Therefore, the determination of PAHs in environmental samples is of high importance. Since PAHs exist in traces in environmental matrices, a preconcentration technique is normally required. Figure 1 shows the chemical structures of common PAHs.

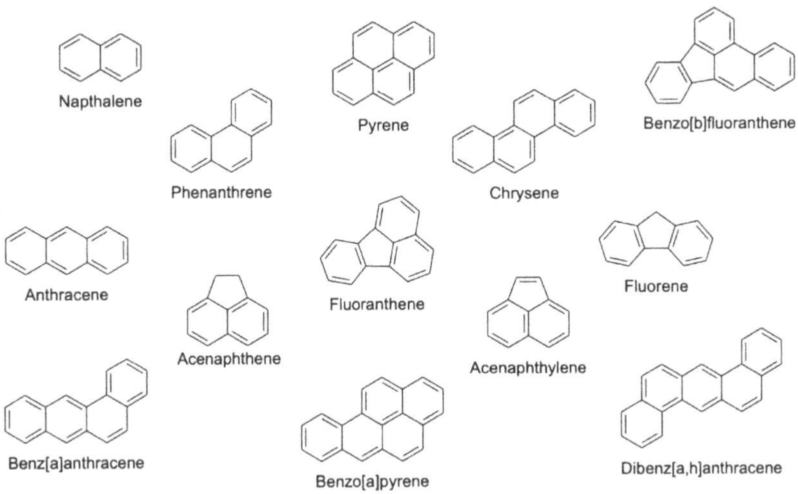

Figure 1. Chemical structures of common polycyclic aromatic hydrocarbons (PAHs).

Currently, the most widely used methods for analyzing these pollutants in environmental matrices are gas chromatography (GC), high performance liquid chromatography (HPLC) and ultra-high pressure liquid chromatography (UHPLC) [3,5]. Various detection systems, including ultraviolet detectors (UV) [6], diode array detectors (DAD) [3], fluorescence detectors (FLD) [1], mass detectors (MS) coupled with HPLC and UHPLC and flame ionization detectors (FID) [7], MS detectors [5] and tandem MS detectors (MS/MS) [8] coupled with GC have been used. Due to the enhanced sensitivity in the determination of PAHs that results in lower LODs, mass detectors and tandem MS detectors are generally preferred.

Solid-phase extraction (SPE) and liquid–liquid extraction (LLE) are two major sample preparation techniques that have been widely used for the extraction and preconcentration of a wide variety of analytes from environmental samples. However, both conventional techniques tend to have many fundamental drawbacks, since they include complicated, time-consuming steps, while they require large amounts of sample and organic solvents. Moreover, in both techniques there are difficulties in automation [9,10].

In order to overcome these drawbacks, different microextraction techniques have been proposed as an efficient alternative to classical extraction techniques, since the introduction of solid-phase microextraction (SPME) by the research group of Pawliszyn [11]. Liquid-phase microextraction (LPME) was introduced a few years later by Liu and Dasgupta, by using organic droplets suspended from the tip of a microsyringe [12]. Those microextraction techniques are widely used today, and they offer certain benefits compared to the conventional sample preparation techniques. Microextraction techniques require a significantly lower sample amount, number of extraction steps, sample preparation time and organic solvent consumption, and they comply with Green Analytical Chemistry principles [10,13].

Typical examples of miniaturized sample preparation techniques include dispersive solid phase extraction (d-SPE) [14], magnetic solid-phase extraction (MSPE) [15], pipette tip solid-phase extraction (PT-SPE) [16], fabric phase sorptive extraction (FPSE) [17], stir bar sorptive extraction (SBSE) [18] etc. In recent years, a wide variety of novel sorbents, including molecular imprinted polymers (MIPs), graphene, graphene oxide (GO), carbon nanotubes (CNTs), metal organic frameworks (MOFs), covalent organic frameworks (COFs) and zeolitic imidazole frameworks (ZIFs) have been successfully coupled with miniaturized extraction techniques and microextraction techniques [14,19–21].

Until now, a plethora of publications discuss the different analytical procedures for the extraction of PAHs from liquid, solid and air matrices [22–29]. To the best of our knowledge, in the last five years there were no review papers regarding the extraction techniques of PAHs from environmental samples with the novel miniaturized extraction and microextraction techniques. Herein, we aim to discuss the recent advances in the extraction of PAHs from environmental samples. Emphasis will be given to the miniaturized sample preparation approaches, as well as the novel sorbents and other materials that have been successfully coupled with various microextraction techniques.

2. Extraction of PAHs from Environmental Matrices

A plethora of novel sample preparation techniques have been employed for the extraction of PAHs. Figure 2 summarizes the recent advances in sorptive extraction techniques that have been employed for the determination of PAHs from environmental samples.

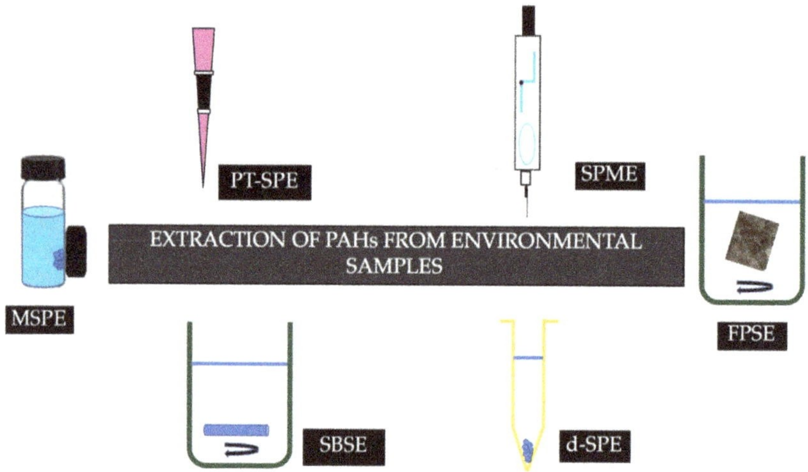

Figure 2. Recent advances in sorptive extraction techniques for the determination of PAHs from environmental samples.

2.1. Dispersive Solid-Phase Extraction of PAHs from Environmental Matrices.

Dispersive solid-phase extraction (d-SPE), is a form of SPE in which the desired sorbent is added directly into the sample aqueous solution followed by dispersion. This technique is taking advantage of the contact between the adsorbent and the target analytes. Once the extraction process is completed, the sorbent with the adsorbed analytes is separated from the sample by a mechanical process, such as centrifugation or filtration. Compared to the conventional SPE process, the main benefit of d-SPE is the reduction of sample preparation time, as well as its simplicity, adaptability and easy handling. A wide variety of sorbents have been utilized for the d-SPE of PAHs from environmental samples [30,31].

This technique gained popularity when Anastassiades et al. [32] introduced the QuEChERS (Quick, Easy, Cheap, Effective, Rugged and Safe) approach for the determination of pesticide residue in food of plant origin. The initial method consists of acetonitrile extraction and addition of a mixture of salts, followed by a dispersive clean-up step with a primary–secondary amine (PSA) as extraction sorbent. QuEChERS was quickly applied for the determination of other analytes in a variety of sample matrices. Cvetkovic et al. [33] developed a QuEChERS extraction procedure of PAHs in soil prior to their determination by GC-MS. The researchers evaluated different solvent systems (acetonitrile/water and hexane/water) and sorbents (PSA, C_{18}, Florisil, diatomaceous earth and clinoptilolite). Among the tested parameters, the best results were obtained with acetonitrile/water, as the extraction solvent and diatomaceous earth as the d-SPE extraction adsorbent.

Until today, a wide variety of novel extraction sorbents have been evaluated for the d-SPE of PAHs from environmental matrices. Among them, metal-organic frameworks and zeolitic imidazole frameworks are currently the most popular d-SPE adsorbents. MOFs became widely known in 1995, when Yaghi and Li [34] reported the hydrothermal synthesis of a MOF material with large rectangular channels. Metal-organic frameworks are a class of hybrid organic-inorganic supramolecular materials, which are based on the coordination of metal ions or clusters with bi- or multidentate organic linkers. What makes MOFs materials so attractive is their unique properties, such as high surface areas (up to 14,600 $m^2 \cdot g^{-1}$) [35], pore size tunability, structure flexibility, luminosity, thermal stability, charge transfer ability from the ligand to the metal or from the metal to the ligands, etc [21,36–39]. As a result, MOFs have gained attention in a plethora of applications, such as gas storage and separation [40], catalysis [41], sensors [42], detoxification [43] and drug delivery [44]. In analytical chemistry, MOFs have been evaluated as stationary phases for GC [45,46] and HPLC [47,48] analysis. However, today, the most popular field of applications of MOFs in analytical chemistry is sample preparation [21,36].

Xia et al. [49] synthesized a JUC-48 (Jilin University China 48) from cadmium nitrate tetrahydrate and biphenyl-4,4′-dicarboxylic acid, and used it for the d-SPE of PAHs from environmental matrices prior to their determination by HPLC-FLD. The novel sorbent exhibited good characteristics regarding its stability and morphology, high surface area and open adsorption site. The researchers observed highly negative relationship between extraction capacity and molecule size. As a result, the JUC-48 adsorbent was used for the extraction of only light PAHs.

Amiri et al. [50] synthesized hybrid nanocomposites prepared from MOF-199 and graphene or fullerene. MOF-199 was prepared from copper nitrate trihydrate and 1,3,5-benzenetricarboxylic acid. Among the examined nanocomposites (i.e., MOF-199, MOF-199/graphene and MOF-199/fullerene), the MOF-199/graphene nanocomposite exhibited the highest adsorption affinity towards PAHs, probably due to the high porosity, surface area and adsorption capacity of graphene.

Zeolitic imidazole frameworks (ZIFs) are a sub-family of metal organic frameworks that favors the benefits of both zeolites and MOFs. ZIFs are composed of imidazolate linkers and metal ions. Their structures are similar to conventional aluminosilicate zeolites. Due to their intrinsic porous characteristics and abundant functionalities, as well as exceptional thermal and chemical stabilities, ZIFs have a wide range of potential applications [37,51].

Liang et al. [52] evaluated the application of a novel cellulose/zeolitic imidazolate frameworks-8 composite for the d-SPE of PAHs from environmental water samples. ZIF-8 is composed of zinc as metal ion and 2-methyl imidazole as an organic linker, and it exhibits good characteristics for the extraction of PAHs due to strong hydrophobic and π–π interaction, as well as permanent porosity and high surface area. However, ZIF-8 suffers from tiny inner pores and strong hydrophobicity. In order to overcome this drawback, ZIF-8 was modified with cellulose microspheres (CMs) generated from natural cellulose. CMs exhibit macro/mesoporous structures, high surface area and abundance of hydroxyl groups. Due to the combination of the benefits of both ZIF-8 and the CMs, the hybrid material exhibited good extraction characteristics.

Other d-SPE sorbents that have been applied for the extraction of PAHs from environmental samples include graphene/sepiolite [53] and *N*-acetyl-L-cysteine modified CdS quantum dots [54].

2.2. Magnetic Solid-Phase Extraction of PAHs from Environmental Matrices.

Magnetic solid-phase extraction (MSPE) is a form of d-SPE in which a magnetic nanomaterial is added into an aqueous sample solution to adsorb the target analytes. After the adsorption of the analyte, an external magnetic field is applied to collect the sorbent and the supernatant solution is discarded. Subsequently, elution of the adsorbed analytes is achieved with the addition of an appropriate solvent, and magnetic separation is performed once again to collect the eluent, which is further analyzed by a suitable analytical technique. Compared to the conventional SPE procedure, in MSPE there is no need for sorbent packing into cartridges, thus avoiding limitations of column blocking and high pressure. Meanwhile, sample and organic solvent consumption is significantly decreased compared to the classic SPE and LLE formats. Finally, the sorbent separation with a magnet is a simple and rapid process, compared to the time-consuming centrifugation and filtration steps that are required in conventional d-SPE [55–57].

Magnetic nanoparticles (MNPs) are characterized by the general formula MFe_2O_4 (M = Fe, Co, Cu, Mn, etc.), and they can be produced by a variety of methods, such as co-precipitation, solvothermal, hydrothermal etc. The most common magnetic nanoparticles that have been used in order to fabricate magnetic sorbents for MSPE are Fe_3O_4 nanoparticles. Iron oxides have been widely used in MSPE due to their super paramagnetism, their low toxicity, their high magnetic saturation, their simple preparation process and their low price [7]. The application of other magnetic nanoparticles, such as $MnFe_2O_4$, have been also reported [58]. However, the utilization of MNPs in sample preparation has some drawbacks, since their selectivity is low. Moreover, MNPs exhibit low stability in strong acidic aqueous media and low dispersion ability in many sample matrices. Therefore, surface modification of magnetic nanoparticles is usually required to enhance their stability and selectivity by the introduction of special functional groups [7]. As seen in Table 1, a wide variety of chemical compounds including carbon-based materials, polymeric materials, MOFs and other molecules have been employed for this purpose.

2.2.1. Metal-Organic Framework Magnetic Nanocomposites for the MSPE of PAHs

A wide variety of metal-organic frameworks have been evaluated as sorbents for the MSPE of PAHs from water samples, after functionalization with magnetic nanoparticles. HKUST-1 (Hong Kong University of Science and Technology), was prepared from copper (II) nitrate hemipentahydrate and benzene-1,3,5-tricarboxylic acid, and has been applied for the MSPE of PAHs from water and fruit tea infusion samples, prior to their determination by UHPLC-FLD. The proposed method provided adequate quantitation and reproducibility of heavy PAHs; however, extraction efficiencies were relatively low and no sorbent reusability was reported [59].

MIL-101(Cr) is another MOF that has been used for the extraction of PAHs after its in situ magnetization [60]. MIL-101 was prepared from chromium nitrate nonahydrate and terephthalic acid, and the magnetic nanocomposite was prepared by the in-situ magnetization of MIL-101 and silica-coated magnetite nanoparticles ($Fe_3O_4@SiO_2$). The novel sorbent exhibited good extraction characteristics, as well as a low sorbent quantity requirement. MIL-101 (Cr) modified zero valent iron nano-particles have been also evaluated as an MSPE adsorbent for the extraction of PAHs [61]. Nanoscale zero valent iron exhibits increased specific surface area and more reactive sites on the surface, compared to Fe_3O_4 nanoparticles. Moreover, it is very easy to prepare core-shell $Fe@SiO_2$ through a one-step approach under normal temperature and pressure, which directly provides the SiO_2 layer. The novel nanocomposite exhibited high extraction efficiency and stability, since it was found to be reusable at least 10 times.

MIL-100(Fe) has been also evaluated as a MSPE sorbent for the extraction of PAHs from water samples [62]. The MIL-100(Fe) sorbent was prepared form iron (III) chloride hexahydrate and 1,3,5-benzenetricarboxylic acid, and it was functionalized with mercaptoacetic acid modified magnetite nanoparticles. The novel sorbent was successfully applied for the MSPE of PAHs from river, well, pond and tap water samples. Although satisfactory extraction recoveries were reported (>80%), no data regarding sorbent reusability were provided. Huo et al. [63] prepared a magnetic MIL-100(Fe) nanocomposite through an pyrolytic in situ magnetization approach. In this case, MIL-100 (Fe) was placed in a tube furnace and it was calcined from 300 °C to 700 °C under nitrogen. The obtained material exhibited high porosity and magnetic characteristics and it was found to be recyclable and reusable (up to 10 times).

Zhang et al. [64] prepared a zinc benzimidazolate (ZIF-7) nanomaterial anchored onto polydopamine-coated Fe_3O_4 nanoparticles. Due to the hydrophobic and $\pi-\pi$ interaction between ZIF and the target analytes, good extraction efficiency was reported. Moreover, the novel nanocomposite was found to be reusable for at least 10 times.

Table 1. Application of different sorbents for the magnetic solid-phase extraction (MSPE) of PAHs from environmental samples.

Sorbent	Matrix	Analytical Technique	Sorbent Mass (mg)	Time (min)	LODs (ng L^{-1})	Extraction Recovery (%)	Reusability	Ref.
HKUST-1	Water	UHPLC-FLD	5 Fe_3O_4/20 HKUST-1	10	0.8–12	39–59	NA	[59]
MIL-101(Cr)	Water	HPLC-PDA	1 Fe_3O_4@SiO_2/0.6 MIL-101	20	2.8–27.2	NA	NA	[60]
Fe@MIL-101(Cr)	Water	HPLC-DAD	50	50	44–64	>80	At least 10 times	[61]
MIL-100(Fe)	Water	HPLC-FLD	10	10	32–2110	>80	NA	[62]
MIL-100(Fe)	Water	GC-FID	12.5	15	4.6–8.9	73–96	Up to 10 times	[63]
Fe_3O_4@ polydopamine/ZIF-7	Water, particulate matter	GC-MS	3 Fe_3O_4@PDA 15 ZIF-7	40	0.71–5.79	>82	At least 10 times	[64]
TpPa-1 COF	Water	HPLC-FLD	5	21	0.24–1.01	73–110	NA	[65]
COF-LZU1@PEI@Fe_3O_4	Water, soil	HPLC-FLD	5	33	0.2–20	NA	At least 6 times	[66]
G/CNF	Water	GC-FID	20	10	4–30	63.0–84.5	Up to 6 times	[67]
Fe_3O_4/C	Water	HPLC-FLD	50	30	0.2–0.6	76–110	At least 10 times	[68]
Hydrophilic Fe_3O_4/C	Water	GC-MS	10	30	15–335	NA	NA	[5]
CNF	Water	GC-FID	10	12	8–30	NA	At least 10 times	[7]
G/Fe_3O_4@PT	Water	GC-FID	20	10	9–20	83–107	At least 17 times	[69]
GO	Water	HPLC-UV	40	16	90–190	76.8–103.2	NA	[6]
GO-Fe_3O_4@PS	Water	GC-FID	15	10	3–10	69.5–88.7	NA	[70]
Poly(Py-co-Ani)@GO-Fe_3O_4	Water	GC-FID	35		3–10	50.4-78.3	At least 20 times	[71]
CNTs	Water	UHPLC-FLD	5	10	25–73	76.4–106.5	Up to 3 times	[72]
mag-MIP	Water	HPLC-PDA	20	55	1.3–969	46–100	At least 3 times	[73]
mag-MIP	Water	GC-MS	5–20	17	30–750	>76	NA	[74]
RAFT-MIP	Water	GC-MS	10	9	1–100	4.5–97	NA	[8]
PDA	Water	HPLC-FLD	20	5	0.5–1.9	76.4–107	NA	[75]
PPy	Water	GC-MS	20	3	0.38–5.01	27.4- 115.7	NA	[76]
PANI/Alginate	Water	HPLC-FLD	400	20	10	86.0–97.8	Up to 6 times	[77]
PoT	Water	GC-FID	60	15	0.3–5.5	NA	Up to 15 times	[58]
IL-MNPs	Water	GC-MS	30	8	40–1111	75–102	Up to 10 times	[78]
MNP@CN/IL	Leachate, sludge	HPLC-DAD	30	35	400–590	89.50–110.2	NA	[79]
MNP-PANI-DICAT	Water, sludge, soil	GC-MS	15	40	0.8–208.6	80.2–111.9	Up to 5 times	[80]
Fe_3O_4@IL@MO	Water	HPLC-FLD	18	26	0.1–2	NA	NA	[81]
Fe_3O_4@SiO_2@Nap	Water	HPLC-FLD	40	12	0.04–0.12	>90	At least 10 times	[1]
PC	Water, milk	HPLC-FLD	100	10	0.2–0.6	>90	NA	[82]
Fe_3O_4-DVB-SO_3-	Water	GC-MS	50	10	0.6–2.1	79.9–115.3	NA	[83]
MPNP	Water	UHPLC-DAD	200	15	10.83–18.53 nM	75.7–106.4	At least 5 times	[3]
Fe_3O_4/SiO_2/TPA	Water	HPLC-FLD	50	15	0.04–37.5	NA	NA	[84]
C_{18}	Water	GC-MS	50	6	0.8–36 × 103	35–99	NA	[85]
C_{10}–C_{18} carboxylates	Water	HPLC-FLD	200	18	0.1–0.25	>90	Up to 5 times	[86]
n-octadecylphosphonic acid	Water	GC-MS	50	1	14.1–70.0 × 103	61.9–119.1	NA	[87]
Nylon 6	Water	HPLC-PDA	40	30	0.05–0.58 × 103	36.2–87.0	NA	[88]
CTAB	Water	UHPLC-FLD	100 Fe_3O_4/50 CTAB	30	0.4–10.3	59.23–87.95	NA	[89]
Palm fatty acid	Leachate, sludge	HPLC-DAD	15	25	10–50	>81.1	Up to 5 times	[90]
TBCD	Water	HPLC-FLD	80	15	0.03–1.2	>80	NA	[91]
TCT	Water, urine	HPLC-FLD	40	13	0.09–0.15	89–93	At least 30 times	[92]
C_{16}-HO	Water	HPLC-UV	30	24	0.14–0.31	88–95	Up to 4 times	[93]

Covalent organic frameworks (COFs) are novel materials, structurally related to MOFs, consisting of light elements (H, O, C, N, B, Si) connected with organic monomers through strong covalent bonds. COFs comprise a class of ordered crystalline porous polymers that are characterized from superior properties, including low crystal density, high specific surface area, pore size tunability and good thermal stability [36,94,95]. He et al. [65] fabricated a bouquet-shaped magnetic porous nanocomposite via grafting a COF prepared from 1,3,5-triformylphloroglucinol and p-phenylenediamine (TpPa-1) onto the surface-modified magnetite nanoparticles. The novel sorbent contained clusters of core–shell magnetic nanoparticles and interconnected porous COF nanofibers. Therefore, it exhibited high specific surface area, high porosity, and supermagnetism.

Another covalent organic framework that has been evaluated as adsorbent for PAHs was COF-LZU1 (= Lan Zhou University-1), after immobilization onto polyethyleneimine-functionalized magnetic nanoparticles (COF-LZU1@PEI@Fe_3O_4) [66]. For this purpose, the examined COF was prepared from 1,3,5-triformylbenzene and 1,4-diaminobenzene through a Schiff base reaction. The novel sorbent exhibited good stability and reusability. Moreover, due to the strong π-stacking and hydrophobic interaction, the novel sorbent was successfully applied for the MSPE of PAHs from water and soil samples.

2.2.2. Carbon-Based Magnetic Sorbents for the MSPE of PAHs

Magnetic carbon-based nanomaterials exhibit superior extraction characteristics, due to the combination of magnetic nanoparticles with carbon-based materials that pose many interesting properties including good thermal stability and tunable miscibility, as well as good extraction capability [67].

Carbon coated Fe_3O_4 nanoparticles were used for the extraction of PAHs from environmental water samples prior to their determination by HPLC-FLD. Due to the large surface area of the nanomaterial and strong adsorption ability of carbon material, high adsorption capacity and extraction efficiency were obtained [68].

A hydrophilic carbon-ferromagnetic nanocomposite was also evaluated for the extraction of PAHs. The sorbent was designed with a hydrophobic sublayer and a hydrophilic surface. As a result, the sorbent exhibited good adsorption efficiency, due to the hydrophobic sublayer of the material, while good compatibility with the aqueous matrices was reported due to its hydrophilic surface [5].

Carbon nanofibers (CNFs) decorated with Fe_3O_4 nanoparticles have been also evaluated as adsorbent for the MSPE of PAHs from water samples. Carbon nanofibers are hydrophobic materials with the ability to establish π–π interaction. Moreover, they exhibit high specific area and flexibility, as well as high mechanical strength [7]. Magnetized graphene (G) layers immobilized on CNFs have been also investigated for the MSPE of PAHs. Graphene is a single layer of carbon atoms that was discovered in 2004 by Geim et al. [96]. In graphene, the carbon atoms are densely packed in a honeycomb crystal lattice, which can be viewed as exfoliated "graphite sheets". Graphene exhibits superior properties, such as a high specific surface area, as well as good chemical stability and thermal stability [6]. The combination of graphene and CNFs provided good adsorption of PAHs that can be attributed to hydrophobic and π–π interactions [67]. A graphene/Fe_3O_4@polythiophene (G/Fe_3O_4@PT) nanocomposite has been also used for the MSPE of PAHs. PT was employed used to enhance the adsorption capacity of magnetic graphene, and also to provide long service life and stability [69].

Graphene oxide is the chemical compound with a form similar to graphene, which consists of one-atom-thick two-dimensional layers of sp^2-bonded carbon and is rich in oxygen-containing groups including hydroxyl, carboxyl and epoxy groups. Graphene oxide exhibits high hydrophilicity and dispersibility, as well as good thermal and mechanical stability and high surface area. GO interacts with organic molecules through strong π-stacking, hydrophobic interaction and hydrogen bonding [57].

Magnetic graphene oxide has been used for the extraction of PAHs from water samples. The composite sorbent exhibited ease in separation and satisfactory recoveries [6]. Polystyrene (PS) modified magnetic GO (GO-Fe_3O_4@PS) was fabricated to enhance the extraction efficiency. Polystyrene is an inexpensive, high porous, environmentally friendly polymer that contains phenyl and alkyl groups [70]. Another polymeric material that has been employed for the functionalization of magnetic GO is poly(pyrrole-co-aniline) in order to increase its efficiency. Both polyaniline (PANI) and polypyrrole (PPy) are promising materials, due to their good chemical stability and ease of synthesis by the chemical and electrochemical method of polymerization. The PANI and PPy copolymer combines the superior properties of the two conducting polymers and its application for the modification of Fe_3O_4/GO can result in a highly efficient sorbent [71].

Carbon nanotubes (CNTs) functionalized magnetite nanoparticles have been also employed for the extraction of PAHs from water samples [72]. CNTs are interesting sorbents for volatile and semi-volatile organic compounds due to hydrogen bonding, π-stacking and hydrophobic interactions. CNTs are divided into single-walled (SWCNTs) and multi-walled carbon nanotubes (MWCNTs), based on whether they consist of one or more sealed tube-shaped layers of graphene, respectively. Adsorption of target analytes can take place in their easily accessible walls, as well as in their interstitial sites [97]. MWCNTs functionalized Fe_3O_4 nanoparticles showed good extraction efficiency and selectivity towards PAHs [72].

2.2.3. Molecularly Imprinted Polymers Magnetic Nanocomposites for the MSPE of PAHs

Molecularly Imprinted Polymers (MIPs) are synthetic polymeric materials that are composed of imprinted sites complementary to a specific chemical molecule. MIPs show high affinity towards the target analytes with analogous molecular structure, resulting in high extraction ability [55]. Molecular recognition is attributed to a combination of shape and size and hydrogen bonding, as well as electrostatic and hydrophobic interactions. MIPs are stable at a wide pH and temperature ranges and in most organic solvents, while there is normally no requirement for special storage conditions [98,99].

MIPs can be prepared through covalent, non-covalent and semi-covalent approaches, and their preparation is based on the polymerization of a functional monomer and a cross-linker around a template molecule. The template molecule should be able to interact with the functional monomer to develop complexes that further interact with the cross-linker during the polymerization reaction. Subsequently, the template molecule is removed and a MIP with imprinted sites complementary to the molecular structure and the functional groups of the template is created. Normally, after the synthesis of MIPs, extensive washing is required to remove any residual molecules of template. However, even after many washing steps, bleeding of the template has been reported. In order to overcome this drawback, MIPs can be prepared from templates with chemical structure that is analogue to the target molecules [98,100].

Magnetic MIPs combine the benefits of ease in separation of magnetic materials with the high molecular recognition of MIPs. Many researchers reported the synthesis of magnetic MIPs as adsorbents of PAHs from environmental samples.

Villar-Navarro et al. [73] used a commercially available magnetic MIP (mag-MIP) by NanoMyP® (Granada, Spain) that was able to extract the 16 EPA-regulatd PAHs. The examined sorbent exhibited good extraction efficiencies (98.8–100%) for the more lipophilic PAHs, however, for the less lipophilic PAHs, extraction recovery rates were lower (46–60%). Finally, mag-MIP was found to be reusable at least three times. Benedetti et al. [74] managed to improve the extraction efficiencies by implementing a Plackett-Burman experimental design. In this case, all recovery values were higher than 76%.

Azizi et al. [8] reported the synthesis of a MIP sorbent by surface polymerization onto magnetic $Fe_3O_4@SiO_2$ nanoparticles through reversible addition fragmentation chain transfer (RAFT) polymerization. For this purpose, methacrylic acid and isopropyl acrylamide were used as functional monomers. In this case, extraction recoveries ranged from 4.5% to 97%, thus extraction of "heavy" PAHs was found to be more efficient compared with the extraction of "lighter" PAHs.

2.2.4. Polymer-Modified Magnetic Nanoparticles for the MSPE of PAHs

Polymers are promising materials in separation science, due to the fact that their network can be chemically anchored around the magnetic core, thus providing large π-conjugated structure, hydrophobicity and polar functional groups for adsorption [58]. Various polymers including polydopamine [75], polypyrrole [76], poly(o-toluidine) [58] and polyaniline [77]. Polyaniline and its derivatives are one of the widely used polymers for the extraction of various types of organic compounds, because of their good stability in high temperature, air and different solvents [58].

Polydopamine (PDA) functionalized magnetic nanoparticles have been also employed for the extraction of PAHs. PDA exhibit several significant advantages, including biocompatibility, good dispersibility in water, as well as multifunctional groups (amino and catechol groups) that enhance the extraction efficiency of PAHs due to π–stacking interaction [75].

Polypopyrole [76], due to its hydrophobicity, large π-conjugated structure, hydrogen bonding and ion exchange properties, exhibits good extraction efficiency and selectivity towards PAHs.

Polyaniline-coated magnetite nanoparticles incorporated in alginate beads have also been employed for the MSPE of PAHs in water samples. In this case, the alginate beads assisted in the increase of the surface area for polyaniline coating, and the novel nanocomposite showed satisfactory extraction efficiency towards the PAH analytes, due to the π–π interactions between the polyaniline moieties [77].

Poly(o-toluidine) (PoT) coated $MnFe_2O_4$ magnetic nanoparticles have been also successfully employed as a magnetic adsorbent for the MSPE of PAHs from water samples [58]. Due to the π–π interactions between the active sites of PoT and PAHs, a satisfactory extraction efficiency was reported.

2.2.5. Ionic liquids Modified Magnetic Nanoparticles for the MSPE of PAHs

Ionic liquids (ILs) are a green alternative to conventional organic solvents that have gained a lot of attention lately. ILs are generally composed of bulky, non-symmetrical organic cations (i.e., imidazolium, pyrrolidinium, pyridinium, ammonium, phosphonium etc.) and different inorganic or organic anions (i.e., tetrafluoroborate anions, bromide anions etc.) [101–103]. The increased popularity of ILs may be attributed to their extraordinary properties, such as a negligible vapor pressure, good thermal stability, as well as tunable viscosity and miscibility with water and organic solvents. By choosing the cationic or the anionic constituent of ILs, their polarity, hydrophobicity, viscosity and other chemical and physical properties can be tuned, in order to prepare ILs with the desired characteristics. Due to their special structures, ionic liquids also exhibit good extractability for various organic compounds and metal ions [103]. As a result, ionic liquids have been successfully coupled with various extraction techniques for the extraction of PAHs.

Galán-Cano et al. [78] prepared ionic liquid-coated magnetic nanoparticles (IL-MNP) and used them for the MSPE of PAHs from water samples, prior to their determination by GC-MS. For this purpose, methylimidazolium-chloride was used to modify Fe_3O_4@SiO_2 magnetic nanoparticles by a simple metathesis reaction, using potassium hexafluorophosphate as a reagent. Due to the covalent immobilization of the ionic liquid onto the surface of silica coated Fe_3O_4 nanoparticles, the developed sorbent exhibited good stability, and it was found to be reusable for up to 10 times.

Cyano-ionic liquid functionalized magnetic nanoparticles (MNP@CN/IL) have been also employed for the MSPE of PAHs from environmental water samples, prior to their determination by HPLC-DAD [79]. For this purpose, 1-benzyl-3-(trimethoxysilylpropyl)imidazolium chloride was chosen as IL, and it was used to modify the cyano-functionalized magnetic nanoparticle. The combination of the cyano group and IL on the surface of the MNPs provided good extraction efficiency, probably due to the combination of π−π and electrostatic interaction.

Magnetic nanoparticles coated with a polyaniline-di-cationic ionic liquid (MNP-PANI-DICAT) were prepared and used for the MSPE of PAHs in environmental samples [80]. The novel sorbent combined the benefits of PANI and ionic liquids. Due to the π–π interaction between polyaniline shell and di-cationic IL with PAHs compounds, high extraction efficiency was observed. Moreover, the novel MSPE sorbent exhibited satisfactory stability, since it was found to be reusable up to five times.

Liu et al. [81] fabricated a magnetic ionic liquid functionalized methyl orange nanoparticles (Fe_3O_4@IL@MO), by self-assembly of the 1-octadecyl-3-methylimidazolium bromide and methyl orange on the surface of Fe_3O_4 silica magnetic nanoparticles. The novel sorbent was successfully applied for the MSPE of PAHs from environmental water samples. Due to the presence of benzene rings of methyl orange and the hydrocarbon chains of ILs, the sorbent provided adsorption sites for organic pollutants through π–π and hydrophobic interactions. The silica modified Fe_3O_4 nanoparticles were found to be reusable after washing, however, assembling with IL and methyl orange was also required.

2.2.6. Other Magnetic Nanocomposites for the MSPE of PAHs

Naphthyl functionalized magnetic nanoparticles (Fe_3O_4@SiO_2@Nap) have been used for the extraction of PAHs from river waters, prior to their determination by HPLC-FLD. Due to the condensed cyclic structure and the hydrophobic property of naphthyl, the novel sorbent exhibited satisfactory selectivity and extraction efficiency through π–π interactions [1].

Phosphatidylcholine (PC) is a phospholipid consisting of a long double carbon chain, and the zwitterions' pair headgroup is composed of phosphate and choline. The extraction of PAHs can be assisted, due to the middle hydrocarbon chains that provide adsorption sites due to hydrophobic interactions. Meanwhile, the zwitterions pair headgroups endow the outer surface of the substrate with perfect hydrophilicity and biocompatibility [82].

Another example of composite material exhibiting both hydrophilicity and hydrophobicity is the magnetic nanocomposite prepared from nanoparticles functionalized with divinylbenzene (DVB) and sulfonate moieties (Fe_3O_4-DVB-SO_3^-). In this case, the hydrophobic DVB moieties were dedicated for extraction, while the hydrophilic sulfonate groups were introduced to enhance dispersion of the magnetic sorbent in the aqueous sample solution [83]. A hydrophilic–hydrophobic magnetic Fe_3O_4-doped polymeric nanoparticle (MPNP), prepared from highly charged poly(styrene-divinylbenzene-co-4-vinylbenzenesulfonic acid sodium salt) has been also employed for the determination of PAHs in environmental matrices [3].

Triphenylamine (TPA) has been employed for the functionalization of magnetic nanoparticles (Fe_3O_4/SiO_2/TPA). Due to the strong π–π conjugate effect between the three benzene rings of TPA and PAHs, satisfactory extraction efficiency and selectivity were observed [84].

Other examples of chemical compounds and groups that have been employed for the functionalization of magnetic nanoparticle for the MSPE of PAHs from environmental samples are C_{18} [85], hemimicelles of alkyl (C_{10}-C_{18}) carboxylates [86], n-octadecylphosphonic acid [87], nylon 6 [88], cetyltrimethylammonium bromide (CTAB) [89], palm fatty acid [90], 1,4,7,10-tetrabenzyl-1,4,7,10-tetraazacyclododecane (TBCD) [91], tetraazacalix[2]arence[2]triazine (TCT) [92] and *n*-hexadecylsilanol-diol (C_{16}-HO) [93].

2.3. *Solid-Phase Microextraction of PAHs from Environmental Matrices.*

Solid-phase microextraction is a sample preparation microextraction technique in which the analytes are directly extracted and preconcentrated at the outer coating of a fused-silica fiber [104]. There are two approaches of SPME that can be used to extract analytes, the headspace (HS-SPME), where the fiber is exposed to the gas phase above the sample, and direct immersion (DI-SPME), where the fiber is directly immersed into the sample solution [105]. After the extraction, desorption takes place either thermally in the injection port of a gas chromatograph or by the addition of an organic solvent [104,105].

Until now, there are many different commercial SPME coated fibers, such as polydimethylsiloxane (PDMS), polydimethylsiloxane/divinylbenzene (PDMS/DVB), polyacrylate (PA), carboxen/polydimethylsiloxane(CAR/PDMS) and divinylbenzene/carboxen/polydimethylsiloxane (DVB/CAR/PDMS) [105]. However, most of them more or less have some disadvantages, such as low selectivity, ease of fiber breakage, a short lifetime and swelling in organic solvents. In order to overcome them, various new coatings have been prepared and evaluated [106].

Ionic liquids and polymeric ionic liquids (PILs) have been successfully employed as SPME coatings, due to their simplicity of synthesis and their high tuneability, that enable the preparation of highly selective fibers [107]. PILs are polymers prepared from IL monomers. Compared with conventional ILs, PILs exhibit a number of advantages when used as coatings in SPME. Generally, PILs often have solid nature and good thermal and mechanical strength, while extraction selectivity is similar with ILs. As a result, they have proved to be more stable coatings [107,108]. Various PILs including poly(1-vinyl-3-hexadecylimidazolium) bis[(trifluoromethyl)sulfonyl]imide [109], poly(1-4-vinylbenzyl)-3-hexadecylimidazolium bis[(trifluoromethyl)sulfonyl]imide [110], poly (1-vinyl-3-octylimidazolium) 2-naphthalene-sulfonate [111] and poly(1-(4-vinylbenzyl)-3-hexadecylimidazolium bis[(trifluoromethyl)sulfonyl]imide [112] have been successfully used as coatings used for the SPME of PAHs. In most cases, the PIL was initially prepared and diluted in a volatile solvent (i.e., acetone or chloroform), and a bare fiber (usually made from stainless steel) was immersed in the solution, followed by slow removal and air drying to remove any excess of solvent that may contribute to high background signals in gas chromatography [109,110,112]. However, in situ polymerization of the IL and creation of the SPME coating on the surface of a stainless steel wire has also been reported [111].

Graphene [113,114], graphene oxide [115–117], MWCNTs [118,119] and other carbon based materials [120] have also been successfully employed as SPME sorbent coatings, either neat or combined with other materials in order to generate more efficient composites. These materials exhibit high chemical, thermal and mechanical stability, as well as great affinity towards PAHs. Additionally, due to their unique structures and sufficient surface areas, rapid extraction and desorption of the target analytes can be achieved [107]. Various techniques for the preparation of the coated fibers including the chemical bonding [113,115,117], electrophoretic deposition [119] and sol-gel approaches [120] have been also evaluated.

Metal-organic frameworks (MOFs) are also promising sorbents for SPME coatings [107,121]. A wide variety of MOF materials, such as HKUST-1 [122], MOF-199 [123], ZIF-8 [124], TMU-6 (Tarbiat Modares University) prepared from N^1,N^4-bis((pyridin-4-yl)methylene)-benzene-1,4-diamine and 4,4′-oxybisbenzoic acid) [125], MOF-177 prepared from zinc nitrate hexahydrate and 1,3,5-tris(4-carboxyphenyl)-benzene [126], UiO-6 (University of Oslo) prepared from zirconium chloride and terephthalic acid [127], an ytterbium based MOF prepared from aminoisophthalic acid and 2,2′-bipyridine [128], as well as bio-MOF-1 prepared from 4,4′-biphenyl dicarboxylic acid (BPDC), zinc acetate dihydrate and adenine [129].

The main considerations when preparing MOFs for extraction purposes are their stability under extraction conditions, and their ability to establish interactions with the target analytes. Moreover, when fabricating MOFs as coatings for SPME fibers the mechanical and thermal stability should be carefully evaluated. The main procedures that can be used to obtain MOF coatings include physical adhesion, immersion in a suspension of MOF, in situ growth and electrodeposition [107].

A metal-azolate framework 66 (MAF-66) has been also used for the SPME of PAHs from environmental samples. Metal-azolate frameworks are a subfamily of MOFs that have recently gained attention. MAF-66 was prepared from 3-amino-1,2,4-triazole and zinc hydroxide and was used to fabricate a SPME coating by a layer-by-layer deposition method, providing enhancement factors of 123-3108 [130].

Examples of other materials that have been employed to fabricate SPME coatings for the extraction of PAHs include polyaminithiophenol (PATP) with Au coating [131], poly(3,4-ethylenedioxythiophene)@gold nanoparticles[132], crosslinked methyl methacrylate–polyhedral oligomeric silsesquioxane hybrid polymeric coating [106], nanostructured octadecyl silica [133] and polythiophene/carboxylic acid modified multi-walled carbon nanotube composite [134]. In the latter approach, the researchers developed a novel SPME technology, in which the features of heating the sample, cooling the sorbent and extraction under vacuum condition were combined [134].

PAL (Prep And Load solution) SPME Arrow technique [135] has been also evaluated for the extraction of PAHs from environmental samples. This technique is based on the use of a robust stainless-steel backbone, carrying the connection to the PAL sampler, the coating and an arrow-shaped tip for septum penetration. SPME Arrow combines the benefits of conventional SPME with the larger sorption phase volumes that are used in stir bar sorptive extraction (SBSE). At the same time, the disadvantages of both techniques include the difficulty in automation for SBSE, and the small volume of sorption phase, as well as the low robustness of classical SPME fibers. The results indicated that extraction efficiency significantly benefited from the larger sorption phase volume.

2.4. Stir Bar Sorptive Extraction (SBSE) and Stir Rod Sorptive Extraction (SRSE) of PAHs from Environmental Matrices.

Stir bar sorptive extraction was initially introduced by Baltussen et al. in 1999 [136]. In SBSE, a coated stir bar is placed into the vial together with the aqueous sample solution. Extraction of the target analytes takes place under rigorous stirring. When equilibrium is reached, the stir bar is removed and elution of the adsorbed analytes takes place either by the addition of an organic solvent or thermally [18]. SBSE by nature is an equilibrium technique, and for water samples the extraction of the target analytes into the extraction medium is controlled by the partitioning coefficient of the solutes between the coating phase and the aqueous phase [137]. Polydimethylsiloxane (PDMS) is the most used commercially available coating phase for stir bars, however, the synthesis and application of many novel coating materials has been reported.

PDMS coated SBSE bars have been successfully used for the extraction of PAHs resulting in good recoveries and low detection limits [138–142]. Apart from the conventional PDMS stir bars, various novel coating materials have been evaluated for the extraction of PAHs from environmental samples. Typical examples are polymeric materials that have been evaluated, as stir bar coatings are polypyrrole and polyaniline copolymer (PPy-PAN) [143] and (octyl methacrylate- ethylene dimethacrylate) copolymer [144]. Poly (ethylene glycol)-grafted multi-walled carbon nanotubes have been also evaluated for the extraction of PAHs from environmental samples [145]. In this case, the extraction efficiency was favored by the superior characteristics of MWCNTs and the ease in operation of the SBSE technique.

Hu et al. [146] prepared polydimethylsiloxane/metal–organic framework (PDMS/Al-MIL-53-NH_2) coated stir bars and used them for the extraction of PAHs. The novel PDMS/MOFs-coated stir bars achieved higher extraction efficiencies for PAHs than the commercial PDMS-coated stir bar. In this case, the reported MOF was prepared from terephthalic acid and aluminum chloride. In contrast to the conventional PDMS coating, in which extraction is based only on hydrophobic interactions, the PDMS/MOF composite also takes advantage of the $\pi-\pi$ conjugations.

Benede et al. [147] used stir bar dispersive liquid microextraction (SBDLME) for the extraction of PAHs from natural waters. SBDLME is a hybrid approach that combines the benefits of SBSE and DLLME. For this purpose, a neodymium stir bar was magnetically coated with a magnetic ionic liquid. At high stirring rates, the ionic liquid was dispersed into the solution, followed by being magnetically retrieved onto the stir bar when the extraction was completed. Subsequently, thermal desorption took place and PAHs were determined in a GC-MS system. The novel method exhibited high sensitivity and low LOD values.

Luo et al. [148] developed a stir rod sorptive extraction (SRSE) method for the preconcentration PAHs from water samples, prior to their determination by GC-MS. SRSE is an improved format of SBSE, which has been proposed to avoid the friction loss of extraction coatings that exist in conventional SBSE procedure. In SBSE, a stir rod (glass insert with coating) is immersed in the sample solution to adsorb the target analytes [149]. For the extraction of PAHs, a graphene-polymer composite was evaluated. Due to the high specific surface area and π–stacking properties of the hybrid coating, good extraction efficiency was reported [148].

Automated stir plate sorptive extraction (SPSE) coupled with HPLC-FLD has been also evaluated for the extraction of PAHs. For this purpose, automatic extraction, desorption and sample loading, was controlled by a programmable flow injection system, and extraction of PAHs took place on the surface of a PDMS/β-cyclodextrin/divinylbenzene (PDMS/β-CD/DVB) coated plates. The researchers investigated three different operation modes, including static, circular flow and continuous flow SPSE. It was found that extraction efficiencies with continuous flow SPSE were slightly better than circular and manual SBSE, probably due to the continuous introduction of new sample solutions [150].

2.5. Liquid-Phase Microextraction of PAHs from Environmental Matrices.

Liquid-phase microextraction (LPME) is a miniaturized version of classical LLE, which is characterized by minimum consumption of solvents. LPME can be divided into three main categories, the single-drop microextraction (SDME), the hollow fiber liquid-phase microextraction (HF-LPME) and the dispersive liquid-liquid microextraction (DLLME), with the latter being the most widely used LPME form [151].

HF-LPME is usually based on the use of disposable propylene porous hollow fibers that are filled with a small amount of extracting solvent (acceptor phase). In order to extract the target analytes, the fibers are immersed into the aqueous sample solution (donor phase) [151,152].

On the other hand, in single-drop microextraction (SDME), a drop acts as the acceptor phase for the extraction. SDME can be divided into two main categories i.e., the direct-immersion single-drop microextraction (DI-SDME), in which the drop is directly immersed into the sample, and the headspace single-drop microextraction (HS-SDME), in which the drop is suspending over the sample [153].

DLLME is based on the initial fast injection of a suitable mixture of two solvents, an extraction solvent and a dispersive solvent, into an aqueous sample solution with the assistance of a syringe, followed by the formation of a cloudy solution that contains droplets of the extraction solvent dispersed into the sample. After phase separation due to the difference in density of the two phases (e.g., by centrifugation), the extraction phase can be removed and analyzed. In the conventional form of DLLME, the extraction phase is accumulated at the bottom of the extraction container [151,154]. DLLME is considered to be simple, cheap and environmentally friendly, while it provides high enrichment. The proper selection of the extraction and dispersive solvents are two critical factors for the optimization of DLLME procedure. Therefore, the dispersive solvent has to be immiscible with the extraction solvent and the aqueous sample, in order to generate a cloudy solution that increases the interaction between the two phases, in order to increase the extraction efficiency [155].

Until today, many different DLLME methods have been proposed for the extraction of PAHs. In 2006, Rezaee et al. [156] developed a DLLME method for the determination of PAHs in surface water, using tetrachloroethylene and acetone as extraction and disperser solvent, respectively. In order to decrease the extraction time, the main drawback of DLLME is the requirement for organic solvents with a density higher than water, in order to be sedimented at the centrifugation step. Guo et al. [157] proposed a low-density solvent-based solvent demulsification DLLME approach. For this purpose, after the formation of the emulsion of the aqueous sample, the disperser solvent (acetone) and the extraction solvent (*n*-hexane), more acetone was added as demulsification solvent, to break up the emulsion. Since no centrifugation was required, the whole procedure was very rapid (2–3 min). A special extraction cell has been also employed for the DLLME of PAHs, based on the use of low-density solvent. Hosseini et al. [158] evaluated the application of air flotation to assist the phase separation, using toluene as an extraction solvent. In this case, the organic solvent was collected at the top part of the designed cell, and no centrifugation was required in this procedure.

The combination of ultra-sound radiation [159] and vortex radiation [160] with DLLME has been evaluated in order to assist the extraction process, and to achieve high extraction efficiency in a short period of time [161]. With the assistance of radiation, the mass transfer of target analytes from the sample to the extraction solvent was facilitated, due to the shorter diffusion distance and larger interfacial area [160].

Ultrasound-assisted emulsification microextraction (USAEME) is a similar approach with ultra-sound assisted DLLME, however, in USUAEME, the dispersion of the extraction solvent in the aqueous sample is attributed to ultra-sound radiation, and no disperser solvent is added. In this case, ultrasonic radiation is used to overcome the drawbacks of disperser solvent, such as the decrease of partition coefficient of analytes into an extraction solvent al [161,162]. USAEME methods for the extraction of PAHs from environmental samples, using toluene [161], chloroform [162], iso-octane [163] and cyclohexane [164] as extraction solvents have been reported. The combination of ultrasound radiation and iso-octane as extraction solvent enabled the handling of high volumes of sample, thus, after concentrating the organic solvent phase, enhancement factors of up to 100,000 were reported. Moreover, the phase separation of iso-octane and sample was assisted by the addition of sodium chloride (NaCl), and no centrifugation was required [163].

An interesting approach to DLLME and USAEME that can be employed for the separation of the extraction solvent is the solidification of floating organic droplet (SFO). In this case, the sample solution is transferred into an ice bath after the extraction process and the floating organic droplet is allowed to solidify. When solidified, the droplet is removed and placed in a vessel to melt prior to the instrumental analysis [165,166]. Yousefi et al. [167] evaluated the use of a deep eutectic solvent (DES) tetra-n-butyl ammonium bromide (TBAB) carboxylic acids for the extraction of PAHs from water samples based on solidification of floating organic droplet. DESs are eco-friendly solvents, which are usually prepared from a hydrogen bond acceptor (e.g., choline chloride salt) and a hydrogen bond donor (e.g., urea, glycerol, carboxylic acids etc.) that can be associated with each other with hydrogen bond interactions. Among the benefits of DESs is their easy preparation from biocompatible, nontoxic and biodegradable chemical compounds [167,168].

The application of ionic liquids in DLLME has been also studied in order to replace the environmentally hazardous volatile organic solvents that are commonly used. Pena et al. [169] developed a DLLME procedure using 1-octyl-3-methylimidazolium hexafluorophosphate ([C_8 MiM][PF_6]) IL as extraction solvent. Due to the chemical affinity between the IL and the PAHs, good extraction recoveries and enhancement factors was reported.

In order to enhance the extraction efficiency and the enhancement factors, Liu et al. [170] combined an SPE method based on a macrocyclic polyamine-functionalized silica with an ionic liquid DLLME procedure. For this purpose, after the elution of PAHs from the SPE column with acetone, distilled water and 1,3-dibutylimidazolium bis[(trifluoromethyl)sulfonyl]imide ([BBIM][Tf_2N]) as extraction solvents were rapidly injected for the DLLME procedure.

In order to enhance the sensitivity of the determination of PAHs in environmental matrices, Shamsipur and Hashemi [171] combined the SBSE with dispersive liquid–liquid microextraction, based on the solidification of floating organic drop. For this purpose, the target analytes were extracted on the surface of a PDMS coated stir bar, which was further placed in a glass vial containing methanol (disperser solvent). After the elution of PAHs, the stir bar was removed and 1-undecanol was added as an extraction solvent. After centrifugation and cooling in an ice bath, the organic drop was collected. Finally, the drop was again melted at room temperature, mixed with acetonitrile and analyzed in a HPLC-UV system. The combination of the extraction techniques provided good extraction recoveries and low LOD values.

Finally, Fernandez et al. [172] developed a lab on valve DLLME method for the extraction of PAHs, prior to their determination by HPLC-FLD. For this purpose, trichloroethylene was used as the extraction solvent, and acetonitrile was used as the dispersive solvent. The automated instrumentation simplified the extraction process and exhibited satisfactory enhancement factors (86–95). Table 2 summarizes the applications of DLLME and USAEME in the extraction of PAHs from water samples. Even though DLLME is the predominant form of LPME that has been employed for the extraction of PAHs, the use of SDME [173,174] and HF-LPME [175,176] has been also evaluated.

Table 2. Applications of dispersive liquid-liquid microextraction (DLLME) and ultrasound-assisted emulsification microextraction (USAEME) in the extraction of PAHs from water samples.

Matrix	Analytical Technique	Extraction Solvent	Disperser Solvent	Phase Separation	LODs(ng·L^{-1})	EF	Extraction Recovery (%)	Ref.
Surface water	GC-MS	Tetrachloroethylene	Acetone	Centrifugation	7–30	603–1113	-	[156]
Rainwater	GC-MS	*n*-Hexane	Acetone	Addition of demulsification solvent	3.7–39.1	NA	-	[157]
River water	GC-FID	Toluene	Methanol	Air flotation	14–41 × 10^3	NA	-	[158]
Sea water	GC-MS	Tetrachloroethylene	Diethyl Ether	Centrifugation	1–10	722–8133	59.2–90.5	[159]
Sediment	HPLC-FLD	Dichloromethane	Acetonitrile	Centrifugation	2.3–6.8 ng g^{-1}	NA	>94.0	[160]
Tap, sea and spring water	GC-FID	Toluene	-	Centrifugation	20–50	1776–2714	99–103	[161]
Tap, well, surface water etc.	GC-MS	Chloroform	-	Centrifugation	1–36	NA	-	[162]
Tap, spring, surface water etc.	GC-MS	Iso-octane	-	Addition of NaCl	0.001–0.009	Up to 100000	-	[163]
Tap, rain and wastewater	HPLC-FLD	Cyclohexane	-	Centrifugation	0.6–62.5	90–247	95–100	[164]
Well, river, lake water etc.	HPLC-FLD	TBAB/2-decanoic acid DES	-	Centrifugation/Solidification	0.7–6.6	163–198	>80.0	[167]
Tap, bottle, fountain water etc.	HPLC-FLD	[C_8 MiM][PF_6]	Acetone	Centrifugation	0.03–2	301–346	-	[169]
Tap, well, surface water etc.	HPLC-UV	[BBIM][Tf_2N]	Acetone	Centrifugation	2	2768–5409	-	[170]
Tap, rain and surface water	HPLC-FLD	Trichloroethylene	Acetonitrile	-	20–600	86–95	-	[172]

2.6. Fabric Phase Sorptive Extraction of PAHs from Environmental Matrices.

Fabric phase sorptive extraction (FPSE) is a novel sample preparation technique proposed by Kabir and Furton in 2014. FPSE utilizes a natural or synthetic fabric substrate, which is chemically coated in the form of ultra-thin coating with sol-gel organic-inorganic hybrid sorbent as the extraction medium. For the FPSE procedure, the sol-gel sorbent coated FPSE media is immersed into a mixture of appropriate solvents, to remove any undesirable impurities from the material, and rinsed with deionized water to remove residual organic solvents. Subsequently, the FPSE media is submerged into the sample solution placed in a glass vial. A magnet is added into the sample solution, and the sample is magnetically stirred for certain time span for the adsorption of the target analytes. Finally, the FPSE media is removed, and elution of the analytes takes place into another vial containing appropriate elution solvent. Analysis of the eluent can take place after centrifugation and or/filtration [177–180].

FPSE provides high primary contact surface area, thus, rapid and efficient analyte extraction can be easily achieved. Moreover, FPSE is also characterized by low organic solvent consumption, ease in operation, reusability and good selectivity towards the target analytes, which is based directly on the different nature of the fabric substrates and the sol-gel coating. Until today, the use of various fabric substrates, including cellulose, fiber glass and polyester, as well as several sol-gel coatings, including polyethylene glycol, polytetrahydrofuran and polydimethyldiphenylsiloxane have been evaluated as adsorbents for a wide variety of analytes [17,178,181,182].

A trace-level determination of selected PAHs in environmental water samples using FPSE prior to their determination by HPLC-FLD has been reported. For this purpose, a non-polar sol-gel C_{18} coated FPSE media was prepared and conditioned in a mixture of methanol and acetonitrile for 5 min, and then rinsed with deionized water. The extraction of PAHs took place in a glass vial containing 10 mL of the aqueous sample solution, in which the sol-gel C_{18} coated FPSE media was directly immersed. After 30 min under constant stirring at 1000 rpm, the FPSE media was removed and PAHs were eluted with acetonitrile under ultrasonic radiation for 5 min. The developed FPSE-HPLC-FLD protocol was proved to simple, efficient, fast, sensitive, green, economical and reliable for trace level determination of environmentally important PAHs [183].

Recently, fabric-phase sorptive extraction was coupled with ion mobility spectrometry (IMS) for on-site rapid detection of PAHs in aquatic environment [184]. Ion mobility spectrometry is a rapid and sensitive gas-phase analytical technique, which can be employed for the in the field testing of various chemical compounds, due to its fast analysis and compact size [185]. For the fabrication of the FPSE media, PDMS was coated on the glass fiber cloth through a sol-gel reaction. Glass wool was chosen based on the inlet temperature of IMS, since the thermal desorption of the PAHs was performed after inserting the FPSE media the inlet port of the IMS instrument directly after analyte extraction. Under optimum conditions, low LODs and satisfactory recoveries were obtaining, thus enabling the on-site monitoring water quality.

2.7. Other Extraction Techniques for the Determination of PAHs in Environmental Matrices.

Yang et al. [186] synthesized MOF-5 from terephthalic acid and zinc acetate, and evaluated it as a sorbent for the SPE of PAHs from environmental samples. Therefore, 300 mg of MOF-5 was packed into SPE cartridges. Although MOF-5 is known to be unstable when exposed to water, the researchers reported that the derived material of MOF-5 still demonstrated good extraction characteristics. This was attributed to the π–π conjugate effect between the terephthalic acid molecules of the framework and the PAHs and the π-complexation between PAHs and the central zinc ions. The analytes were further separated and detected in a HPLC-UV system. The method exhibited satisfactory extraction ability and low LODs (0.4–4.0 ng L^{-1}), however, no sorbent reusability was reported.

Hu et al. [187] evaluated two zeolitic imidazolate frameworks for the SPE disk extraction of PAHs in aid of filter membrane of PAHs from environmental water samples. The studied ZIFs were both composed of the same metal ion (zinc) and organic linker (benzimidazole), thus differing in spatial structures with one in cube (ZIF-7), while the other was in rhombic dodecahedron (ZIF-11). ZIF-11 with markedly better extraction efficiencies due to its unique spatial structure with large cages and its molecular composition that was composed of abundant benzyl groups and metal sites on the surface.

ZIF-8 [188] has been also evaluated for the porous membrane-protected micro-solid-phase extraction (μ-SPE) of PAHs. For this purpose, the sorbent was packed in a sealed porous polypropylene membrane envelope. The novel extraction devices exhibited good extraction characteristics and decreased consumption of the organic solvent.

A molecularly imprinted polymer has been also applied for the SPE of 16 PAHs from seawater, prior to their determination by GC-MS [189]. MIPs were prepared by using the 16 PAHs mixture through non-covalent polymerization as a template based on sol-gel surface imprinting. The developed sorbent exhibited good affinity towards the target analytes. Other examples of novel SPE sorbents that have been implemented for the extraction and preconcentration of PAHs from environmental samples include a cyclodextrin-silica microporous composite [190,191], aminopropyl imidazole-modified silica gel [192], tetraazacalix[2]arene[2]triazine [193], titanium oxide nanotubes [194] and a titanate nanotube array modified by cetyltrimethylammonium bromide [195].

Krupadam et al. [196] prepared MIP microspheres in a continuous segmented flow microfluidic reactor, and used them as packing material for microtraps for the selective extraction of benzo[a]pyrene from environmental water samples. For this purpose, the pumping of monodisperse droplets of acetonitrile containing methacrylic acid as the functional monomer took place, benzo[a]pyrene was used as a template, and ethylene glycol dimethacrylate as cross-linking monomer into the microchannels of the microfluidic reactor. The obtained microspheres exhibited high extraction efficiency and selectivity towards benzo[a]pyrene. In comparison with commercially available activated carbon, the novel microspheres showed 300% higher adsorption capacity.

A portable system for the in situ extraction of PAHs was proposed by Foan et al. [197]. The researchers designed a microfluidic device for the fast extraction of PAHs using low volume samples. The work was performed on a lab-on-a-chip, made of a silicon/glass microfluidic device functionalized with PDMS. Among the benefits of the novel technique was the low organic solvent consumption and the portability. A comparison of the novel device with SBSE showed approximately 50 times less sample preparation time for the high molecular weight PAHs. However, for the lightest PAHs, the performance of the microchip required improvement.

Flow injection solid-phase extraction (FI-SPE) of PAHs from environmental samples with novel extraction sorbents has been also proposed by the research groups of Wu [198] and Zhou [199]. The first group used a multi-walled carbon nanotubes (MWCNTs)-packed micro-column for the extraction of PAHs, prior to their determination by GC-MS, while the second group prepared a copper(II) isonicotinate coordination polymer packed in a pre-column for the extraction of PAHs, prior to their determination by HPLC-DAD. Both methods exhibited good extraction characteristics. In the case of GC-MS detection, after the FI-SPE process, the eluates were collected, and manual injection was performed, while for the HPLC-DAD analysis elution of the adsorbed analytes was also performed on-line in the backflush mode by the HPLC mobile phase directly into the chromatographic column, thus minimizing the required analysis steps.

In-syringe solid-phase extraction of PAHs has been also proposed for the on-site sampling of water samples. In-syringe SPE is characterized by portability, simplicity in use, low cost and short extraction time. Zhang et al. [200] evaluated the application of MIL-101 as a novel sorbent, due to its good thermal and mechanical stability, as well as its resistance towards organic solvent and waters. The proposed technique exhibited excellent adsorption performance, since the analytes could be completely adsorbed during one adsorption cycle, thus reducing the extraction time. Moreover, it was found that the adsorbed analytes remained stable on the in-syringe device for at least 7 days.

3. Conclusions

A wide variety of novel microextraction techniques and miniaturized extraction techniques have been developed and applied for the extraction of PAHs from environmental samples. Most of the novel extraction techniques are variations of the conventional SPE approach, however LLE based extraction approaches have been also developed.

Unequivocally, a lot of progress has been made in the field of sorbent development for micro and miniaturized solid-phase extraction sample preparation techniques. Metal-organic frameworks, covalent organic frameworks, zeolitic imidazole frameworks, graphene, graphene oxide and carbon nanotubes are some typical examples of novel extraction sorbents that have been implemented as such, or after surface modification for the extraction of PAHs from environmental samples. At the same time, ionic liquids, polymeric ionic liquids and deep eutectic solvents have been implemented in liquid phase microextraction approaches as extraction solvents and in solid-phase extraction approaches for the surface modification of a plethora of adsorbents.

The selection of the extraction technique, as well as the extraction sorbent can be based on the needs of the analysis (e.g., selectivity) and the laboratory equipment. MSPE and d-SPE are some examples of simple, rapid and environmentally friendly extraction procedures, that have recently gained a lot of attention, due to their convenience in sample handling. The application of these techniques has rapidly increased, and a wide variety of sorbents have been evaluated. Other novel miniaturized extraction forms, including SBSE, PT-SPE and FPSE have also become popular during the last years. These techniques have been evaluated in less extend compared to SPE, MSPE and d-SPE methods, however, due to their ease in operation, they are considered useful alternatives that enrich the toolbox of analytical chemists.

Proper attention should be also given to the stability of the sorbent in the case of solid-phase extraction, since physical and chemical stability in aqueous solutions is important to provide sorbent reusability, which significantly reduces the cost of the analysis. An important technical problem that has to be overcome is the limited stability of some SPE sorbents (e.g., MOFs) in an aqueous environment that limits their applications, their potential regeneration and recyclability. Another bottleneck of many sample preparation techniques is the lack of compatibility with desorption chambers for HPLC and GC applications.

Therefore, when designing novel sorbents for the extraction of PAHs from environmental matrices, their stability should be carefully examined. Moreover, it is generally preferred to prepare sorbents rich in π-electron moieties. In this case, the extraction of PAHs can be assisted through π-stacking, as well as hydrophobic and hydrogen interactions. Additionally, developing materials with an improved affinity towards PAHs, such as MIPs, has also proved to enhance the selectivity and sensitivity of PAHs extraction. Finally, sorbent functionalization with appropriate materials, such as polymers, ionic liquids etc., can also be beneficial for the overall extraction process.

Coupling and combination of different extraction approaches has been also reported for the efficient extraction and enhancement of PAHs from environmental matrices. For this purpose, a solid-phase extraction technique (e.g., SBSE, d-SPE) can be followed by an LPME technique, to further increase the extraction sensitivity.

Future perspectives in the field of sample preparation of environmental samples for PAHs extraction, should focus on developing robust extraction approaches of the on-site extraction and the determination of the environmental contaminants. For this kind of approach, several sorbents and extraction techniques can be examined in order to develop accurate, sensitive and selective sample preparation methods.

Author Contributions: Conceptualization, N.M. and G.A.Z.; writing—original draft preparation, N.M. and G.A.Z.; writing—review and editing, N.M. and G.A.Z.; supervision, G.A.Z. All authors have read and agreed to the published version of the manuscript.

Funding: The research work was supported by the Hellenic Foundation for Research and Innovation (HFRI) under the HFRI PhD Fellowship grant (Fellowship Number: 138).

Conflicts of Interest: The authors declare no conflict of interest.

References

1. Cai, Y.; Yan, Z.H.; Wang, N.Y.; Cai, Q.Y.; Yao, S.Z. Preparation of naphthyl functionalized magnetic nanoparticles for extraction of polycyclic aromatic hydrocarbons from river waters. *RSC Adv.* **2015**, *5*, 56189–56197. [CrossRef]
2. Pérez, R.A.; Albero, B.; Tadeo, J.L.; Fraile, M.V.; Sánchez-Brunete, C. Determination of PAHs in soil leachates by magnetic solid-phase extraction using nanoparticles and gas chromatography-tandem mass spectrometry. *Anal. Methods* **2014**, *6*, 1941–1950. [CrossRef]
3. Zhang, X.; Xie, S.; Paau, M.C.; Zheng, B.; Yuan, H.; Xiao, D.; Choi, M.M.F. Ultrahigh performance liquid chromatographic analysis and magnetic preconcentration of polycyclic aromatic hydrocarbons by Fe3O4-doped polymeric nanoparticles. *J. Chromatogr. A* **2012**, *1247*, 1–9. [CrossRef] [PubMed]
4. Boffetta, P.; Jourenkova, N.; Gustavsson, P. Cancer risk from occupational and environmental exposure to polycyclic aromatic hydrocarbons. *Cancer Causes Control* **1997**, *8*, 444–472. [CrossRef]
5. Bai, L.; Mei, B.; Guo, Q.Z.; Shi, Z.G.; Feng, Y.Q. Magnetic solid-phase extraction of hydrophobic analytes in environmental samples by a surface hydrophilic carbon-ferromagnetic nanocomposite. *J. Chromatogr. A* **2010**, *1217*, 7331–7336. [CrossRef]
6. Han, Q.; Wang, Z.; Xia, J.; Chen, S.; Zhang, X.; Ding, M. Facile and tunable fabrication of Fe_3O_4/graphene oxide nanocomposites and their application in the magnetic solid-phase extraction of polycyclic aromatic hydrocarbons from environmental water samples. *Talanta* **2012**, *101*, 388–395. [CrossRef]
7. Sarafraz-Yazdi, A.; Rokhian, T.; Amiri, A.; Ghaemi, F. Carbon nanofibers decorated with magnetic nanoparticles as a new sorbent for the magnetic solid phase extraction of selected polycyclic aromatic hydrocarbons from water samples. *New J. Chem.* **2015**, *39*, 5621–5627. [CrossRef]
8. Azizi, A.; Shahhoseini, F.; Bottaro, C.S. Magnetic molecularly imprinted polymers prepared by reversible addition fragmentation chain transfer polymerization for dispersive solid phase extraction of polycyclic aromatic hydrocarbons in water. *J. Chromatogr. A* **2020**, *1610*, 460534. [CrossRef]
9. Manousi, N.; Raber, G.; Papadoyannis, I. Recent Advances in Microextraction Techniques of Antipsychotics in Biological Fluids Prior to Liquid Chromatography Analysis. *Separations* **2017**, *4*, 18. [CrossRef]
10. Filippou, O.; Bitas, D.; Samanidou, V. Green approaches in sample preparation of bioanalytical samples prior to chromatographic analysis. *J. Chromatogr. B Anal. Technol. Biomed. Life Sci.* **2017**, *1043*, 44–62. [CrossRef]
11. Arthur, C.L.; Pawliszyn, J. Solid Phase Microextraction with Thermal Desorption Using Fused Silica Optical Fibers. *Anal. Chem.* **1990**, *62*, 2145–2148. [CrossRef]
12. Liu, H.; Dasgupta, P.K. Analytical chemistry in a drop. solvent extraction in a microdrop. *Anal. Chem.* **1996**, *68*, 1817–1821. [CrossRef] [PubMed]
13. Armenta, S.; Garrigues, S.; de la Guardia, M. The role of green extraction techniques in Green Analytical Chemistry. *TrAC - Trends Anal. Chem.* **2015**, *71*, 2–8. [CrossRef]
14. Manousi, N.; Gomez-Gomez, B.; Madrid, Y.; Deliyanni, E.A.; Zachariadis, G.A. Determination of rare earth elements by inductively coupled plasma-mass spectrometry after dispersive solid phase extraction with novel oxidized graphene oxide and optimization with response surface methodology and central composite design. *Microchem. J.* **2020**, *152*. [CrossRef]
15. Filippou, O.; Deliyanni, E.A.; Samanidou, V.F. Fabrication and evaluation of magnetic activated carbon as adsorbent for ultrasonic assisted magnetic solid phase dispersive extraction of bisphenol A from milk prior to high performance liquid chromatographic analysis with ultraviolet detection. *J. Chromatogr. A* **2017**, *1479*, 20–31. [CrossRef]
16. Chen, Z.; Yu, C.; Xi, J.; Tang, S.; Bao, T.; Zhang, J. A hybrid material prepared by controlled growth of a covalent organic framework on amino-modified MIL-68 for pipette tip solid-phase extraction of sulfonamides prior to their determination by HPLC. *Microchim. Acta* **2019**, *186*, 393. [CrossRef]
17. Karageorgou, E.; Manousi, N.; Samanidou, V.; Kabir, A.; Furton, K.G. Fabric phase sorptive extraction for the fast isolation of sulfonamides residues from raw milk followed by high performance liquid chromatography with ultraviolet detection. *Food Chem.* **2016**, *196*, 428–436. [CrossRef]
18. Nazyropoulou, C.; Samanidou, V. Stir bar sorptive extraction applied to the analysis of biological fluids. *Bioanalysis* **2015**, *7*, 2241–2250. [CrossRef]

19. Li, N.; Jiang, H.L.; Wang, X.; Wang, X.; Xu, G.; Zhang, B.; Wang, L.; Zhao, R.S.; Lin, J.M. Recent advances in graphene-based magnetic composites for magnetic solid-phase extraction. *TrAC - Trends Anal. Chem.* **2018**, *102*, 60–74. [CrossRef]
20. Riahi-Zanjani, B.; Balali-Mood, M.; Asoodeh, A.; Es'haghi, Z.; Ghorani-Azam, A. Developing a new sensitive solid-phase microextraction fiber based on carbon nanotubes for preconcentration of morphine. *Appl. Nanosci.* **2018**, *8*, 2047–2056. [CrossRef]
21. Manousi, N.; Zachariadis, G.A.; Deliyanni, E.A.; Samanidou, V.F. Applications of metal-organic frameworks in food sample preparation. *Molecules* **2018**, *23*, E2896. [CrossRef] [PubMed]
22. Lau, E.V.; Gan, S.; Ng, H.K. Extraction Techniques for Polycyclic Aromatic Hydrocarbons in Soils. *Int. J. Anal. Chem.* **2010**, *2010*, 398381. [CrossRef] [PubMed]
23. Pandey, S.K.; Kim, K.H.; Brown, R.J.C. A review of techniques for the determination of polycyclic aromatic hydrocarbons in air. *TrAC - Trends Anal. Chem.* **2011**, *30*, 1716–1739. [CrossRef]
24. Plaza-Bolaños, P.; Frenich, A.G.; Vidal, J.L.M. Polycyclic aromatic hydrocarbons in food and beverages. Analytical methods and trends. *J. Chromatogr. A* **2010**, *1217*, 6303–6326. [CrossRef]
25. Moret, S.; Conte, L.S. Polycyclic aromatic hydrocarbons in edible fats and oils: Occurrence and analytical methods. *J. Chromatogr. A* **2000**, *882*, 245–253. [CrossRef]
26. Locatelli, M.; Forcucci, L.; Sciascia, F.; Cifelli, R.; Ferrone, V.; Carlucci, G. Extraction and Detection Techniques for PAHs Determination in Beverages: A Review. *Curr. Chromatogr.* **2014**, *1*, 122–138. [CrossRef]
27. Poster, D.L.; Schantz, M.M.; Sander, L.C.; Wise, S.A. Analysis of polycyclic aromatic hydrocarbons (PAHs) in environmental samples: A critical review of gas chromatographic (GC) methods. *Anal. Bioanal. Chem.* **2006**, *386*, 859–881. [CrossRef]
28. Mahgoub, H.A. Extraction Techniques for Determination of Polycyclic Aromatic Hydrocarbons in Water Samples. *Int. J. Sci. Res.* **2016**, *5*, NOV152648.
29. Ncube, S.; Madikizela, L.; Cukrowska, E.; Chimuka, L. Recent advances in the adsorbents for isolation of polycyclic aromatic hydrocarbons (PAHs) from environmental sample solutions. *TrAC - Trends Anal. Chem.* **2018**, *99*, 101–116. [CrossRef]
30. Islas, G.; Ibarra, I.S.; Hernandez, P.; Miranda, J.M.; Cepeda, A. Dispersive Solid Phase Extraction for the Analysis of Veterinary Drugs Applied to Food Samples: A Review. *Int. J. Anal. Chem.* **2017**, *2017*, 8215271. [CrossRef]
31. Han, L.; Sapozhnikova, Y.; Lehotay, S.J. Streamlined sample cleanup using combined dispersive solid-phase extraction and in-vial filtration for analysis of pesticides and environmental pollutants in shrimp. *Anal. Chim. Acta* **2014**, *827*, 40–46. [CrossRef] [PubMed]
32. Anastassiades, M.; Lehotay, S.J.; Stajnbaher, D.; Schenck, F. Fast and Easy Multiresidue Method Employing Acetonitrile. *J. AOAC Int.* **2003**, *86*, 412–431. [CrossRef] [PubMed]
33. Cvetkovic, J.S.; Mitic, V.D.; Stankov Jovanovic, V.P.; Dimitrijevic, M.V.; Petrovic, G.M.; Nikolic-Mandic, S.D.; Stojanovic, G.S. Optimization of the QuEChERS extraction procedure for the determination of polycyclic aromatic hydrocarbons in soil by gas chromatography-mass spectrometry. *Anal. Methods* **2016**, *8*, 1711–1720. [CrossRef]
34. Yaghi, O.M.; Li, H. Hydrothermal Synthesis of a Metal-Organic Framework Containing Large Rectangular Channels. *J. Am. Chem. Soc.* **1995**, *117*, 10401–10402. [CrossRef]
35. Farha, O.K.; Eryazici, I.; Jeong, N.C.; Hauser, B.G.; Wilmer, C.E.; Sarjeant, A.A.; Snurr, R.Q.; Nguyen, S.T.; Yazaydin, A.Ö.; Hupp, J.T. Metal-organic framework materials with ultrahigh surface areas: Is the sky the limit? *J. Am. Chem. Soc.* **2012**, *134*, 15016–15021. [CrossRef]
36. Manousi, N.; Giannakoudakis, D.A.; Rosenberg, E.; Zachariadis, G.A. Extraction of metal ions with metal–organic frameworks. *Molecules* **2019**, *24*, E4605. [CrossRef]
37. Vardali, S.C.; Manousi, N.; Barczak, M.; Giannakoudakis, D.A. Novel approaches utilizing metal-organic framework composites for the extraction of organic compounds and metal traces from fish and seafood. *Molecules* **2020**, *25*, E513. [CrossRef]
38. Giliopoulos, D.; Zamboulis, A.; Giannakoudakis, D.; Bikiaris, D.; Triantafyllidis, K. Polymer/metal organic framework (MOF) nanocomposites for biomedical applications. *Molecules* **2020**, *25*, 185. [CrossRef]
39. Giannakoudakis, D.A.; Hu, Y.; Florent, M.; Bandosz, T.J. Smart textiles of MOF/g-C3N4 nanospheres for the rapid detection/detoxification of chemical warfare agents. *Nanoscale Horizons* **2017**, *2*, 356–364. [CrossRef]

40. Li, H.; Li, L.; Lin, R.-B.; Zhou, W.; Zhang, Z.; Xiang, S.; Chen, B. Porous metal-organic frameworks for gas storage and separation: Status and challenges. *EnergyChem* **2019**, *1*, 100006. [CrossRef]
41. Yang, D.; Gates, B.C. Catalysis by Metal Organic Frameworks: Perspective and Suggestions for Future Research. *ACS Catal.* **2019**, *9*, 1779–1798. [CrossRef]
42. Achmann, S.; Hagen, G.; Kita, J.; Malkowsky, I.M.; Kiener, C.; Moos, R. Metal-Organic frameworks for sensing applications in the gas phase. *Sensors* **2009**, *9*, 1574–1589. [CrossRef] [PubMed]
43. Rojas, S.; Baati, T.; Njim, L.; Manchego, L.; Neffati, F.; Abdeljelil, N.; Saguem, S.; Serre, C.; Najjar, M.F.; Zakhama, A.; et al. Metal-Organic Frameworks as Efficient Oral Detoxifying Agents. *J. Am. Chem. Soc.* **2018**, *140*, 9581–9586. [CrossRef] [PubMed]
44. Wang, L.; Zheng, M.; Xie, Z. Nanoscale metal-organic frameworks for drug delivery: A conventional platform with new promise. *J. Mater. Chem. B* **2018**, *6*, 707–717. [CrossRef]
45. Chen, B.; Liang, C.; Yang, J.; Contreras, D.S.; Clancy, Y.L.; Lobkovsky, E.B.; Yaghi, O.M.; Dai, S. A microporous metal-organic framework for gas-chromatographic separation of alkanes. *Angew. Chemie - Int. Ed.* **2006**, *9*, 1418–1421. [CrossRef]
46. Gu, Z.Y.; Jiang, J.Q.; Yan, X.P. Fabrication of isoreticular metal-organic framework coated capillary columns for high-resolution gas chromatographic separation of persistent organic pollutants. *Anal. Chem.* **2011**, *83*, 5093–5100. [CrossRef]
47. Yang, C.X.; Yan, X.P. Metal-organic framework MIL-101(Cr) for high-performance liquid chromatographic separation of substituted aromatics. *Anal. Chem.* **2011**, *83*, 7144–7150. [CrossRef]
48. Yang, C.X.; Chen, Y.J.; Wang, H.F.; Yan, X.P. High-performance separation of fullerenes on metal-organic framework MIL-101(Cr). *Chem. - A Eur. J.* **2011**, *42*, 11734–11737. [CrossRef]
49. Xia, L.; Liu, L.; Dou, Y.; Guo, L.; Li, G.; Sun, Z.; You, J. A stable mesoporous metal-organic framework as highly efficient sorbent of dispersive micro solid-phase extraction for the determination of polycyclic aromatic hydrocarbons by HPLC. *J. Sep. Sci.* **2018**, *41*, 4331–4339. [CrossRef]
50. Amiri, A.; Ghaemi, F.; Maleki, B. Hybrid nanocomposites prepared from a metal-organic framework of type MOF-199(Cu) and graphene or fullerene as sorbents for dispersive solid phase extraction of polycyclic aromatic hydrocarbons. *Microchim. Acta* **2019**, *186*, 1–8. [CrossRef]
51. Chen, B.; Yang, Z.; Zhu, Y.; Xia, Y. Zeolitic imidazolate framework materials: Recent progress in synthesis and applications. *J. Mater. Chem. A* **2014**, *2*, 16811–16831. [CrossRef]
52. Liang, X.; Liu, S.; Zhu, R.; Xiao, L.; Yao, S. Highly sensitive analysis of polycyclic aromatic hydrocarbons in environmental water with porous cellulose/zeolitic imidazolate framework-8 composite microspheres as a novel adsorbent coupled with high-performance liquid chromatography. *J. Sep. Sci.* **2016**, *39*, 2806–2814. [CrossRef] [PubMed]
53. Mateos, R.; Vera-López, S.; Saz, M.; Díez-Pascual, A.M.; San Andrés, M.P. Graphene/sepiolite mixtures as dispersive solid-phase extraction sorbents for the anaysis of polycyclic aromatic hydrocarbons in wastewater using surfactant aqueous solutions for desorption. *J. Chromatogr. A* **2019**, *1596*, 30–40. [CrossRef] [PubMed]
54. Yang, X.P.; Luo, N.; Zong, Y.Y.; Jia, Z.H.; Liao, X.J. Quantum dots extraction coupled with high-performance liquid chromatography for the determination of polycyclic aromatic hydrocarbons in water. *Appl. Ecol. Environ. Res.* **2017**, *15*, 171–186.
55. Hemmati, M.; Rajabi, M.; Asghari, A. Magnetic nanoparticle based solid-phase extraction of heavy metal ions: A review on recent advances. *Microchim. Acta* **2018**, *185*, 160. [CrossRef]
56. Giakisikli, G.; Anthemidis, A.N. Magnetic materials as sorbents for metal/metalloid preconcentration and/or separation. A review. *Anal. Chim. Acta* **2013**, *789*, 1–16. [CrossRef]
57. Manousi, N.; Rosenberg, E.; Deliyanni, E.; Zachariadis, G.A.; Samanidou, V. Magnetic Solid-Phase Extraction of Organic Compounds Based on Graphene Oxide Nanocomposites. *Molecules* **2020**, *25*, 1148. [CrossRef]
58. Hesampour, M.; Ali Taher, M.; Behzadi, M. Synthesis, characterization and application of a MnFe2O4@poly(:O-toluidine) nanocomposite for magnetic solid-phase extraction of polycyclic aromatic hydrocarbons. *New J. Chem.* **2017**, *41*, 12910–12919. [CrossRef]
59. Rocío-Bautista, P.; Pino, V.; Ayala, J.H.; Pasán, J.; Ruiz-Pérez, C.; Afonso, A.M. A magnetic-based dispersive micro-solid-phase extraction method using the metal-organic framework HKUST-1 and ultra-high-performance liquid chromatography with fluorescence detection for determining polycyclic aromatic hydrocarbons in waters and fruit tea. *J. Chromatogr. A* **2016**, *1426*, 42–50. [CrossRef]

60. Huo, S.H.; Yan, X.P. Facile magnetization of metal-organic framework MIL-101 for magnetic solid-phase extraction of polycyclic aromatic hydrocarbons in environmental water samples. *Analyst* **2012**, *137*, 3445–3451. [CrossRef]
61. Zhou, Q.; Lei, M.; Wu, Y.; Yuan, Y. Magnetic solid phase extraction of typical polycyclic aromatic hydrocarbons from environmental water samples with metal organic framework MIL-101 (Cr) modified zero valent iron nano-particles. *J. Chromatogr. A* **2017**, *1487*, 22–29. [CrossRef] [PubMed]
62. Du, F.; Qin, Q.; Deng, J.; Ruan, G.; Yang, X.; Li, L.; Li, J. Magnetic metal–organic framework MIL-100(Fe) microspheres for the magnetic solid-phase extraction of trace polycyclic aromatic hydrocarbons from water samples. *J. Sep. Sci.* **2016**, *12*, 2356–2364. [CrossRef] [PubMed]
63. Huo, S.H.; An, H.Y.; Yu, J.; Mao, X.F.; Zhang, Z.; Bai, L.; Huang, Y.F.; Zhou, P.X. Pyrolytic in situ magnetization of metal-organic framework MIL-100 for magnetic solid-phase extraction. *J. Chromatogr. A* **2017**, *1517*, 18–25. [CrossRef] [PubMed]
64. Zhang, S.; Yao, W.; Ying, J.; Zhao, H. Polydopamine-reinforced magnetization of zeolitic imidazolate framework ZIF-7 for magnetic solid-phase extraction of polycyclic aromatic hydrocarbons from the air-water environment. *J. Chromatogr. A* **2016**, *1452*, 18–26. [CrossRef]
65. He, S.; Zeng, T.; Wang, S.; Niu, H.; Cai, Y. Facile synthesis of magnetic covalent organic framework with three-dimensional bouquet-like structure for enhanced extraction of organic targets. *ACS Appl. Mater. Interfaces* **2017**, *9*, 2959–2965. [CrossRef]
66. Wang, R.; Chen, Z. A covalent organic framework-based magnetic sorbent for solid phase extraction of polycyclic aromatic hydrocarbons, and its hyphenation to HPLC for quantitation. *Microchim. Acta* **2017**, *184*, 3867–3874. [CrossRef]
67. Rezvani-Eivari, M.; Amiri, A.; Baghayeri, M.; Ghaemi, F. Magnetized graphene layers synthesized on the carbon nanofibers as novel adsorbent for the extraction of polycyclic aromatic hydrocarbons from environmental water samples. *J. Chromatogr. A* **2016**, *1465*, 1–8. [CrossRef]
68. Zhang, S.; Niu, H.; Hu, Z.; Cai, Y.; Shi, Y. Preparation of carbon coated Fe_3O_4 nanoparticles and their application for solid-phase extraction of polycyclic aromatic hydrocarbons from environmental water samples. *J. Chromatogr. A* **2010**, *1217*, 4757–4764. [CrossRef]
69. Mehdinia, A.; Khodaee, N.; Jabbari, A. Fabrication of graphene/Fe_3O_4@polythiophene nanocomposite and its application in the magnetic solid-phase extraction of polycyclic aromatic hydrocarbons from environmental water samples. *Anal. Chim. Acta* **2015**, *868*, 1–9. [CrossRef]
70. Amiri, A.; Baghayeri, M.; Sedighi, M. Magnetic solid-phase extraction of polycyclic aromatic hydrocarbons using a graphene oxide/Fe_3O_4@polystyrene nanocomposite. *Microchim. Acta* **2018**, *185*, 393. [CrossRef]
71. Amiri, A.; Baghayeri, M.; Hamidi, E. Poly(pyrrole-: Co -aniline)@graphene oxide/Fe_3O_4 sorbent for the extraction and preconcentration of polycyclic aromatic hydrocarbons from water samples. *New J. Chem.* **2018**, *42*, 16744–16751. [CrossRef]
72. Corps Ricardo, A.I.; Guzmán Bernardo, F.J.; Zougagh, M.; Rodríguez Martín-Doimeadios, R.C.; Ríos, Á. Magnetic nanoparticles—carbon nanotubes hybrid composites for selective solid-phase extraction of polycyclic aromatic hydrocarbons and determination by ultra-high performance liquid chromatography. *Anal. Bioanal. Chem.* **2017**, *409*, 5125–5132. [CrossRef] [PubMed]
73. Villar-Navarro, M.; Martín-Valero, M.J.; Fernández-Torres, R.M.; Callejón-Mochón, M.; Bello-López, M.Á. Easy, fast and environmental friendly method for the simultaneous extraction of the 16 EPA PAHs using magnetic molecular imprinted polymers (mag-MIPs). *J. Chromatogr. B Anal. Technol. Biomed. Life Sci.* **2017**, *1044–1045*, 63–69. [CrossRef] [PubMed]
74. Benedetti, B.; Di Carro, M.; Magi, E. Multivariate optimization of an extraction procedure based on magnetic molecular imprinted polymer for the determination of polycyclic aromatic hydrocarbons in sea water. *Microchem. J.* **2019**, *145*, 1199–1206. [CrossRef]
75. Wang, Y.; Wang, S.; Niu, H.; Ma, Y.; Zeng, T.; Cai, Y.; Meng, Z. Preparation of polydopamine coated Fe_3O_4 nanoparticles and their application for enrichment of polycyclic aromatic hydrocarbons from environmental water samples. *J. Chromatogr. A* **2013**, *1283*, 20–26. [CrossRef]
76. Xu, S.N.; Zhao, Q.; He, H.B.; Yuan, B.F.; Feng, Y.Q.; Yu, Q.W. Rapid determination of polycyclic aromatic hydrocarbons in environmental water based on magnetite nanoparticles/polypyrrole magnetic solid-phase extraction. *Anal. Methods* **2014**, *6*, 7046–7053. [CrossRef]

77. Nurerk, P.; Kanatharana, P.; Bunkoed, O. Polyaniline-coated magnetite nanoparticles incorporated in alginate beads for the extraction and enrichment of polycyclic aromatic hydrocarbons in water samples. *Int. J. Environ. Anal. Chem.* **2017**, *2*, 145–158. [CrossRef]
78. Galán-Cano, F.; Del Carmen Alcudia-León, M.; Lucena, R.; Cárdenas, S.; Valcárcel, M. Ionic liquid coated magnetic nanoparticles for the gas chromatography/mass spectrometric determination of polycyclic aromatic hydrocarbons in waters. *J. Chromatogr. A* **2013**, *1300*, 134–140. [CrossRef]
79. Bakhshaei, S.; Kamboh, M.A.; Nodeh, H.R.; Md Zain, S.; Mahmad Rozi, S.K.; Mohamad, S.; Mohammed Mohialdeen, I.A. Magnetic solid phase extraction of polycyclic aromatic hydrocarbons and chlorophenols based on cyano-ionic liquid functionalized magnetic nanoparticles and their determination by HPLC-DAD. *RSC Adv.* **2016**, *6*, 77047–77058. [CrossRef]
80. Shahriman, M.S.; Ramachandran, M.R.; Zain, N.N.M.; Mohamad, S.; Manan, N.S.A.; Yaman, S.M. Polyaniline-dicationic ionic liquid coated with magnetic nanoparticles composite for magnetic solid phase extraction of polycyclic aromatic hydrocarbons in environmental samples. *Talanta* **2018**, *178*, 211–221. [CrossRef] [PubMed]
81. Liu, X.; Lu, X.; Huang, Y.; Liu, C.; Zhao, S. Fe_3O_4@ionic liquid@methyl orange nanoparticles as a novel nano-adsorbent for magnetic solid-phase extraction of polycyclic aromatic hydrocarbons in environmental water samples. *Talanta* **2014**, *119*, 341–347. [CrossRef] [PubMed]
82. Zhang, S.; Niu, H.; Zhang, Y.; Liu, J.; Shi, Y.; Zhang, X.; Cai, Y. Biocompatible phosphatidylcholine bilayer coated on magnetic nanoparticles and their application in the extraction of several polycyclic aromatic hydrocarbons from environmental water and milk samples. *J. Chromatogr. A* **2012**, *1238*, 38–45. [CrossRef] [PubMed]
83. Xue, S.W.; Tang, M.Q.; Xu, L.; Shi, Z. guo Magnetic nanoparticles with hydrophobicity and hydrophilicity for solid-phase extraction of polycyclic aromatic hydrocarbons from environmental water samples. *J. Chromatogr. A* **2015**, *1411*, 9–16. [CrossRef] [PubMed]
84. Long, Y.; Chen, Y.; Yang, F.; Chen, C.; Pan, D.; Cai, Q.; Yao, S. Triphenylamine-functionalized magnetic microparticles as a new adsorbent coupled with high performance liquid chromatography for the analysis of trace polycyclic aromatic hydrocarbons in aqueous samples. *Analyst* **2012**, *137*, 2716–2722. [CrossRef] [PubMed]
85. Liu, Y.; Li, H.; Lin, J.M. Magnetic solid-phase extraction based on octadecyl functionalization of monodisperse magnetic ferrite microspheres for the determination of polycyclic aromatic hydrocarbons in aqueous samples coupled with gas chromatography-mass spectrometry. *Talanta* **2009**, *77*, 1037–1042. [CrossRef]
86. Ballesteros-Gómez, A.; Rubio, S. Hemimicelles of alkyl carboxylates chemisorbed onto magnetic nanoparticles: Study and application to the extraction of carcinogenic polycyclic aromatic hydrocarbons in environmental water samples. *Anal. Chem.* **2009**, *81*, 9012–9020. [CrossRef]
87. Ding, J.; Gao, Q.; Luo, D.; Shi, Z.G.; Feng, Y.Q. n-Octadecylphosphonic acid grafted mesoporous magnetic nanoparticle: Preparation, characterization, and application in magnetic solid-phase extraction. *J. Chromatogr. A* **2010**, *1217*, 7351–7358. [CrossRef]
88. Reyes-Gallardo, E.M.; Lucena, R.; Cárdenas, S.; Valcárcel, M. Magnetic nanoparticles-nylon 6 composite for the dispersive micro solid phase extraction of selected polycyclic aromatic hydrocarbons from water samples. *J. Chromatogr. A* **2014**, *1345*, 43–49. [CrossRef]
89. Wang, H.; Zhao, X.; Meng, W.; Wang, P.; Wu, F.; Tang, Z.; Han, X.; Giesy, J.P. Cetyltrimethylammonium Bromide-Coated Fe_3O_4 Magnetic Nanoparticles for Analysis of 15 Trace Polycyclic Aromatic Hydrocarbons in Aquatic Environments by Ultraperformance, Liquid Chromatography With Fluorescence Detection. *Anal. Chem.* **2015**, *87*, 7667–7675. [CrossRef]
90. Rozi, S.K.M.; Nodeh, H.R.; Kamboh, M.A.; Manan, N.S.A.; Mohamad, S. Novel palm fatty acid functionalized magnetite nanoparticles for magnetic solid-phase extraction of trace polycyclic aromatic hydrocarbons from environmental samples. *J. Oleo Sci.* **2017**, *66*, 771–784. [CrossRef]
91. Zou, Y.; Chen, Y.; Yan, Z.; Chen, C.; Wang, J.; Yao, S. Magnetic solid-phase extraction based on tetrabenzyl modified Fe_3O_4 nanoparticles for the analysis of trace polycyclic aromatic hydrocarbons in environmental water samples. *Analyst* **2013**, *138*, 5904–5912. [CrossRef] [PubMed]

92. Zhang, W.; Zhang, Y.; Jiang, Q.; Zhao, W.; Yu, A.; Chang, H.; Lu, X.; Xie, F.; Ye, B.; Zhang, S. Tetraazacalix[2]arence[2]triazine Coated Fe_3O_4/SiO_2 Magnetic Nanoparticles for Simultaneous Dispersive Solid Phase Extraction and Determination of Trace Multitarget Analytes. *Anal. Chem.* **2016**, *88*, 10523–10532. [CrossRef]
93. Al-rashdi, A.A. Double-functionalized magnetic nanoparticles for preconcentration and determination of polycyclic aromatic hydrocarbons in water samples. *Anal. Chem. Res.* **2016**, *10*, 9–17. [CrossRef]
94. Feng, X.; Ding, X.; Jiang, D. Covalent organic frameworks. *Chem. Soc. Rev.* **2012**, *41*, 6010–6022. [CrossRef] [PubMed]
95. Li, N.; Du, J.; Wu, D.; Liu, J.; Li, N.; Sun, Z.; Li, G.; Wu, Y. Recent advances in facile synthesis and applications of covalent organic framework materials as superior adsorbents in sample pretreatment. *TrAC - Trends Anal. Chem.* **2018**, *108*, 154–166. [CrossRef]
96. Novoselov, K.S.; Geim, A.K.; Morozov, S.V.; Jiang, D.; Zhang, Y.; Dubonos, S.V.; Grigorieva, I.V.; Firsov, A.A. Electric field in atomically thin carbon films. *Science* **2004**, *306*, 666–669. [CrossRef]
97. González-Fuenzalida, R.A.; López-García, E.; Moliner-Martínez, Y.; Campíns-Falcó, P. Adsorbent phases with nanomaterials for in-tube solid-phase microextraction coupled on-line to liquid nanochromatography. *J. Chromatogr. A* **2016**, *1432*, 17–25. [CrossRef]
98. Bitas, D.; Samanidou, V. Molecularly imprinted polymers as extracting media for the chromatographic determination of antibiotics in milk. *Molecules* **2018**, *32*, E316. [CrossRef]
99. Queiroz, M.E.C.; Melo, L.P. Selective capillary coating materials for in-tube solid-phase microextraction coupled to liquid chromatography to determine drugs and biomarkers in biological samples: A review. *Anal. Chim. Acta* **2014**, *826*, 1–11. [CrossRef]
100. Tamayo, F.G.; Turiel, E.; Martín-Esteban, A. Molecularly imprinted polymers for solid-phase extraction and solid-phase microextraction: Recent developments and future trends. *J. Chromatogr. A* **2007**, *1152*, 32–40. [CrossRef]
101. Kissoudi, M.; Samanidou, V. Recent advances in applications of ionic liquids in miniaturized microextraction techniques. *Molecules* **2018**, *23*, E1437. [CrossRef]
102. Chatzimitakos, T.; Stalikas, C. Carbon-based nanomaterials functionalized with ionic liquids for microextraction in sample preparation. *Separations* **2017**, *4*, 14. [CrossRef]
103. Han, D.; Row, K.H. Recent applications of ionic liquids in separation technology. *Molecules* **2010**, *15*, 2405–2426. [CrossRef] [PubMed]
104. Kataoka, H.; Lord, H.L.; Pawliszyn, J. Applications of solid-phase microextraction in food analysis. *J. Chromatogr. A* **2000**, *880*, 35–62. [CrossRef]
105. Aulakh, J.S.; Malik, A.K.; Kaur, V.; Schmitt-Kopplin, P. A review on solid phase micro extraction - High performance liquid chromatography (SPME-HPLC) analysis of pesticides. *Crit. Rev. Anal. Chem.* **2005**, *35*, 71–85. [CrossRef]
106. Chen, C.; Liang, X.; Wang, J.; Yang, S.; Yan, Z.; Cai, Q.; Yao, S. Development of a highly robust solid phase microextraction fiber based on crosslinked methyl methacrylate-polyhedral oligomeric silsesquioxane hybrid polymeric coating. *Anal. Chim. Acta* **2013**, *792*, 45–51. [CrossRef]
107. Rocío-Bautista, P.; Pacheco-Fernández, I.; Pasán, J.; Pino, V. Are metal-organic frameworks able to provide a new generation of solid-phase microextraction coatings? – A review. *Anal. Chim. Acta* **2016**, *939*, 26–41. [CrossRef]
108. Yu, H.; Ho, T.D.; Anderson, J.L. Ionic liquid and polymeric ionic liquid coatings in solid-phase microextraction. *TrAC - Trends Anal. Chem.* **2013**, *45*, 219–232. [CrossRef]
109. López-Darias, J.; Pino, V.; Anderson, J.L.; Graham, C.M.; Afonso, A.M. Determination of water pollutants by direct-immersion solid-phase microextraction using polymeric ionic liquid coatings. *J. Chromatogr. A* **2010**, *1217*, 1236–1243. [CrossRef]
110. Meng, Y.; Anderson, J.L. Tuning the selectivity of polymeric ionic liquid sorbent coatings for the extraction of polycyclic aromatic hydrocarbons using solid-phase microextraction. *J. Chromatogr. A* **2010**, *1217*, 6143–6152. [CrossRef]
111. Feng, J.; Sun, M.; Li, J.; Liu, X.; Jiang, S. A novel aromatically functional polymeric ionic liquid as sorbent material for solid-phase microextraction. *J. Chromatogr. A* **2012**, *1227*, 54–59. [CrossRef] [PubMed]

112. López-Darias, J.; Pino, V.; Meng, Y.; Anderson, J.L.; Afonso, A.M. Utilization of a benzyl functionalized polymeric ionic liquid for the sensitive determination of polycyclic aromatic hydrocarbons; parabens and alkylphenols in waters using solid-phase microextraction coupled to gas chromatography-flame ionization detection. *J. Chromatogr. A* **2010**, *1217*, 7189–7197. [PubMed]
113. Zhang, S.; Du, Z.; Li, G. Layer-by-layer fabrication of chemical-bonded graphene coating for solid-phase microextraction. *Anal. Chem.* **2011**, *83*, 7531–7541. [CrossRef] [PubMed]
114. Wang, F.; Zheng, Y.; Qiu, J.; Liu, S.; Tong, Y.; Zhu, F.; Ouyang, G. Graphene-based metal and nitrogen-doped carbon composites as adsorbents for highly sensitive solid phase microextraction of polycyclic aromatic hydrocarbons. *Nanoscale* **2018**, *10*, 10073–10078. [CrossRef]
115. Xu, L.; Feng, J.; Liang, X.; Li, J.; Jiang, S. C_{18} functionalized graphene oxide as a novel coating for solid-phase microextraction. *J. Sep. Sci.* **2012**, *35*, 1531–1537. [CrossRef]
116. Sun, M.; Feng, J.; Bu, Y.; Duan, H.; Wang, X.; Luo, C. Development of a solid-phase microextraction fiber by the chemical binding of graphene oxide on a silver-coated stainless-steel wire with an ionic liquid as the crosslinking agent. *J. Sep. Sci.* **2014**, *37*, 3691–3698. [CrossRef]
117. Xu, L.; Feng, J.; Li, J.; Liu, X.; Jiang, S. Graphene oxide bonded fused-silica fiber for solid-phase microextraction-gas chromatography of polycyclic aromatic hydrocarbons in water. *J. Sep. Sci.* **2012**, *35*, 93–100. [CrossRef]
118. Behzadi, M.; Noroozian, E.; Mirzaei, M. A novel coating based on carbon nanotubes/poly-ortho-phenylenediamine composite for headspace solid-phase microextraction of polycyclic aromatic hydrocarbons. *Talanta* **2013**, *108*, 66–73. [CrossRef]
119. Maghsoudi, S.; Noroozian, E. HP-SPME of volatile polycyclic aromatic hydrocarbons from water using multiwalled carbon nanotubes coated on a steel fiber through electrophoretic deposition. *Chromatographia* **2012**, *75*, 913–921. [CrossRef]
120. Zhang, X.; Zang, X.H.; Wang, J.T.; Wang, C.; Wu, Q.H.; Wang, Z. Porous carbon derived from aluminum-based metal organic framework as a fiber coating for the solid-phase microextraction of polycyclic aromatic hydrocarbons from water and soil. *Microchim. Acta* **2015**, *182*, 2353–2359. [CrossRef]
121. Gutiérrez-Serpa, A.; Pacheco-Fernández, I.; Pasán, J.; Pino, V. Metal–organic frameworks as key materials for solid-phase microextraction devices—a review. *Separations* **2019**, *6*, 47. [CrossRef]
122. Sun, S.; Huang, L.; Xiao, H.; Shuai, Q.; Hu, S. In situ self-transformation metal into metal-organic framework membrane for solid-phase microextraction of polycyclic aromatic hydrocarbons. *Talanta* **2019**, *202*, 145–151. [CrossRef] [PubMed]
123. Zhang, G.; Zang, X.; Li, Z.; Wang, C.; Wang, Z. Polydimethylsiloxane/metal-organic frameworks coated fiber for solid-phase microextraction of polycyclic aromatic hydrocarbons in river and lake water samples. *Talanta* **2014**, *129*, 600–605. [CrossRef] [PubMed]
124. Kong, J.; Zhu, F.; Huang, W.; He, H.; Hu, J.; Sun, C.; Xian, Q.; Yang, S. Sol–gel based metal-organic framework zeolite imidazolate framework-8 fibers for solid-phase microextraction of nitro polycyclic aromatic hydrocarbons and polycyclic aromatic hydrocarbons in water samples. *J. Chromatogr. A* **2019**, *1603*, 92–101. [CrossRef] [PubMed]
125. Amanzadeh, H.; Yamini, Y.; Masoomi, M.Y.; Morsali, A. Nanostructured metal-organic frameworks, TMU-4, TMU-5, and TMU-6, as novel adsorbents for solid phase microextraction of polycyclic aromatic hydrocarbons. *New J. Chem.* **2017**, *20*, 12035–12043. [CrossRef]
126. Wang, G.; Lei, Y.; Song, H. Exploration of metal-organic framework MOF-177 coated fibers for headspace solid-phase microextraction of polychlorinated biphenyls and polycyclic aromatic hydrocarbons. *Talanta* **2015**, *144*, 369–374. [CrossRef]
127. Gao, J.; Huang, C.; Lin, Y.; Tong, P.; Zhang, L. In situ solvothermal synthesis of metal-organic framework coated fiber for highly sensitive solid-phase microextraction of polycyclic aromatic hydrocarbons. *J. Chromatogr. A* **2016**, *1436*, 1–8. [CrossRef]
128. Li, Q.L.; Wang, X.; Chen, X.F.; Wang, M.L.; Zhao, R.S. In situ hydrothermal growth of ytterbium-based metal-organic framework on stainless steel wire for solid-phase microextraction of polycyclic aromatic hydrocarbons from environmental samples. *J. Chromatogr. A* **2015**, *1415*, 11–19. [CrossRef]
129. Huo, S.H.; Yu, J.; Fu, Y.Y.; Zhou, P.X. In situ hydrothermal growth of a dual-ligand metal-organic framework film on a stainless steel fiber for solid-phase microextraction of polycyclic aromatic hydrocarbons in environmental water samples. *RSC Adv.* **2016**, *6*, 14042–14048. [CrossRef]

130. Liu, M.; Liu, J.; Guo, C.; Li, Y. Metal azolate framework-66-coated fiber for headspace solid-phase microextraction of polycyclic aromatic hydrocarbons. *J. Chromatogr. A* **2019**, *1584*, 57–63. [CrossRef]
131. Mehdinia, A.; Mohammadi, A.A.; Davarani, S.S.H.; Banitaba, M.H. Application of self-assembled monolayers in the preparation of solid-phase microextraction coatings. *Chromatographia* **2011**, *5–6*, 421–427. [CrossRef]
132. Yang, L.; Zhang, J.; Zhao, F.; Zeng, B. Electrodeposition of self-assembled poly(3,4-ethylenedioxythiophene) @gold nanoparticles on stainless steel wires for the headspace solid-phase microextraction and gas chromatographic determination of several polycyclic aromatic hydrocarbons. *J. Chromatogr. A* **2016**, *1471*, 80–86. [CrossRef] [PubMed]
133. Harati, F.; Ghiasvand, A.; Dalvand, K.; Haddad, P.R. Fused-silica capillary internally modified with nanostructured octadecyl silica for dynamic in-tube solid-phase microextraction of polycyclic aromatic hydrocarbons from aqueous media. *Microchem. J.* **2020**, *155*, 104672. [CrossRef]
134. Ghiasvand, A.; Yazdankhah, F.; Paull, B. Heating-, Cooling- and Vacuum-Assisted Solid-Phase Microextraction (HCV-SPME) for Efficient Sampling of Environmental Pollutants in Complex Matrices. *Chromatographia* **2020**, *83*, 531–540. [CrossRef]
135. Kremser, A.; Jochmann, M.A.; Schmidt, T.C. PAL SPME Arrow - Evaluation of a novel solid-phase microextraction device for freely dissolved PAHs in water. *Anal. Bioanal. Chem.* **2016**, *408*, 943–952. [CrossRef]
136. Baltussen, E.; Sandra, P.; David, F.; Cramers, C. Stir bar sorptive extraction (SBSE), a novel extraction technique for aqueous samples: Theory and principles. *J. Microcolumn Sep.* **1999**, *10*, 737–747. [CrossRef]
137. David, F.; Sandra, P. Stir bar sorptive extraction for trace analysis. *J. Chromatogr. A* **2007**, *1152*, 54–69. [CrossRef]
138. García-Falcón, M.S.; Cancho-Grande, B.; Simal-Gándara, J. Stirring bar sorptive extraction in the determination of PAHs in drinking waters. *Water Res.* **2004**, *38*, 1679–1684. [CrossRef]
139. Popp, P.; Bauer, C.; Wennrich, L. Application of stir bar sorptive extraction in combination with column liquid chromatography for the determination of polycyclic aromatic hydrocarbons in water samples. *Anal. Chim. Acta* **2001**, *436*, 1–9. [CrossRef]
140. Popp, P.; Bauer, C.; Hauser, B.; Keil, P.; Wennrich, L. Extraction of polycyclic aromatic hydrocarbons and organochloride compounds from water: A comparison between solid-phase microextraction and stir bar sorptive extraction. *J. Sep. Sci.* **2003**, *9–10*, 961–967. [CrossRef]
141. Bourdat-Deschamps, M.; Daudin, J.J.; Barriuso, E. An experimental design approach to optimise the determination of polycyclic aromatic hydrocarbons from rainfall water using stir bar sorptive extraction and high performance liquid chromatography-fluorescence detection. *J. Chromatogr. A* **2007**, *1167*, 143–153. [CrossRef] [PubMed]
142. Mao, X.; Hu, B.; He, M.; Fan, W. Stir bar sorptive extraction approaches with a home-made portable electric stirrer for the analysis of polycyclic aromatic hydrocarbon compounds in environmental water. *J. Chromatogr. A* **2012**, *1260*, 16–24. [CrossRef] [PubMed]
143. Mollahosseini, A.; Rokue, M.; Mojtahedi, M.M.; Toghroli, M.; Kamankesh, M.; Motaharian, A. Mechanical stir bar sorptive extraction followed by gas chromatography as a new method for determining polycyclic aromatic hydrocarbons in water samples. *Microchem. J.* **2016**, *126*, 431–437. [CrossRef]
144. Huang, X.; Yuan, D. Preparation of stir bars for sorptive extraction based on monolithic material. *J. Chromatogr. A* **2007**, *1154*, 152–157. [CrossRef] [PubMed]
145. Ekbatani Amlashi, N.; Hadjmohammadi, M.R. Sol–gel coating of poly(ethylene glycol)-grafted multiwalled carbon nanotubes for stir bar sorptive extraction and its application to the analysis of polycyclic aromatic hydrocarbons in water. *J. Sep. Sci.* **2016**, *39*, 3445–3456. [CrossRef] [PubMed]
146. Hu, C.; He, M.; Chen, B.; Zhong, C.; Hu, B. Sorptive extraction using polydimethylsiloxane/metal-organic framework coated stir bars coupled with high performance liquid chromatography-fluorescence detection for the determination of polycyclic aromatic hydrocarbons in environmental water samples. *J. Chromatogr. A* **2014**, *1356*, 45–53. [CrossRef]
147. Benedé, J.L.; Anderson, J.L.; Chisvert, A. Trace determination of volatile polycyclic aromatic hydrocarbons in natural waters by magnetic ionic liquid-based stir bar dispersive liquid microextraction. *Talanta* **2018**, *176*, 253–261. [CrossRef]

148. Luo, Y.B.; Cheng, J.S.; Ma, Q.; Feng, Y.Q.; Li, J.H. Graphene-polymer composite: Extraction of polycyclic aromatic hydrocarbons from water samples by stir rod sorptive extraction. *Anal. Methods* **2011**, *3*, 92–98. [CrossRef]
149. Luo, Y.B.; Ma, Q.; Feng, Y.Q. Stir rod sorptive extraction with monolithic polymer as coating and its application to the analysis of fluoroquinolones in honey sample. *J. Chromatogr. A* **2010**, *1217*, 3583–3589. [CrossRef]
150. Yu, C.; Hu, B. Automated stir plate (bar) sorptive extraction coupled to high-performance liquid chromatography for the determination of polycyclic aromatic hydrocarbons. *J. Sep. Sci.* **2010**, *33*, 2176–2183. [CrossRef]
151. Rutkowska, M.; Płotka-Wasylka, J.; Sajid, M.; Andruch, V. Liquid–phase microextraction: A review of reviews. *Microchem. J.* **2019**, *149*, 103989. [CrossRef]
152. Bello-López, M.Á.; Ramos-Payán, M.; Ocaña-González, J.A.; Fernández-Torres, R.; Callejón-Mochón, M. Analytical Applications of Hollow Fiber Liquid Phase Microextraction (HF-LPME): A Review. *Anal. Lett.* **2012**, *45*, 804–830. [CrossRef]
153. Kokosa, J.M. Recent trends in using single-drop microextraction and related techniques in green analytical methods. *TrAC - Trends Anal. Chem.* **2015**, *71*, 194–204. [CrossRef]
154. Kocúrová, L.; Balogh, I.S.; Šandrejová, J.; Andruch, V. Recent advances in dispersive liquid-liquid microextraction using organic solvents lighter than water. A review. *Microchem. J.* **2012**, *102*, 11–17. [CrossRef]
155. Rykowska, I.; Ziemblińska, J.; Nowak, I. Modern approaches in dispersive liquid-liquid microextraction (DLLME) based on ionic liquids: A review. *J. Mol. Liq.* **2018**, *259*, 319–339. [CrossRef]
156. Rezaee, M.; Assadi, Y.; Milani Hosseini, M.R.; Aghaee, E.; Ahmadi, F.; Berijani, S. Determination of organic compounds in water using dispersive liquid-liquid microextraction. *J. Chromatogr. A* **2006**, *1116*, 1–9. [CrossRef]
157. Guo, L.; Lee, H.K. Low-density solvent-based solvent demulsification dispersive liquid-liquid microextraction for the fast determination of trace levels of sixteen priority polycyclic aromatic hydrocarbons in environmental water samples. *J. Chromatogr. A* **2011**, *1218*, 5040–5046. [CrossRef]
158. Hosseini, M.H.; Rezaee, M.; Akbarian, S.; Mizani, F.; Pourjavid, M.R.; Arabieh, M. Homogeneous liquid-liquid microextraction via flotation assistance for rapid and efficient determination of polycyclic aromatic hydrocarbons in water samples. *Anal. Chim. Acta* **2013**, *762*, 54–60. [CrossRef]
159. Song, X.; Li, J.; Liao, C.; Chen, L. Ultrasound-assisted dispersive liquid-liquid microextraction combined with low solvent consumption for determination of polycyclic aromatic hydrocarbons in seawater by GC-MS. *Chromatographia* **2011**, *74*, 89–98. [CrossRef]
160. Leng, G.; Lui, G.; Chen, Y.; Yin, H.; Dan, D. Vortex-assisted extraction combined with dispersive liquid-liquid microextraction for the determination of polycyclic aromatic hydrocarbons in sediment by high performance liquid chromatography. *J. Sep. Sci.* **2012**, *35*, 2796–2804. [CrossRef]
161. Saleh, A.; Yamini, Y.; Faraji, M.; Rezaee, M.; Ghambarian, M. Ultrasound-assisted emulsification microextraction method based on applying low density organic solvents followed by gas chromatography analysis for the determination of polycyclic aromatic hydrocarbons in water samples. *J. Chromatogr. A* **2009**, *1216*, 6673–6679. [CrossRef] [PubMed]
162. Ozcan, S.; Tor, A.; Aydin, M.E. Determination of polycyclic aromatic hydrocarbons in waters by ultrasound-assisted emulsification-microextraction and gas chromatography-mass spectrometry. *Anal. Chim. Acta* **2010**, *665*, 193–199. [CrossRef] [PubMed]
163. Avino, P.; Notardonato, I.; Perugini, L.; Russo, M.V. New protocol based on high-volume sampling followed by DLLME-GC-IT/MS for determining PAHs at ultra-trace levels in surface water samples. *Microchem. J.* **2017**, *133*, 251–257. [CrossRef]
164. Cheng, J.; Matsadiq, G.; Liu, L.; Zhou, Y.W.; Chen, G. Development of a novel ultrasound-assisted surfactant-enhanced emulsification microextraction method and its application to the analysis of eleven polycyclic aromatic hydrocarbons at trace levels in water. *J. Chromatogr. A* **2011**, *1218*, 2476–2482. [CrossRef] [PubMed]
165. Mansour, F.R.; Danielson, N.D. Solidification of floating organic droplet in dispersive liquid-liquid microextraction as a green analytical tool. *Talanta* **2017**, *170*, 22–35. [CrossRef]
166. Guiñez, M.; Martinez, L.D.; Fernandez, L.; Cerutti, S. Dispersive liquid–liquid microextraction based on solidification of floating organic drop and fluorescence detection for the determination of nitrated polycyclic aromatic hydrocarbons in aqueous samples. *Microchem. J.* **2017**, *131*, 1–8. [CrossRef]

167. Yousefi, S.M.; Shemirani, F.; Ghorbanian, S.A. Hydrophobic Deep Eutectic Solvents in Developing Microextraction Methods Based on Solidification of Floating Drop: Application to the Trace HPLC/FLD Determination of PAHs. *Chromatographia 2018 81*, 1201–1211. [CrossRef]
168. Shishov, A.; Bulatov, A.; Locatelli, M.; Carradori, S.; Andruch, V. Application of deep eutectic solvents in analytical chemistry. A review. *Microchem. J.* **2017**, *135*, 33–38. [CrossRef]
169. Pena, M.T.; Casais, M.C.; Mejuto, M.C.; Cela, R. Development of an ionic liquid based dispersive liquid-liquid microextraction method for the analysis of polycyclic aromatic hydrocarbons in water samples. *J. Chromatogr. A* **2009**, *1216*, 6356–6364. [CrossRef]
170. Liu, L.; He, L.; Jiang, X.; Zhao, W.; Xiang, G.; Anderson, J.L. Macrocyclic polyamine-functionalized silica as a solid-phase extraction material coupled with ionic liquid dispersive liquid-liquid extraction for the enrichment of polycyclic aromatic hydrocarbons. *J. Sep. Sci.* **2014**, *37*, 1004–1011. [CrossRef]
171. Shamsipur, M.; Hashemi, B. Extraction and determination of polycyclic aromatic hydrocarbons in water samples using stir bar sorptive extraction (SBSE) combined with dispersive liquid-liquid microextraction based on the solidification of floating organic drop (DLLME-SFO) followed by. *RSC Adv.* **2015**, *5*, 20339–20345. [CrossRef]
172. Fernández, M.; Clavijo, S.; Forteza, R.; Cerdà, V. Determination of polycyclic aromatic hydrocarbons using lab on valve dispersive liquid-liquid microextraction coupled to high performance chromatography. *Talanta* **2015**, *138*, 190–195. [CrossRef] [PubMed]
173. Pena-Pereira, F.; Costas-Mora, I.; Lavilla, I.; Bendicho, C. Rapid screening of polycyclic aromatic hydrocarbons (PAHs) in waters by directly suspended droplet microextraction-microvolume fluorospectrometry. *Talanta* **2012**, *89*, 217–222. [CrossRef] [PubMed]
174. Santos, L.O.; dos Anjos, J.P.; Ferreira, S.L.C.; de Andrade, J.B. Simultaneous determination of PAHS, nitro-PAHS and quinones in surface and groundwater samples using SDME/GC-MS. *Microchem. J.* **2017**, *133*, 431–440. [CrossRef]
175. Sibiya, P.; Cukrowska, E.; Jönsson, J.Å.; Chimuka, L. Hollow-fibre liquid-phase microextraction for the determination of polycyclic aromatic hydrocarbons in Johannesburg Jukskei River, South Africa. *Chromatographia* **2013**, *76*, 427–436. [CrossRef]
176. Ratola, N.; Alves, A.; Kalogerakis, N.; Psillakis, E. Hollow-fibre liquid-phase microextraction: A simple and fast cleanup step used for PAHs determination in pine needles. *Anal. Chim. Acta* **2008**, *618*, 70–78. [CrossRef]
177. Aznar, M.; Alfaro, P.; Nerin, C.; Kabir, A.; Furton, K.G. Fabric phase sorptive extraction: An innovative sample preparation approach applied to the analysis of specific migration from food packaging. *Anal. Chim. Acta* **2016**, *936*, 97–107. [CrossRef]
178. Zilfidou, E.; Kabir, A.; Furton, K.G.; Samanidou, V. Fabric phase sorptive extraction: Current state of the art and future perspectives. *Separations* **2018**, *5*, 40. [CrossRef]
179. Kabir, A.; Furton, K.G. Sample preparation in food analysis: Practices, problems and future outlook. In *Analytical Chemistry: Developments, Applications and Challenges in Food Analysis*; Nova Science Publishers, Inc.: Hauppauge, NY, USA, 2017; ISBN 9781536122824.
180. Locatelli, M.; Kabir, A.; Innosa, D.; Lopatriello, T.; Furton, K.G. A fabric phase sorptive extraction-High performance liquid chromatography-Photo diode array detection method for the determination of twelve azole antimicrobial drug residues in human plasma and urine. *J. Chromatogr. B Anal. Technol. Biomed. Life Sci.* **2017**, *1040*, 192–198. [CrossRef]
181. Samanidou, V.; Galanopoulos, L.D.; Kabir, A.; Furton, K.G. Fast extraction of amphenicols residues from raw milk using novel fabric phase sorptive extraction followed by high-performance liquid chromatography-diode array detection. *Anal. Chim. Acta* **2015**, *855*, 41–50. [CrossRef]
182. Samanidou, V.; Michaelidou, K.; Kabir, A.; Furton, K.G. Fabric phase sorptive extraction of selected penicillin antibiotic residues from intact milk followed by high performance liquid chromatography with diode array detection. *Food Chem.* **2017**, *224*, 131–138. [CrossRef] [PubMed]
183. Saini, S.S.; Kabir, A.; Rao, A.L.J.; Malik, A.K.; Furton, K.G. A novel protocol to monitor trace levels of selected polycyclic aromatic hydrocarbons in environmental water using fabric phase sorptive extraction followed by high performance liquid chromatography-fluorescence detection. *Separations* **2017**, *4*, 22. [CrossRef]
184. Sun, T.; Wang, D.; Tang, Y.; Xing, X.; Zhuang, J.; Cheng, J.; Du, Z. Fabric-phase sorptive extraction coupled with ion mobility spectrometry for on-site rapid detection of PAHs in aquatic environment. *Talanta* **2019**, *195*, 109–116. [CrossRef] [PubMed]

185. Armenta, S.; Alcala, M.; Blanco, M. A review of recent, unconventional applications of ion mobility spectrometry (IMS). *Anal. Chim. Acta* **2011**, *703*, 114–123. [CrossRef] [PubMed]
186. Yang, S.; Chen, C.; Yan, Z.; Cai, Q.; Yao, S. Evaluation of metal-organic framework 5 as a new SPE material for the determination of polycyclic aromatic hydrocarbons in environmental waters. *J. Sep. Sci.* **2013**, *36*, 1283–1290. [CrossRef] [PubMed]
187. Hu, H.; Liu, S.; Chen, C.; Wang, J.; Zou, Y.; Lin, L.; Yao, S. Two novel zeolitic imidazolate frameworks (ZIFs) as sorbents for solid-phase extraction (SPE) of polycyclic aromatic hydrocarbons (PAHs) in environmental water samples. *Analyst* **2014**, *139*, 5818–5826. [CrossRef]
188. Ge, D.; Lee, H.K. Water stability of zeolite imidazolate framework 8 and application to porous membrane-protected micro-solid-phase extraction of polycyclic aromatic hydrocarbons from environmental water samples. *J. Chromatogr. A* **2011**, *1218*, 8490–8495. [CrossRef]
189. Song, X.; Li, J.; Xu, S.; Ying, R.; Ma, J.; Liao, C.; Liu, D.; Yu, J.; Chen, L. Determination of 16 polycyclic aromatic hydrocarbons in seawater using molecularly imprinted solid-phase extraction coupled with gas chromatography-mass spectrometry. *Talanta* **2012**, *99*, 75–82. [CrossRef]
190. Mauri-Aucejo, A.; Amorós, P.; Moragues, A.; Guillem, C.; Belenguer-Sapiña, C. Comparison of the solid-phase extraction efficiency of a bounded and an included cyclodextrin-silica microporous composite for polycyclic aromatic hydrocarbons determination in water samples. *Talanta* **2016**, *32*, 659–665. [CrossRef]
191. Soler-Seguí, S.; Belenguer-Sapiña, C.; Amorós, P.; Mauri-Aucejo, A. Evaluation of a cyclodextrin-silica hybrid microporous composite for the solid-phase extraction of polycyclic aromatic hydrocarbons. *Anal. Sci.* **2016**, *32*, 659–665. [CrossRef]
192. Wang, N.; Guo, Y.; Wang, L.; Liang, X.; Liu, S.; Jiang, S. Preparation of an aminopropyl imidazole-modified silica gel as a sorbent for solid-phase extraction of carboxylic acid compounds and polycyclic aromatic hydrocarbons. *Analyst* **2014**, *139*, 2531–2537. [CrossRef] [PubMed]
193. Zhao, W.; Yang, L.; He, L.; Zhang, S. Simultaneous Enrichment of Polycyclic Aromatic Hydrocarbons and Cu^{2+} in Water Using Tetraazacalix[2]arene[2]triazine as a Solid-Phase Extraction Selector. *J. Agric. Food Chem.* **2016**, *64*, 6233–6239. [CrossRef] [PubMed]
194. Kefi, B.B.; El Atrache, L.L.; Kochkar, H.; Ghorbel, A. TiO_2 nanotubes as solid-phase extraction adsorbent for the determination of polycyclic aromatic hydrocarbons in environmental water samples. *J. Environ. Sci.* **2011**, *23*, 860–867. [CrossRef]
195. Huang, Y.; Zhou, Q.; Xie, G. Development of micro-solid phase extraction with titanate nanotube array modified by cetyltrimethylammonium bromide for sensitive determination of polycyclic aromatic hydrocarbons from environmental water samples. *J. Hazard. Mater.* **2011**, *183*, 82–89. [CrossRef] [PubMed]
196. Krupadam, R.J.; Korde, B.A.; Ashokkumar, M.; Kolev, S.D. Novel molecularly imprinted polymeric microspheres for preconcentration and preservation of polycyclic aromatic hydrocarbons from environmental samples. *Anal. Bioanal. Chem.* **2014**, *406*, 5313–5321. [CrossRef] [PubMed]
197. Foan, L.; Ricoul, F.; Vignoud, S. A novel microfluidic device for fast extraction of polycyclic aromatic hydrocarbons (PAHs) from environmental waters – comparison with stir-bar sorptive extraction (SBSE). *Int. J. Environ. Anal. Chem.* **2015**, *95*, 1171–1185. [CrossRef]
198. Wu, H.; Wang, X.; Liu, B.; Lu, J.; Du, B.; Zhang, L.; Ji, J.; Yue, Q.; Han, B. Flow injection solid-phase extraction using multi-walled carbon nanotubes packed micro-column for the determination of polycyclic aromatic hydrocarbons in water by gas chromatography-mass spectrometry. *J. Chromatogr. A* **2010**, *1217*, 2911–2917. [CrossRef]
199. Zhou, Y.Y.; Yan, X.P.; Kim, K.N.; Wang, S.W.; Liu, M.G. Exploration of coordination polymer as sorbent for flow injection solid-phase extraction on-line coupled with high-performance liquid chromatography for determination of polycyclic aromatic hydrocarbons in environmental materials. *J. Chromatogr. A* **2006**, *1116*, 172–178. [CrossRef]
200. Zhang, X.; Wang, P.; Han, Q.; Li, H.; Wang, T.; Ding, M. Metal–organic framework based in-syringe solid-phase extraction for the on-site sampling of polycyclic aromatic hydrocarbons from environmental water samples. *J. Sep. Sci.* **2018**, *41*, 1856–1863. [CrossRef]

molecules

Perspective

Layered Double Hydroxides as Rising-Star Adsorbents for Water Purification: A Brief Discussion

Brígida Maria Villar da Gama [1], Rangabhashiyam Selvasembian [2], Dimitrios A. Giannakoudakis [3,*], Konstantinos S. Triantafyllidis [3], Gordon McKay [4] and Lucas Meili [1,*]

1 Laboratory of Processes, Center of Technology, Federal University of Alagoas, Maceió 57072-900, AL, Brazil; brigidavillar@gmail.com
2 Department of Biotechnology, School of Chemical and Biotechnology, SASTRA Deemed University, Thanjavur 613401, Tamilnadu, India; rambhashiyam@gmail.com
3 Department of Chemistry, Aristotle University of Thessaloniki, 54124 Thessaloniki, Greece; ktrianta@chem.auth.gr
4 Division of Sustainable Development, College of Science and Engineering, Hamad Bin Khalifa University, Qatar Foundation, P.O. Box 5825 Doha, Qatar; gmckay@hbku.edu.qa
* Correspondence: dagchem@gmail.com (D.A.G.); lucas.meili@ctec.ufal.br (L.M.)

Abstract: Within the frame of this article, briefly but comprehensively, we present the existing knowledge, perspectives, and challenges for the utilization of Layered Double Hydroxides (LDHs) as adsorbents against a plethora of pollutants in aquatic matrixes. The use of LDHs as adsorbents was established by considering their significant physicochemical features, including their textural, structural, morphological, and chemical composition, as well as their method of synthesis, followed by their advantages and disadvantages as remediation media. The utilization of LDHs towards the adsorptive removal of dyes, metals, oxyanions, and emerging pollutants is critically reviewed, while all the reported kinds of interactions that gather the removal are collectively presented. Finally, future perspectives on the topic are discussed. It is expected that this discussion will encourage researchers in the area to seek new ideas for the design, development, and applications of novel LDHs-based nanomaterials as selective adsorbents, and hence to further explore the potential of their utilization also for analytic approaches to detect and monitor various pollutants.

Keywords: layered double hydroxides (LDHs); removal; pollutants; wastewater treatment

Citation: Gama, B.M.V.d.; Selvasembian, R.; Giannakoudakis, D.A.; Triantafyllidis, K.S.; McKay, G.; Meili, L. Layered Double Hydroxides as Rising-Star Adsorbents for Water Purification: A Brief Discussion. *Molecules* **2022**, *27*, 4900. https://doi.org/10.3390/molecules27154900

Academic Editor: Gavino Sanna

Received: 23 June 2022
Accepted: 28 July 2022
Published: 31 July 2022

1. General Background

Freshwater quality has been threatened over the few last decades due to rapid urbanization. To meet the enhanced demand for products, the expansion of industrial and agricultural activities is taking place at an accelerated pace, which ends up releasing a large number of contaminants into water bodies [1,2]. In turn, these activities can lead to water contamination with various types of chemical substances, which are constantly and well reported in the specialized literature; namely dyes, metals, refining substances, pharmaceuticals, fertilizers, personal care products, pathogenic bacteria, and more. The most common between of them are:

Dyes: Various colored substances are widely utilized as dyes by the textile, food, paper, and pharmaceutical industries. Among these sectors, the textile industry releases 10 to 15% of the dyes produced in the environment. These elements, even in low concentrations, can pose a threat to the environment and living beings. The presence of dyes in water resources reduces the penetration of light through the water, affecting photosynthesis and dissolved oxygen levels, and harming the aquatic biota. In addition, some dyes can degrade to produce highly toxic and carcinogenic compounds [3–5].

Metals: Effluents from industries such as battery manufacturing, electroplating, mining, smelting, and many other industrial activities consist of heavy metals mixtures (such as Cd, Pb, Hg, Cu, Ni, Cr, As). The presence of these metals in water bodies is harmful to human health, as they are bioaccumulative and toxic. Among the negative effects of these elements in living beings, even at a low concentration are diarrhea, stomach pain, headaches, chronic bronchitis, and lung cancer [3,6,7].

Pharmaceuticals: Drugs are widely used in human medicines, livestock, aquaculture, beekeeping, and poultry to aid species growth. Currently, the release of these elements into the environment is not always absolutely regulated/controlled; hence, it is difficult to monitor their effects on living beings. The discharge of drugs into water resources occurs through the excretion of compounds not metabolized by living beings and industrial and hospital effluents. The continued presence of these pharmaceutical pollutants can cause oxidative stress and negative effects on reproduction (sperm motility and abnormal fetal development), osmoregulation, and altered immune functions in aquatic biotics [8,9].

Emerging pollutants: In general, emerging pollutants are chemical or microbiological products that are not commonly monitored or regulated and can cause problems to the environment as well as to the health of living beings. Except for the abovementioned, substances that can end up acting as pollutants are usually found in personal care products, pharmaceuticals, industrial additives, pesticides, plasticizers, and solvents. Among these substances, exposure to emerging organic compounds such as phenol, benzene, toluene, and xylenes can cause serious health problems such as gastrointestinal disorders, lung and kidney damage, heart attacks, and cancer. Yet, the increased use of pesticides and personal care products has increased dramatically in recent years. Although these products appear in low concentrations in water bodies, they can negatively influence the environment due to their resistance to biodegradation and accumulation in tissues [3,10].

Among the several technologies applied for water treatment, adsorption using layered double hydroxides (LDHs) will be highlighted in this article. Adsorption is a unitary operation in which the separation process occurs due to contact between a fluid phase—in this case, liquid—containing one or more contaminants (adsorbate) to be adsorbed/removed, and a solid (adsorbent). Due to the imbalance of forces (attraction or repulsion) for adsorption, the contaminant is attracted to the solid surface by physical or chemical interactions. This mass transfer process occurs until the balance between the adsorbed contaminant and what remains in the liquid phase, the residual, is reached. This is termed equilibrium. Adsorption is recognized as an effective process for water treatment due to certain characteristics, such as high efficiency, feasibility, low cost, flexibility, simplicity, wide processing range, cost-effective applications for water treatment, easy operation and implementation, great availability, and the possibility of adsorbents' regeneration. A very important aspect is that the adsorption-based processes do not result in the formation of hazardous substances/by-products [1,3,11–13].

Around 70% of the adsorption operation cost is related to the adsorbents [14]. Therefore, scientists involved in this area play a fundamental role in the development, characterization, and optimization of novel adsorbents. Hence, we seek materials with low cost and ease of production, high efficiency and selectivity towards the removal of the targeted contaminant, large surface area and volume of accessible pores, and good mechanical resistance. It is also desirable for the adsorbents to be regenerated/reused for numerous sufficient cycles to justify the costs.

Currently, and within this context, layered double hydroxides (LDHs), also known as anionic clays, have drawn great attention in their application as adsorbents. Several researchers have proposed the use of LDHs as efficient adsorbents against a plethora of organic and inorganic contaminants [4,15–18]. Furthermore, recent research demonstrates that biochar/LDH co-blends are highly promising, sustainable, and eco-economic materials for water treatment. Conventional adsorbents such as activated carbon have specific limitations such as high cost, low reuse performance, and low selectivity for water treatment. On the other hand, biochar/LDH blends can have a low cost, high surface area, an elevated amount of active adsorption sites, inherent interchangeability, a considerable increase in anions, and less toxicity for the removal of organic substances [19–23]. A wide variety of works have focused on evaluating the adsorptive capacity of LDHs against various pollutants such as dyes, drugs, arsenic, rare earth, radioactive substances, phosphate, metals, among other pollutants. Table 1 collects the most important advantages and disadvantages of using LDHs for water treatment compared to other materials [19–26].

Table 1. Advantages and disadvantages of using LDH in water treatment.

Advantages	Disadvantages
Low cost	Few studies regarding their toxicity in the environment
Sustainable nature	Current methods limit the amount of LDHs produced
Can be engineered for specific purposes	Few studies on its application in real wastewater
Excellent thermal stability	Functional groups preferences for anionic dyes
High removal efficiency	Can be exfoliated during synthesis
Extensive specific surface area	Cannot be easily regenerated/reused
High number of active sites	
Easy to prepare	
Memory effect	
High anion exchange capacities	
Chemical stability	

Considering the enhanced surface chemical heterogeneity and the nanostructured nature of LDHs, their elevated remediation efficiency against different substances is linked in the literature to plenty of physical and/or chemical interactions, with the most reported ones collected in Figure 1 [15,27]. Among all the interactions/mechanisms involved in the removal of various pollutants by Layered Double Hydroxides (LDHs), the most predominantly reported are:

(1) Physical adsorption. LDHs can have a high specific surface area, and hence present high adsorption capacities due to the presence and availability of active adsorption sites. Furthermore, the specific surface area of LDHs can be increased through calcinating or modifying/depositing on supports with three-dimensional structures.
(2) Ion exchange. Strongly negative molecules can be easily changed for the original anions in LDHs. In addition, positive ions can also be exchanged in the intermediate layer of LDHs, if pre-interleaved by some chelators.
(3) Interleaving. This starts from a preparation process, such as co-precipitation. Furthermore, the capture of molecules via the intercalation process is faster and more complete than ion exchange.

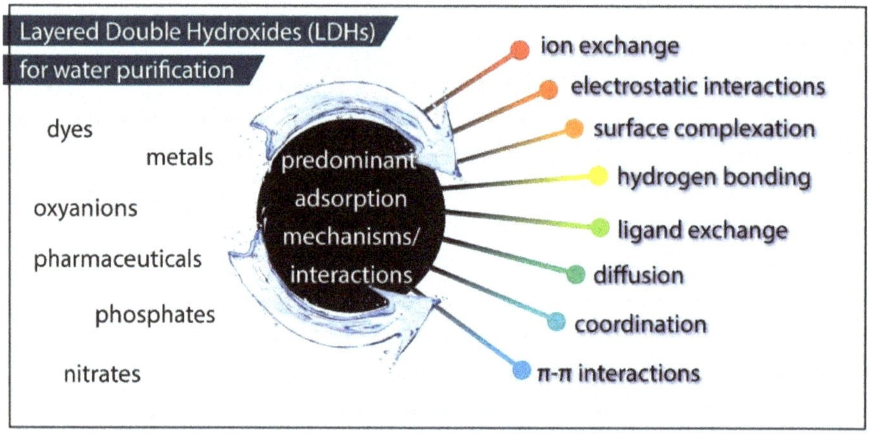

Figure 1. All the reported interactions/mechanisms involved in the removal of various pollutants by Layered Double Hydroxides (LDHs).

2. LDHs Physicochemical Characteristics

For the design and utilization of a proper adsorbent, certain characteristics are essential, such as low production cost, thermal, mechanical, and chemical stabilities, desirable physicochemical characteristics (such as elevated textural properties and high surface functional groups availability), high efficiency and adsorption capacity, rapid kinetics, and regeneration/reuse potential [28]. Some of the mentioned properties can only be verified when applying the material in a specific process due to the dependence on operational conditions and adsorbate characteristics. Of course, it is very challenging to develop materials that possess all the abovementioned characteristics. Thus, scientists involved in the water purification area of research have developed and tested a broad range of materials. Among them, layered double hydroxides (LDHs) are prominent and meet many demands, considering the sustainability of the approach, high anion exchange capacity, high specific surface area, high ion exchange capacities, and regenerative adsorptivity [28]. In addition to their attractive properties as adsorbents, LDHs have applications in various fields such as drug carriers, catalysis, flame retardants, pharmaceutical transport systems, electrocatalytic water separation, additives for polymers, photocatalytic degradation, and medicine [6,18,29].

LDHs are two-dimensional (2D) nanostructures composed of stacked layers consisting of mixed hydroxides of di- and trivalent cations with hydrated anions in the spaces between the positively charged *lamellae* (Figure 2). LDHs are also called hydrotalcite compounds, due to the rigid layer structure derived from brucite with edge-sharing $M(OH)_6$, similar to flexible graphene oxide nanosheets [30]. In general, such materials are represented by the formula: $[M^{2+}{}_{1-x}M^{3+}{}_{x}(OH)_2]^{x+}$. $A^{m-}{}_{x/m}.nH_2O$ [30], where M^{2+} represents a divalent metal cation; M^{3+}, the trivalent cation; A^{m-}, an anion intercalated with charge m; X, the ratio between divalent (M^{2+}) and trivalent cations (M^{3+}), for which the value of the molar ratio can be between 2.0 and 4.0; and *n*, the number of soft water molecules. Many divalent (such as Mg^{2+}, Fe^{2+}, Ca^{2+}, Co^{2+}, Cu^{2+}, Ni^{2+}, and Zn^{2+}) and trivalent (such as Al^{3+}, Cr^{3+}, Ga^{3+}, Mn^{3+}, and Fe^{3+}) metal cations can be used for the preparation of LDHs. Except for bivalent and trivalent metal ions, a wide range of monovalent and tetravalent metal ions (such as Li^{+}, Ti^{4+}, Sn^{4+}, or Zr^{4+}) may also be inserted in the octahedral sites [31]. In addition, the electroneutrality of LDHs can be due to the presence of hydrated organic or inorganic anions (CO_3^{2-}, NO_3^{-}, SO_4^{2-}, OH^{-}, Cl^{-}, Br^{-}) in the interlamellar space [18]. Because of these and the different combinations between the cations that participate in the formation of the layers, LDHs can have numerous structural varieties. However, this combination must consider the octahedral coordination and the ionic radius (preferably between 0.50 and 0.74 Å), as distortions may occur with the use of cations. In addition, it is necessary to consider the relationship between the size and charge of the interlayer anion so that it is possible to balance the positive charges of the layer homogeneously [32].

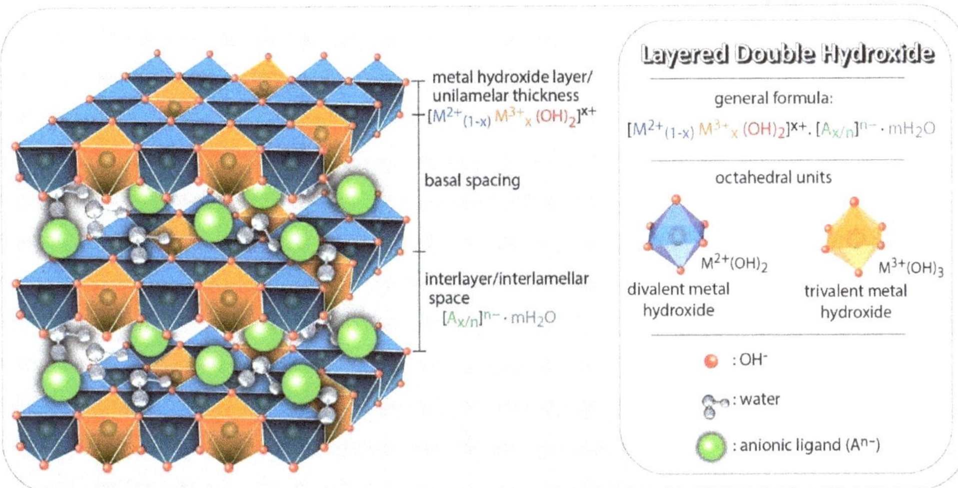

Figure 2. A schematic presentation of the layered double hydroxides (LDHs) chemical composition and structure. Figure reproduced from reference [30]. Copyrights Elsevier, 2022.

For LDHs, the precisely controlled chemical composition of the lamellar layers and the interlayer composition provide a unique supramolecular nanometric structure with the ability to disperse active sites on an atomic scale, in addition to facilitating morphological manipulation [33,34]. Thus, the versatility of the chemical composition and nanostructure of LDHs establish them as promising adsorbents against a wide variety of pollutants. The most attractive properties of LDHs are the chemical compositions according to the metals used, the space between layers, and the high surface capable of adsorbing bioactive substances, which are dominated by the synthesis conditions [32]. All the above can be tuned on demand based on the synthetic approach/protocol. Another important feature of LDHs is the ease, low cost, and variety of available synthesis methods, which can be classified as direct and indirect methods:

(i) Direct methods: The preparation of LDHs occurs via direct precipitation from the addition of tri- and divalent cations, in a solution in alkaline pH with the main methods of coprecipitation, salt–oxide, sol–gel, induced hydrolysis, and hydrothermal synthesis.
(ii) Indirect methods: involve replacing an interlamellar anion from a previously produced precursor LDH. Examples of this substitution method are ion exchange in solution, ion exchange in acidic medium, double phase replacement, and regeneration through the delaminate precursor [35–37]. Therefore, the supramolecular structure, the facile manipulation of adsorption sites at the atomic scale, the versatility of compositions, in addition to the possibility of morphological manipulation create the possibility of tuning the amount and accessibility of the active adsorption sites, and hence the adsorption kinetics, as well as the efficiency for a specifically targeted pollutant [34]. As in any case, LDHs have some specific characteristics that can complicate their use as adsorbents. The low mechanical resistance is a problem for continuous water treatment units and in certain regeneration processes, as LDHs can be exfoliated. Therefore, there is a series of studies in the literature proposing to support LDHs in larger and recalcitrant particles [38–40]. In addition, in acidic media, the removal capacity of LDHs is compromised due to low structural stability at low pH [26]. In

Table 2, we collected characteristics of methods of synthesis which can be followed for the preparation of pure LDHs, as well as their composites and hybrids [18,25,41,42].

Table 2. Provides details of some of the methods used for the synthesis of LDH.

Methods of Synthesis	Characteristics
Coprecipitation	This method is based on the controlled and slow addition of a base (such as sodium hydroxide and/or bicarbonate, sodium carbonate or ammonium hydroxide) to a solution containing simultaneous divalent and trivalent metal cations. Since more than two cations can precipitate simultaneously, the process must be carried out under supersaturation conditions. It is recommended that the pH of the reaction medium be kept constant in the range of 7–10. Subsequently, the suspension is subjected to hydrothermal treatment to increase the yield or crystallinity.
Salt-oxide	This method was developed by Boehm in 1977 to prepare zinc and chromium LDHs, using an aqueous suspension of ZnO to react with excess $CrCl_3$ in an aqueous solution. The salt–oxide method, in short, is a solid–liquid reaction in which the aqueous solution of the excess trivalent ion chloride salt is treated with an aqueous suspension of the divalent metal oxide.
Sol-gel	This synthetic protocol is widely used for the preparation of a plethora of metal oxides due to the possible high efficiency and purity of the final material. One important advantage of this method is the variety of compositions obtained through temperature adjustment. This process consists of the constant agitation of the component that transforms sol to gel. This sol–gel transformation occurs during the strong acid hydrolysis of metallic precursors, predominately using a strong acid such chloric acid or nitric acid. After the formation of the gel, the material is filtered and washed with distilled water, and later with ethanol.
Hydrothermal	The hydrothermal method is generally used when low-affinity anions need to be intercalated into the intermediate layers. This method uses gibbsite and brucite, double-layered hydroxide–deoxycholate intercalation compounds, which are not feasible to obtain easily via other syntheses. An aqueous suspension consists of two oxides, one trivalent metal ion and the other bivalent, which are placed in a pressurized container and subjected to hydrothermal treatment at high temperature for a few days. During this process, the hydrated amorphous precursor crystallizes in the presence of reactive basic oxide.
Ion exchange	This is an indirect method usually applied to pre-synthesized LDHs. This method is used when the anions or the divalent/trivalent metal cations are unstable in the alkaline solution, or when the LDHs have a greater affinity for the guest anions than for the intercalated anions of a pre-synthesized LDH. An aqueous suspension of the LDH precursors/pre-synthesized is mixed with a large excess of the salt of the anion to be intercalated. The reaction is carried out under an inert atmosphere to avoid excess carbonate in the intermediate layers. It is recommended the reaction not occur at pH lower than 4, due to the anion interaction in the LDH layers being weaker and presenting a high temperature in this pH range.
Regeneration/"memory effect"	One of the main properties of LDH is its ability to restructure. After being subjected to heat treatment or calcination (400 to 500 °C), the layered structure of LDH changes to mixed metallic oxides (water, anion, and hydroxyl groups are highlighted). When calcined LDH is placed in a solution containing guest anions, they can recover their original layered structure and form a new LDH phase. This procedure of retrieving its original form (rehydration) is called the "memory effect", and must be carried out in an inert atmosphere, mostly comprised of nitrogen.

3. LDHs as Adsorbents

As already mentioned, LDHs are very promising materials for the removal of a wide variety of pollutants via adsorption. There are many reports in the literature; consequently, we will address some of the most important ones that gather the highest level of research attention for specific pollutants, with emphasis on the utilized LDHs.

Dyes: The presence of dyes in water and wastewater is highly undesirable, and LDHs have been reported as efficient remediation media of a large number of industrial dyes [4]. Dyes are mainly categorized into anionic and cationic dyes. LDHs can remove anionic dyes in the same order of magnitude as cationic dyes in activated carbon [43,44]. Examples of anionic dyes include methyl orange, amaranth, sunset yellow FCF, reactive blue 21, Eriochrome Black T, Congo red, and others. In the literature, it is common to find reports of high adsorption capacities regarding the removal of cationic dyes by LDHs [4,18,45].

El-Abboubi et al. [45] evaluated the double hydroxides of Mg-Al synthesized with dodecyl sulfate (Mg-Al-Ds) and carbonate (Mg-Al-CO_3), through the coprecipitation method to remove methyl orange from aqueous solution. The prepared LDHs showed the intercalary-like structure with (00l) reflections and a nanoscale nature, as well as the presence of the desired anions in the interlayer space. The maximum capacities were 185.06 mg·g^{-1} for Mg-Al-Ds and 97.5 mg·g^{-1} for Mg-Al-CO_3. They also observed the influence of the pH of the solution on the adsorption capacity of the dye in the case of Mg-Al-CO_3. The adsorptive capacity of Mg-Al-Ds was not influenced by the pH of the solution, while Mg-Al-CO_3 showed greater efficiency at pH in the range of 3–7. According to the authors, this behavior may be associated with two mechanisms: (1) anion exchange of carbonate anions for dye anions; (2) association of the positively charged surface groups of LDHs and dye anions.

Kostić et al. [18] synthesized MgCoAl-CO_3-LDH double hydroxides using the coprecipitation method for the removal of dye RB19 from an aqueous solution. The material presented a surface area of around 48 m^2·g^{-1}, high crystallinity and the presence of carbonate, hydroxyl, and metal ions groups within its structure. The proposed adsorption mechanisms involved in the process were electrostatic attraction, physisorption, and chemical bonding. The maximum capacity was found to be equal to 367.93 mg·g^{-1}.

Ahmed et al. [4] prepared Mg/Fe-LDHs nanoparticles through the precipitation method for adsorption of Congo red dye from effluents. The Mg/Fe-LDHs showed high crystallinity and the presence of hydroxyl functional groups. The adsorption process was governed both by physical and chemical interactions, with the maximum capacity reaching above 9000 mg per gram.

Metals: The discharge of heavy metals into water bodies results in harmful effects to the environment and human health due to their toxicity and persistence. The presence of these elements in water systems has caused concern in recent years; hence, the interest in novel and efficient remediation media is still elevated [9,16]. In the literature, it is quoted that the work of Satoshi Fujji et al. [46], from 1992, was the first in which LDHs were used for metal removal (Pb^{2+}, Cu^{2+} and Zn^{2+}) [47]. There is no established mechanism for sorption/adsorption of metals in each LDH; hence, the study of the involved mechanism of interaction is still a major challenge for researchers in the field. However, several possible mechanisms/kinds of interactions are proposed, such as surface complexation, isomorphic substitution, surface precipitation, and electrostatic interactions, and chelation with the binding anion has been reported [16,29,48].

Dinari and Neamati [48] synthesized Ca/Fe double hydroxides (LDHs) modified with 3-aminopropyl triethoxysilane as the silane coupling agent. Polyaniline nanocomposites with 5 and 10% by weight of modified silane Ca/Fe LDH-Cit were also studied. Pure polyaniline and that modified with NCs were used as adsorbent to remove Pb^{2+} ions from an aqueous solution. The results showed that NC10% showed higher adsorptive capacity (110 mg·g^{-1}; 0.52 mmol·g^{-1}) compared to NC5% (56 mg·g^{-1}; 0.27 mmol·g^{-1}) and NIBP (47 mg·g^{-1}; 0.22 mmol·g^{-1}), demonstrating the favoring of adsorption in the presence of Ca/Fe double hydroxides.

Guo et al. [29] synthesized ZnNiCr double-layer hydroxides (ZnNiCr-LDHs) using the microwave hydrothermal method to remove Cr^{6+} from an aqueous solution. The material presented a surface area of 354 m^2·g^{-1}, and the SEM analysis showed an irregular block structure. In this study, it was proposed that predominantly electrostatic interaction between the metal ion and the adsorbent took place with a maximum capacity of 28.2 mg·g^{-1} (0.54 mmol·g^{-1}). The results demonstrate potential application prospects in the removal of Cr^{6+} in wastewater.

Zhang et al. [16] synthesized sodium alginate intercalated with MgAl-LDH (SA-LDH) for adsorption of Cd^{2+}, Pb^{2+}, and Cu^{2+} from an aqueous solution. The characterization results showed characteristic peaks at 2θ = 13.24°, 22.88°, 35.09°, 39.19°, 47.23°, and 60.97°, while CO, COO^- and CH surface functional groups were also detected in its structure. The maximum adsorption capacities of SA-LDH were 60 $mg \cdot g^{-1}$ (0.945 $mmol \cdot g^{-1}$) for Cu^{2+}, 243.66 $mg \cdot g^{-1}$ (1.176 $mmol \cdot g^{-1}$) for Pb^{2+}, and 95.55 $mg \cdot g^{-1}$ (0.850 $mmol \cdot g^{-1}$) for Cd^{2+}. The results showed that the possible mechanisms involved in the adsorption process are: (1) bonding or complexation with SurOH or Sur-O- of SA-LDH; (2) precipitation of metal hydroxides or carbonates; (3) isomorphic substitution, and (4) chelation with COO^- in the interlayers.

Oxyanions: High levels of oxyanions found in the environment are reported all over the world. Oxyanions (e.g., arsenate, chromate, phosphate, selenite, selenate, borate, nitrate, etc.) are considered dangerous to humans and wildlife, even at very low concentrations [17,49]. In the case of oxyanions, the primarily reported mechanisms are ion exchange, electrostatic attraction, and coordination [17,50].

Motandi et al. [51] synthesized zirconium-modified Mg-Al-LDH layered double hydroxides (Zr-LDH) using the coprecipitation method, and then used calcination to obtain an oxide (Zr-LDO), and the materials were used as adsorbents for the removal of phosphate from aqueous solution. The materials showed a well-developed layered structure. In addition, Zr was homogeneously distributed in the adsorbent with a well-defined crystallinity and the presence of brucite, while hydroxyl groups and M–O elongation were detected. The results showed that Zr-LDH and Zr-LDO are excellent adsorbents for phosphate, obtaining maximum adsorption capacities of 99.35 mg/g for Zr-LDH and 80.33 $mg \cdot g^{-1}$ for Zr-LDO.

Jung et al. [17] synthesized Mg-Al double hydroxides (Mg-Al LDHs-FHC) via the one-pot in situ hydrothermal method for the monocomponent and multicomponent adsorption of arsenate and phosphate from aqueous solution. Initially, the influence of the Mg:Al ratio and temperature on the preparation of the adsorbent material was investigated. The results showed that Mg:Al molar ratios and temperature influenced the Mg-Al LDHs-FHC structural properties. The best adsorption efficiency was for Mg:Al molar ratio of 2:1 and temperature of 150 °C. In the one-component system, the maximum adsorption capacities were 56.30 $mg \cdot g^{-1}$ and 33.21 $mg \cdot g^{-1}$ for arsenate and phosphate, respectively. For the multicomponent system, the maximum adsorption capacities were 16.22 $mg \cdot g^{-1}$ for arsenate and 20.26 $mg \cdot g^{-1}$ for phosphate. In the single-component system, a possible adsorption mechanism is the ion exchange between the nitrate interlayer and the arsenate or phosphate groupings. In the multicomponent system, coordinated bonds are also possibly responsible for the competition of arsenate or phosphate in the adsorptive process.

Zhou et al. [50] prepared FeMgMn-LDH via the co-precipitation method for the removal of nitrate from an aqueous solution. The surface area of the material was 47 $m^2 \cdot g^{-1}$; the presence of –OH, N=O, M–OH and M–O–M was detected in its structure. The maximum amount of nitrate adsorption was 10.56 $mg \cdot g^{-1}$. The adsorptive process was spontaneous and exothermic. The possible proposed adsorption mechanisms of nitrate removal using FeMgMn-LDH were an electrostatic attraction and ion exchange.

Emerging pollutants: Plenty of synthetic substances that have recently been detected at low concentrations ($ng \cdot L^{-1}$ or μL^{-1}) are assumed to be emerging pollutants nowadays. These components are used in different industrial processes for the production of pharmaceuticals, personal care products (PCPs), beverages, foods, and others [11,12,15,52], and as a result are discarded in aquatic environments. Although LDHs have been reported as efficient adsorbents for various emerging pollutants, such as sodium diclofenac, caffeine, inorganic endocrine disruptor, hormones, and bisphenol A, among others, there are still few studies in the literature.

Santamaría et al. [53] synthesized by the co-precipitation method Zinc–Titanium–Aluminum (ZnTiAl) double layer hydroxides (LDHs) with a Zn/(Al-Ti) molar ratio of 3:1 and studied them for the adsorption of diclofenac and salicylic acid. The study also evaluated the use of commercial aluminum (Al) and aluminum extracted from saline slag. It was shown that the increase in Ti content negatively affected the crystallinity of the material and that the increase in temperature decreased the surface area of the material due to the increase in amorphous mixed oxides. The results showed a great potential of the synthesized hydrotalcites for the adsorption of diclofenac (409 μmol/g) compared to salicylic acid (80 $\mu mol \cdot g^{-1}$)

Kumari et al. [11] used a double layer of Zn-Al hydroxide (LDH) loaded with Bi_2O_3 as an adsorbent for the removal of diclofenac from an aqueous solution. The material presented the main diffraction peaks 003, 006, 012, 041, 111, 241, and 110, and a surface area of 102 $m^2 \cdot g^{-1}$. The Zn-Al-LDH.xBi_2O_3 presented a capacity of 574.71 $mg \cdot g^{-1}$ for the removal of diclofenac. The results indicated that the adsorptive process occurs in a monolayer and that the diffusion of diclofenac occurs mainly on the external surface.

Other kinds of pollutants: There are few reports of the use of LDHs for the adsorptive removal of bacteria and viruses, per- and poly-fluoroalkyl substances, and rare earth and radioactive substances; such as Cs, Sr, and Th (excluding Uranium, which is well reported). Therefore, exploring the use of LDHs in the removal of this type of contaminant is very promising and important for environmental health protection.

Table 3 summarizes some cited articles in which LDH were studied as adsorbents against eclectic organic and inorganic substances from aqueous solutions.

Table 3. Some characteristic works in which LDHs are used as adsorbents for organic and inorganic substances.

Pollutant	LDH	Synthesis Method	q_{max} ($mg \cdot g^{-1}$)	Reference
Dye methyl orange	Mg-Al-Ds Mg-Al-CO_3	Coprecipitation	185.06 97.50	[45]
Dye RB19	MgCoAl-CO_3-LDH	Coprecipitation	367.93	[18]
Dye Congo red	Mg/Fe-LDHs	Precipitation	9127.08	[4]
Pb^{2+}	Ca/Fe LDH-Cit(NC10%) Ca/Fe LDH-Cit(NC5%)	Precipitation	110.00 56.00	[48]
Cr^{6+}	ZnNiCr-LDHs	Hydrothermal	28.20	[29]
Cd^{2+} Pb^{2+} Cu^{2+}	MgAl-LDH (SA-LDH)	Coprecipitation	60.00 243.66 95.55	[16]
Phosphate	Zr-LDH Zr-LDO	Coprecipitation	99.35 80.33	[51]
Arsenate (mono) Arsenate (mult) Phosphate (mono) Phosphate (mult)	Mg-Al LDHs-FHC	Hydrothermal	56.30 16.22 33.21 20.26	[17]
Nitrate	FeMgMn-LDH	Co-precipitation	10.56	[50]
Diclofenac Salicylic acid	ZnTiAl	Co-precipitation	0.07 0.01	[53]
Diclofenac	Zn-Al-LDH.xBi_2O_3	Hydrothermal	574.71	[11]

Utilization in real-life applications: Industrial effluents contain various pollutants simultaneously, such as dyes, metals, pesticides, and antibiotics, which impose a high physicochemical complexity on the real system. Therefore, it is necessary to understand the interaction between the adsorbates and the adsorbent in the water decontamination process [1,3]. Despite the importance of multicomponent adsorption, there are few references in the literature evaluating the use of LDHs to remove dyes, metals, and rare earth elements in multicomponent systems. Consequently, it is necessary to study the multicomponent adsorption, as well as the evaluation of the selectivity or affinity of a given LDH for each adsorbate and the competition between them. Another developing field is for LDH containing hybrids of two types—nanocomposites and organically modified LDH hybrids—which have been recently reported [54] for the removal of metal ions and dyes from wastewater, with particularly high capacities.

Considering the adsorption mode, batch experiments are well reported in the literature. On the contrary, although the design and efficiency are well approached in lab-scale batch studies, the fixed beds procedure has received little attention regarding exploration of the use of LDHs as adsorbents. The fixed bed columns are applied for continuous flow, and are effective for the treatment of large volumes of effluents, allowing application in industries, adsorption–desorption cycles, efficient fluid–particle contact and easy phase separation, which makes the treatment process cheaper and more sustainable [28,55,56]. In this context, the main challenge is to overcome the LDHs' poor mechanical resistance for continuous use, because they can be sprayed or exfoliated [40]. In addition, leaching tests should be always considered when studying LDHs in aquatic matrixes.

Fixed beds are employed primarily for commercial water treatment scale-up. Discovering and proposing new operational modes to improve fluid–particle interaction and phase separation is a key issue. Therefore, a smaller treatment area, faster operation, larger water volumes treated, and cost reduction are important advantages. As a result, there may be benefits such as a smaller treatment area, faster operation, treatment of bigger quantities of water, and cheaper costs. Therein, the study of alternative contacting devices, such as fluidized bed, spouted bed, simulated moving bed, expanded bed, continuous stirred tanks, and multi-batch tanks, is much appreciated and warrants substantially further investigation [28].

4. Discussion

The studies presented and discussed above demonstrate the prosperity of utilizing LDHs as remediation media, and hence, the growing interest of researchers against various pollutants regarding interpretation and proposals for the involved mechanisms and interactions. The studies revealed elevated adsorptive capacities of organic and inorganic pollutants, even compared to benchmark materials such as porous carbons [4,16,45,48]. The high adsorptive capacity of LDHs is attributed to their high specific surface area, high thermal and chemical stability, but, more importantly, to their surface chemical composition. The characterization of LDHs via various analyses/techniques such as the pH of the point of zero charge (pH_{PZC}), Fourier transform infrared spectroscopy (FT-IR), and X-ray diffraction (XRD) helps us to understand the adsorptive mechanism of the process. The most widely proposed interactions/mechanisms involved in the adsorption of organic compounds generally involve electrostatic interaction between the surface of the LDH (positively or negatively charged) with the cationic (positively charged) or anionic (negatively charged) compounds. The adsorptive mechanism associated with the adsorption of metal ions involves the ion exchange between the surface of the material and the metal ions. Other mechanisms are also involved in the adsorption of metal ions in LDHs, such as precipitation and complexation. Among the functional groups present in the structure of LDHs, the hydroxyl group favors the adsorption of both organic and inorganic compounds. Another important parameter in the adsorption process using LDHs is the pH of the solution. It is observed that, in general, the favorable pH for the adsorption of anionic compounds occurs at pH lower than 7 and for cationic compounds at pH above 7.

5. Future Prospects

Layered double hydroxides (LDHs) represent one category of materials with an increasing trend of interest toward the removal of water pollutants. To uplift the direction of LDHs-based adsorbent application on a commercial scale, novel approaches and considerations with emphasis on sustainability and low cost are presented. However, more adsorption investigations of LDHs need to be explored in multi-component systems rather than in the current trend, where most of the studies are limited to mono-component systems and lab-scale batch experiments. The parametric influence on the adsorption process requires an advanced optimization approach to attain maximum removal performance of water pollutants using LDHs-based adsorbents. The major setback of a "one time" use of the adsorbent must be overcome with the proper assessment of desorption strategies, followed by a good number of adsorbent recycle options, treated water re-use, and the adsorbate pollutant recovery and re-use. More mechanistic insights into water pollutant removal need to be understood in multicomponent simulated and real effluent systems to study the antagonistic effects targeting the development of adsorbents with high selectivity for the specific pollutant(s). To understand the complete application of layered double hydroxides in water pollutant elimination, the batch system followed by a continuous system needs to be performed at lab and pilot scale, with real effluents.

The simplicity and low cost of LDHs production (for instance, via straightforward coprecipitation of metal salts at mild basic conditions), in addition to the ability to tailor design their composition and structure, establish them as sustainable and attractive candidates for real-life applications. Moving a step forward regarding the advantages of nanomaterials, designing and synthesizing novel nanocomposites/hybrids should be assumed as a worthwhile effort to be further explored in the field of research. The development of synthesis methods capable of producing LDHs on a commercial scale is one of the current problems for the use of these materials on an industrial scale. In addition, LDHs syntheses are generally limited to the use of MgAl and MgFe, requiring the exploration of new compounds. Within the frame of circular (bio)economy and sustainability, the development of novel and advanced composited of LDHs with biomass-derived biochar/carbon and determining the optimum ratio between the counterparts is a prosperous topic of research, since still only a few studies exist.

Author Contributions: All authors contributed to the study conception and design. Conceptualization, B.M.V.d.G., D.A.G. and L.M.; writing—original draft preparation, B.M.V.d.G. and L.M.; writing—review and editing, B.M.V.d.G., R.S., D.A.G., K.S.T. and G.M. All authors have read and agreed to the published version of the manuscript.

Funding: The authors thank to National Council for Scientific and Technological Development (CNPq/Brazil), Coordination for the Improvement of Higher Education Personnel (CAPES/Brazil) and Foundation for Research Support of the State of Alagoas (FAPEAL/Brazil).

Institutional Review Board Statement: Not applicable.

Informed Consent Statement: Not applicable.

Conflicts of Interest: The authors have no relevant financial or non-financial interest to disclose.

References

1. Nimbalkar, M.N.; Bhat, B.R. Simultaneous adsorption of methylene blue and heavy metals from water using Zr-MOF having free carboxylic group. *J. Environ. Chem. Eng.* **2021**, *9*, 106216. [CrossRef]
2. Yadav, A.; Bagotia, N.; Sharma, A.K.; Kumar, S. Advances in decontamination of wastewater using biomass-basedcomposites: A critical review. *Sci. Total Environ.* **2021**, *784*, 147108. [CrossRef]
3. Yadav, A.; Bagotia, N.; Sharma, A.K.; Kumar, S. Simultaneous adsorptive removal of conventional and emerging contaminants in multi-component systems for wastewater remediation: A critical review. *Sci. Total Environ.* **2021**, *799*, 149500. [CrossRef] [PubMed]
4. Ahmed, D.N.; Naji, L.A.; Faisal, A.; Al-Ansari, N.; Naushad, M. Waste foundry sand/MgFe-layered double hydroxides composite material for efficient removal of Congo red dye from aqueous solution. *Sci. Rep.* **2020**, *10*, 2042. [CrossRef]
5. Solehudin, M.; Sirimahachai, U.; Ali, G.A.; Chong, K.F.; Wongnawa, S. One-pot synthesis of isotype heterojunction g-C_3N_4-MU photocatalyst for effective tetracycline hydrochloride antibiotic and reactive orange 16 dye removal. *Adv. Powder Technol.* **2020**, *31*, 1891–1902. [CrossRef]
6. Mureseanu, M.; Eliescu, A.; Ignat, E.-C.; Carja, G.; Cioatera, N. Different routes of MgAl–LDH synthesis for tailoring the adsorption of Pb(II) pollutant from water. *Comptes Rendus. Chim.* **2022**, *25*, 1–12. [CrossRef]
7. Gayathri, R.; Gopinath, K.; Kumar, P.S. Adsorptive separation of toxic metals from aquatic environment using agro waste biochar: Application in electroplating industrial wastewater. *Chemosphere* **2021**, *262*, 128031. [CrossRef] [PubMed]
8. Nava-Andrade, K.; Carbajal-Arízaga, G.; Obregón, S.; Rodríguez-González, V. Layered double hydroxides and related hybrid materials for removal of pharmaceutical pollutants from water. *J. Environ. Manag.* **2021**, *288*, 112399. [CrossRef]
9. Manjunath, S.; Baghel, R.S.; Kumar, M. Antagonistic and synergistic analysis of antibiotic adsorption on Prosopis juliflora activated carbon in multicomponent systems. *Chem. Eng. J.* **2020**, *381*, 122713. [CrossRef]
10. Somma, S.; Reverchon, E.; Baldino, L. Water Purification of Classical and Emerging Organic Pollutants: An Extensive Review. *ChemEngineering* **2021**, *5*, 47. [CrossRef]
11. Kumari, P.; Pal, B.; Das, R.K. Superior adsorptive removal of eco-toxic drug diclofenac sodium by Zn–Al LDH·xBi_2O_3 layer double hydroxide composites. *Appl. Clay Sci.* **2021**, *208*, 106119. [CrossRef]

12. Cheng, N.; Wang, B.; Wu, P.; Lee, X.; Xing, Y.; Chen, M.; Gao, B. Adsorption of emerging contaminants from water and wastewater by modified biochar: A review. *Environ. Pollut.* **2021**, *273*, 116448. [CrossRef]
13. Quintela, D.U.; Henrique, D.C.; dos Santos Lins, P.V.; Ide, A.H.; Erto, A.; da Silva Duarte, J.L.; Meili, L. Waste of Mytella Falcata shells for removal of a triarylmethane biocide from water: Kinetic, equilibrium, regeneration and thermodynamic studies. *Colloids Surf. B Biointerfaces* **2020**, *195*, 111230. [CrossRef] [PubMed]
14. Li, W.; Mu, B.; Yang, Y. Feasibility of industrial-scale treatment of dye wastewater via bio-adsorption technology. *Bioresour. Technol.* **2019**, *277*, 157–170. [CrossRef] [PubMed]
15. dos Santos, G.E.D.S.; dos Santos Lins, P.V.; de Magalhães Oliveira, L.M.T.; da Silva, E.O.; Anastopoulos, I.; Erto, A.; Giannakoudakis, D.A.; de Almeida, A.R.F.; da Silva Duarte, J.L.; Meili, L. Layered double hydroxides/biochar composites as adsorbents for water remediation applications: Recent trends and perspectives. *J. Clean. Prod.* **2021**, *284*, 124755. [CrossRef]
16. Zhang, X.; Shan, R.; Li, X.; Yan, L.; Ma, Z.; Jia, R.; Sun, S. Effective removal of Cu(II), Pb(II) and Cd(II) by sodium alginate intercalated MgAl-layered double hydroxide: Adsorption properties and mechanistic studies. *Water Sci. Technol.* **2021**, *83*, 975–984. [CrossRef]
17. Jung, K.-W.; Lee, S.Y.; Choi, J.-W.; Hwang, M.-J.; Shim, W.G. Synthesis of Mg–Al layered double hydroxides-functionalized hydrochar composite via an in situ one-pot hydrothermal method for arsenate and phosphate removal: Structural characterization and adsorption performance. *Chem. Eng. J.* **2021**, *420*, 129775. [CrossRef]
18. Kostić, M.; Najdanović, S.; Velinov, N.; Vučić, M.R.; Petrović, M.; Mitrović, J.; Bojić, A. Ultrasound-assisted synthesis of a new material based on MgCoAl-LDH: Characterization and optimization of sorption for progressive treatment of water. *Environ. Technol. Innov.* **2022**, *26*, 102358. [CrossRef]
19. Zubair, M.; Ihsanullah, I.; Aziz, H.A.; Ahmad, M.A.; Al-Harthi, M.A. Sustainable wastewater treatment by biochar/layered double hydroxide composites: Progress, challenges, and outlook. *Bioresour. Technol.* **2021**, *319*, 124128. [CrossRef]
20. Zubair, M.; Manzar, M.S.; Mu'Azu, N.D.; Anil, I.; Blaisi, N.I.; Al-Harthi, M.A. Functionalized MgAl-layered hydroxide intercalated date-palm biochar for Enhanced Uptake of Cationic dye: Kinetics, isotherm and thermodynamic studies. *Appl. Clay Sci.* **2020**, *190*, 105587. [CrossRef]
21. Mu'Azu, N.D.; Zubair, M.; Jarrah, N.; Alagha, O.; Al-Harthi, M.A.; Essa, M.H. Sewage Sludge $ZnCl_2$-Activated Carbon Intercalated MgFe–LDH Nanocomposites: Insight of the Sorption Mechanism of Improved Removal of Phenol from Water. *Int. J. Mol. Sci.* **2020**, *21*, 1563. [CrossRef] [PubMed]
22. Palapa, N.R.; Taher, T.; Rahayu, B.R.; Mohadi, R.; Rachmat, A.; Lesbani, A. CuAl LDH/Rice Husk Biochar Composite for Enhanced Adsorptive Removal of Cationic Dye from Aqueous Solution. *Bull. Chem. React. Eng. Catal.* **2020**, *15*, 525–537. [CrossRef]
23. Alagha, O.; Manzar, M.S.; Zubair, M.; Anil, I.; Mu'Azu, N.D.; Qureshi, A. Comparative Adsorptive Removal of Phosphate and Nitrate from Wastewater Using Biochar-MgAl LDH Nanocomposites: Coexisting Anions Effect and Mechanistic Studies. *Nanomaterials* **2020**, *10*, 336. [CrossRef] [PubMed]
24. Liu, Q.; Ehite, E.; Houston, R.; Li, Y.; Pope, C.; Labbé, N.; Abdoulmoumine, N. Synthesis and evaluation of layered double hydroxide based sorbent for hot gas cleanup of hydrogen chloride. *Mater. Sci. Energy Technol.* **2021**, *4*, 46–53. [CrossRef]
25. Daniel, S.; Thomas, S. *Layered Double Hydroxides: Fundamentals to Applications*; Woodhead Publishing: Sawston, UK, 2020; pp. 1–76. [CrossRef]
26. Santos, L.C.; da Silva, A.F.; dos Santos Lins, P.V.; da Silva Duarte, J.L.; Ide, A.H.; Meili, L. Mg-Fe layered double hydroxide with chloride intercalated: Synthesis, characterization and application for efficient nitrate removal. *Environ. Sci. Pollut. Res.* **2020**, *27*, 5890–5900. [CrossRef]
27. Tang, S.; Yao, Y.; Chen, T.; Kong, D.; Shen, W.; Lee, H.K. Recent advances in the application of layered double hydroxides in analytical chemistry: A review. *Anal. Chim. Acta* **2020**, *1103*, 32–48. [CrossRef] [PubMed]
28. Dotto, G.L.; McKay, G. Current scenario and challenges in adsorption for water treatment. *J. Environ. Chem. Eng.* **2020**, *8*, 103988. [CrossRef]
29. Guo, L.; Zhang, Y.; Zheng, J.; Shang, L.; Shi, Y.; Wu, Q.; Liu, X.; Wang, Y.; Shi, L.; Shao, Q. Synthesis and characterization of ZnNiCr-layered double hydroxides with high adsorption activities for Cr(VI). *Adv. Compos. Hybrid Mater.* **2021**, *4*, 819–829. [CrossRef]
30. Gabriel, R.; de Carvalho, S.H.; da Silva Duarte, J.L.; Oliveira, L.M.; Giannakoudakis, D.A.; Triantafyllidis, K.S.; Soletti, J.I.; Meili, L. Mixed metal oxides derived from layered double hydroxide as catalysts for biodiesel production. *Appl. Catal. A Gen.* **2021**, *630*, 118470. [CrossRef]
31. Shao, M.; Wei, M.; Evans, D.G.; Duan, X. Layered Double Hydroxide Materials in Photocatalysis. *Photofunct. Layer. Mater.* **2015**, *166*, 105–136. [CrossRef]
32. de Sousa, A.L.M.D.; dos Santos, W.M.; de Souza, M.L.; Silva, L.C.P.B.B.; Yun, A.E.H.K.; Aguilera, C.S.B.; de França Chagas, B.; Rolim, L.A.; da Silva, R.M.F.; Neto, P.J.R. Layered Double Hydroxides as Promising Excipients for Drug Delivery Purposes. *Eur. J. Pharm. Sci.* **2021**, *165*, 105922. [CrossRef] [PubMed]
33. Wang, Q.; O'Hare, D. Recent Advances in the Synthesis and Application of Layered Double Hydroxide (LDH) Nanosheets. *Chem. Rev.* **2012**, *112*, 4124–4155. [CrossRef] [PubMed]
34. Guo, X.; Zhang, F.; Evans, D.G.; Duan, X. Layered double hydroxide films: Synthesis, properties and applications. *Chem. Eng. J.* **2014**, *46*, 5197–5210. [CrossRef] [PubMed]

35. Goh, K.-H.; Lim, T.-T.; Dong, Z. Application of layered double hydroxides for removal of oxyanions: A review. *Water Res.* **2008**, *42*, 1343–1368. [CrossRef] [PubMed]
36. Rives, V.; Del Arco, M.; Martin, C. Intercalation of drugs in layered double hydroxides and their controlled release: A review. *Appl. Clay Sci.* **2014**, *88–89*, 239–269. [CrossRef]
37. Conterosito, E.; Gianotti, V.; Palin, L.; Boccaleri, E.; Viterbo, D.; Milanesio, M. Facile preparation methods of hydrotalcite layered materials and their structural characterization by combined techniques. *Inorganica Chim. Acta* **2018**, *470*, 36–50. [CrossRef]
38. Lins, P.V.S.; Henrique, D.C.; Ide, A.H.; da Silva Duarte, J.L.; Dotto, G.L.; Yazidi, A.; Sellaoui, L.; Erto, A.; e Silva Zanta, C.L.D.P.; Meili, L. Adsorption of a non-steroidal anti-inflammatory drug onto MgAl/LDH-activated carbon composite–Experimental investigation and statistical physics modeling. *Colloids Surf. A Physicochem. Eng. Asp.* **2020**, *586*, 124217. [CrossRef]
39. Wan, S.; Wang, S.; Li, Y.; Gao, B. Functionalizing biochar with Mg–Al and Mg–Fe layered double hydroxides for removal of phosphate from aqueous solutions. *J. Ind. Eng. Chem.* **2017**, *47*, 246–253. [CrossRef]
40. Meili, L.; Lins, P.V.; Zanta, C.L.P.S.; Soletti, J.I.; Ribeiro, L.M.O.; Dornelas, C.B.; Silva, T.L.; Vieira, M.G.A. MgAl-LDH/Biochar composites for methylene blue removal by adsorption. *Appl. Clay. Sci.* **2019**, *168*, 11–20. [CrossRef]
41. Harizi, I. Synthèse et Caractérisation des Matériaux à Base de Zéolithe et D'hydroxydes Doubles Lamellaires: Application à L'élimination des Colorants. Ph.D. Thesis, Université Ferhat Abbas, Sétif, El Bez, 2020.
42. Jijoe, P.S.; Yashas, S.R.; Shivaraju, H.P. Fundamentals, synthesis, characterization and environmental applications of layered double hydroxides: A review. *Environ. Chem. Lett.* **2021**, *19*, 2643–2661. [CrossRef]
43. Rojas, R. *Applications of Layered Double Hydroxides on Environmental Remediation*; Nova Science Publishers Inc.: Hauppauge, NY, USA, 2012; pp. 39–71.
44. Daud, M.; Hai, A.; Banat, F.; Wazir, M.; Habib, M.; Bharath, G.; Al-Harthi, M.A. A review on the recent advances, challenges and future aspect of layered double hydroxides (LDH)—Containing hybrids as promising adsorbents for dyes removal. *J. Mol. Liq.* **2019**, *288*, 110989. [CrossRef]
45. El-Abboubi, M.; Taoufik, N.; Mahjoubi, F.; Oussama, A.; Kzaiber, F.; Barka, N. Sorption of methyl orange dye by dodecyl-sulfate intercalated Mg-Al layered double hydroxides. *Mater. Today Proc.* **2021**, *37*, 3894–3897. [CrossRef]
46. Fujii, S.; Sugie, Y.; Kobune, M.; Touno, A.; Touji, J. Uptakes of Cu^{2+}, Pb^{2+} and Zn^{2+} on Synthetic Hydrotalcite in Aqueous Solution. *J. Jpn. Chem. Soc.* **1992**, *1992*, 1504–1507. [CrossRef]
47. Liang, X.; Zang, Y.; Xu, Y.; Tan, X.; Hou, W.; Wang, L.; Sun, Y. Sorption of metal cations on layered double hydroxides. *Colloids Surf. A Physicochem. Eng. Asp.* **2013**, *433*, 122–131. [CrossRef]
48. Dinari, M.; Neamati, S. Surface modified layered double hydroxide/polyaniline nanocomposites: Synthesis, characterization and Pb2+ removal. *Colloids Surf. A Physicochem. Eng. Asp.* **2020**, *589*, 124438. [CrossRef]
49. Li, S.; Ma, X.; Ma, Z.; Dong, X.; Wei, Z.; Liu, X.; Zhu, L. Mg/Al-layered double hydroxide modified biochar for simultaneous removal phosphate and nitrate from aqueous solution. *Environ. Technol. Innov.* **2021**, *23*, 101771. [CrossRef]
50. Zhou, H.; Tan, Y.; Gao, W.; Zhang, Y.; Yang, Y. Selective nitrate removal from aqueous solutions by a hydrotalcite-like absorbent FeMgMn-LDH. *Sci. Rep.* **2020**, *10*, 16126. [CrossRef]
51. Motandi, M.K.; Zhang, Z.; Inkoua, S.; Yan, L. Application of zirconium modified layered double hydroxide and calcination product for adsorptive removal of phosphate from aqueous solution. *Environ. Prog. Sustain. Energy* **2022**, *41*, e13744. [CrossRef]
52. dos Santos, G.E.D.S.; Ide, A.H.; Duarte, J.L.S.; McKay, G.; Silva, A.O.S.; Meili, L. Adsorption of anti-inflammatory drug diclofenac by MgAl/layered double hydroxide supported on Syagrus coronata biochar. *Powder Technol.* **2020**, *364*, 229–240. [CrossRef]
53. Santamaría, L.; López-Aizpún, M.; García-Padial, M.; Vicente, M.; Korili, S.; Gil, A. Zn-Ti-Al layered double hydroxides synthesized from aluminum saline slag wastes as efficient drug adsorbents. *Appl. Clay Sci.* **2020**, *187*, 105486. [CrossRef]
54. Zubair, M.; Daud, M.; McKay, G.; Shehzad, F.; Al-Harthi, M.A. Recent progress in layered double hydroxides (LDH)-containing hybrids as adsorbents for water remediation. *Appl. Clay Sci.* **2017**, *143*, 279–292. [CrossRef]
55. de Sousa, F.W.; Moreira, S.A.; Oliveira, A.G.; Cavalcante, R.M.; Nascimento, R.F.; Rosa, M.F. Uso da casca de coco verde como adsorbente na remoção de metais tóxicos. *Química Nova* **2007**, *30*, 1153–1157. [CrossRef]
56. Aksu, Z.; Gönen, F. Biosorption of phenol by immobilized activated sludge in a continuous packed bed: Prediction of breakthrough curves. *Process Biochem.* **2004**, *39*, 599–613. [CrossRef]

MDPI
St. Alban-Anlage 66
4052 Basel
Switzerland
www.mdpi.com

Molecules Editorial Office
E-mail: molecules@mdpi.com
www.mdpi.com/journal/molecules

www.ingramcontent.com/pod-product-compliance
Lightning Source LLC
Chambersburg PA
CBHW040824250926
51850CB00007BA/1255
9783036588056